Symposium on Foods:

Carbohydrates and Their Roles

Other Oregon State University Symposia

Published by AVI

FOOD ENZYMES—1959 Symposium
  edited by Schultz

LIPIDS AND THEIR OXIDATION—1961 Symposium
  edited by Schultz, Day and Sinnhuber

PROTEINS AND THEIR REACTIONS—1963 Symposium
  edited by Schultz and Anglemier

THE CHEMISTRY AND PHYSIOLOGY OF FLAVORS—1965 Symposium
  edited by Schultz, Day and Libbey

Symposium on Foods:
Carbohydrates and Their Roles

The fifth in a series of Symposia on foods held at Oregon State University

Editor   H. W. Schultz, Ph.D.

Head, Department of Food Science and Technology
Oregon State University
Corvallis, Oregon

Associate Editors   R. F. Cain, Ph.D.

Professor, Department of Food Science and Technology
Oregon State University
Corvallis, Oregon

R. W. Wrolstad, Ph.D.

Assistant Professor, Department of
Food Science and Technology
Oregon State University
Corvallis, Oregon

THE AVI PUBLISHING COMPANY, INC.

Westport, Connecticut

1969

Contributors to this Volume

J. N. BeMILLER, Southern Illinois University, Carbondale, Illinois

G. N. BOLLENBACK, Refined Syrups & Sugars, Inc., Yonkers, New York

S. M. CANTOR, Sidney Cantor Associates, Ardmore, Pennsylvania

D. FRENCH, Iowa State University, Ames, Iowa

E. S. GORDON, University of Wisconsin Medical School, Madison, Wisconsin

A. E. HARPER, University of Wisconsin, Madison, Wisconsin

R. M. HOROWITZ, United States Department of Agriculture, Pasadena, California

D. R. LINEBACK, University of Nebraska, Lincoln, Nebraska

L. M. MASSEY, JR., Cornell University, Geneva, New York

W. A. MITCHELL, General Foods Corp., Tarrytown, New York

R. MONTGOMERY, University of Iowa, Iowa City, Iowa

G. E. MORTIMORE, Hershey Medical Center, Hershey, Pennsylvania

A. NEUBERGER, St. Mary's Hospital, London, England

A. S. PERLIN, McGill University, Montreal, Canada

R. A. PIERINGER, Temple University, Philadelphia, Pennsylvania

T. M. REYNOLDS, CSIRO, Ryde, New South Wales, Australia

T. J. SCHOCH, Cornell University, Ithaca, New York

K. WARD, JR., Institute of Paper Chemistry, Appleton, Wisconsin

R. L. WHISTLER, Purdue University, Lafayette, Indiana

M. L. WOLFROM, Ohio State University, Columbus, Ohio

W. W. ZORBACH, Gulf South Research Institute, New Iberia, Louisiana

This book contains the papers which were prepared at the Fifth Biennial *Symposium on Foods: Carbohydrates and Their Roles* held in the Department of Food Science and Technology at Oregon State University July 23–25, 1968. It is hoped it will complement the books resulting from previous Symposia on enzymes, lipids, proteins, and flavors by providing a comprehensive treatment of the scientific aspects of man's most abundant class of nutrients—the carbohydrates.

The Symposium was planned to be similar to earlier ones. It provided a bridge of communication for scientists who, because of their specific interests, rarely have an opportunity to present their subject directly to certain others whose interests are somewhat remote but nevertheless significantly related through the basic substance with which all are dealing. Thus, the carbohydrate chemists enlightened those studying diseases involving carbohydrates and the food scientists dealing with carbohydrates in food systems. Vice versa, the food scientists and medical scientists discussed their subjects and problems with the chemists. As a consequence, it is believed there has resulted one of the most comprehensive treatments of the entire subject of carbohydrates.

It was clearly evident each participant understood his role and fulfilled it. In the end, the parts of the puzzle have been properly placed to give an up-to-date picture of present knowledge of carbohydrates and their roles. The references at the end of the chapters further extend the usefulness of the book as a reference volume. Each speaker, thereby a contributor of a chapter of this book, deserves highest praise and thanks are hereby extended.

Dr. S. M. Cantor is given special recognition for providing some of the concepts in programming, aiding in the selection of speakers, and introducing and summarizing the Symposium. Deep gratitude has been shown by all concerned to Dr. R. F. Cain who was disinclined to accept responsibility for organizing the Symposium but did indeed do his job well. He deserves heartiest thanks and it is sincerely hoped he is not unrewarded. Dr. Ronald E. Wrolstad very ably assisted Dr. Cain, for which we are extremely grateful.

The staff of the Department of Food Science and Technology, as always, did its part well as host to those in attendance. Special recognition and thanks is due Mrs. Mary Gleicher for bearing the burden of much of the editorial detail.

We are thankful for the financial assistance. This Symposium was supported by U.S. Public Health Service Grant No. UI 00393 from the National Center for Urban and Industrial Health. The grant plus the generosity of the publisher permits purchase of the book at an extremely reasonable price, thereby permitting even those who were not able to attend the Symposium to participate in its benefits.

H. W. SCHULTZ
Corvallis, Oregon

January 1969

Contents

CHAPTER   PAGE

Preface vii

Section I. Introduction

1. INTRODUCTION TO THE SYMPOSIUM, *Sidney M. Cantor* 1

Section II. Physical and Chemical Structures of Carbohydrates

2. MONO- AND OLIGOSACCHARIDES, *Melville L. Wolfrom* 12
3. PHYSICAL AND CHEMICAL STRUCTURE OF STARCH AND GLYCOGEN, *Dexter French* 26
4. CELLULOSE, *Kyle Ward, Jr.* 55
5. PECTINS AND GUMS, *Roy L. Whistler* 73
6. GLYCOLIPIDS, *Ronald A. Pieringer* 100
7. ASPECTS OF THE STRUCTURE OF GLYCOPROTEINS, *A. Neuberger* and *R. D. Marshall* 115

Section III. Advances in Analytical Methodology

8. THE CHEMICAL ANALYSIS OF CARBOHYDRATES, *Rex Montgomery* 133
9. PHYSICAL METHODS, *David R. Linebeck* 147
10. BIOCHEMICAL METHODS OF CARBOHYDRATE ANALYSIS, *J. N. BeMiller* 178

Section IV. Reactions and Interactions of Carbohydrates

11. CARBOHYDRATE OXIDATIONS, *A. S. Perlin* 205
12. NONENZYMIC BROWNING SUGAR-AMINE INTERACTIONS, *Thelma M. Reynolds* 219
13. GLYCOSIDIC PIGMENTS AND THEIR REACTIONS, *Robert M. Horowitz* and *Bruno Gentili* 253
14. IRRADIATION—EFFECTS ON POLYSACCHARIDES, *Louis M. Massey, Jr.,* and *Miklos Faust* 269

Section V.   Carbohydrates in Nutrition and Disease

15. CARBOHYDRATES IN HUMAN NUTRITION, *A. E. Harper* 298
16. THE METABOLIC IMPORTANCE OF CARBOHYDRATE IN OBESITY,
     Edgar S. Gordon 322
17. CARDIAC GLYCOSIDES, *W. Werner Zorbach* 347
18. DIABETES MELLITUS: HORMONAL REGULATION OF HEPATIC
     CARBOHYDRATE METABOLISM, *Glenn E. Mortimore* 359

Section VI.   Role of Carbohydrates in the Food Industry

19. SUGARS, *G. N. Bollenback* 373
20. STARCHES IN FOODS, *Thomas J. Schoch* 395
21. CARBOHYDRATE HYDROPHILIC COLLOID SYSTEMS, *W. A. Mitchell* 421
22. SUMMARY AND PANEL DISCUSSION 444

INDEX 451

Introduction

CHAPTER 1

Sidney M. Cantor

Carbohydrates & Their Roles in Foods
Introduction to the Symposium

INTRODUCTION

As you have heard this is the fifth in a series of Symposia sponsored by the Oregon State University's Department of Food Science and Technology. All of these have been concerned with the role in foods of particular classes of compounds or of products identified functionally with an interrelated set of chemical reactions. These Symposia enjoy an outstanding reputation because a stated objective of them has been to relate the basic chemistry of the subject field to applications in the food industry. Usually this objective is not easily achieved, but the people at this center of excellence seem to have a particular way of organizing programs to encourage interdisciplinary exchanges.

PROGRAM OBJECTIVES

Perhaps a more specific objective of these programs has been to draw the kinds of logical relationships which allow extrapolation. For example, in our case we can ask whether our understanding of carbohydrates individually or in combination can be interrelated, expanded, and projected by being exposed not only to an expert interpretation of their fundamental physical and chemical properties, but also by examining them in an interdisciplinary context which includes their applications. Past experience suggests that the answer can be "yes," and so we hope that in the elaboration of the relationship between fundamental carbohydrate science and applications of carbohydrates in the food field we can find some new insights and be stimulated to a more complete appreciation of our major dietary constituent.

My scientific orientation is applied and my work in a broad sense is concerned with both domestic and international developments related to agribusiness. Hence in introducing this subject I would like to refer

1

to some arbitrarily selected food industry developments which have appeared recently and relate them in my understanding to certain long developing patterns of technologic and economic influence. These patterns, in my experience, appear to be related to the process of development and hence are global. So this Symposium, which is international as regards participants, accentuates the international character of the food problems involved.

MARKETING INFLUENCES

Major changes in the technique of food products development have occurred in the last 20 yr which in turn have had a substantial impact on research and development practices and therefore results. The most important change relates to industry's gradual but reasonably broad acceptance of the so-called "total marketing concept" in product development. This concept says in essence that an identified consumer need (expressed or interpreted) is the chief motivation for determining the characteristics of a product which will be developed, manufactured, and offered for sale. Moreover in meeting the need, the compromises required will be determined on the one hand by the best interests of the consumer and on the other by those of the producing complex.

Generally, the slowly developing appreciation of this process—marketing oriented product development—has resulted in a wide spectrum of practices: from a compromise totally in favor of the producer to one, perhaps not as frequently as desirable, totally in favor of the consumer. Unfortunately, in too many cases, the compromise is in favor of an intermediate agent—producer, sales organization, technology, commodity lobby—but in any case a vested interest or prejudice. Properly respected the process has been remarkably successful in producing useful and saleable products. Ignored, it is responsible for a high rate of frustration and new product failures.

Marketing orientation has resulted from an appreciation of the market as a dynamic social and economic structure. In the United States we have become a substantially urbanized people as a result of our advanced industrialization. More members of a family work and there is an ever-increasing demand for the convenience of buying food preparation. This is reflected in the nature of home consumed foods and the continuing rise in importance of the food service business.

In developing countries the same process is being followed but not always in the same sequence. It is possible with understanding to avoid some unnecessary steps and this is a major but often missed advantage of development which ignores the systematization coming from knowledge of the market.

Product convenience and product formulation go together because the consumer's wants can most readily be expressed in the food technologists' functionally oriented language. For example, flavor, mouthfeel, richness, shelf-life, freeze-thaw stability, antioxidant, nutritional requirements are familiar terms which are immediately translatable into product components by the food technologist.

Developments in commodity processing over the last 20 yr have contributed to this relationship and reflect the unbounded curiosity of scientists whose interest is stimulated in industry by a need for a more meaningful association between the terms of the food technologists and the chemical structures. Gradually, the commodities have been taken apart and their various homogenous fractions identified with one or more functions reflected in product use. Aside from curiosity as a stimulant for such progress, another major motivation is the improved control over product identity which is thus provided and for which both manufacturer and consumer are willing to pay. The manufacturing consumer or product formulator buys today essentially on the basis of function and cost. This has led inevitably to minimizing the importance of source and elevating the importance of function and has opened the door to synthetics. Today the chemical industry is a major source of functional food additives.

The chemical industry has only recently discovered the market which the food formulator represents. An interesting aspect of this discovery is the array of new products which are now regularly directed at the food industry by the chemical industry. The promise for an ever-increasing flow of products because of the research experience and expertise owned by the chemical industry is great. Such facilities have never before been made available to the food industry since it traditionally makes the lowest budgetary commitments of any industry to research and development.

This astonishing flexibility—which the recognition of the critical role of function has generated—is developing because a market exists, but the motivation may initially be more than that, for example, a national advantage. Let me illustrate various motivations.

Calorie-control

The concept of controlled calorie rather than noncalorie formulation, a more rational approach to problems of obesity, has required products to go with noncaloric sweeteners which simulate mouth-feel at low concentrations. More critical has been the interpreted need for non-hygroscopic, nonretrograding (i.e., dissolving to clear solutions) carriers which can be used to simulate the physical characteristics of table sugar. The starch industry's response of new low dextrose-equivalent syrups supplied in a spray-dried form testifies to both the quality and quantity of basic information available on enzyme degraded starches. It would appear now that with proper drying and expansion techniques a wide range of product densities and therefore calorie densities is available. Conversely, high calorie density is sometimes desirable and is currently being formulated into infant foods for developing areas.

Flavor

A greater elaboration of taste mechanisms and motivations is critical to guide the food formulator toward better product acceptance. I am sure we will hear more about this later but the great interest in flavors and flavor enhancers current in the food industry and among its suppliers needs a much broader understanding. It is encouraging, therefore, that more fundamental work is being organized to provide the guidance that this interest is seeking.

The new interdisciplinary institute being organized by Dr. Morley Kare in the University of Pennsylvania which is devoted to studying the complex interrelationships among chemical senses and nutrition promises new insights for understanding flavor systems. I refer to this here because sweetness and sweetness perception are so important in food formulation, and we are just at the beginning of appreciating the significance of variations in response among species and the function of enhancers.

Isomerization of sugars

As much as the market may dominate the product development scene and reflect on research in our consumer culture, the process by which commodities or at least their properties are delivered is becoming as frequently determined by collective needs.

The process for enzyme catalyzed isomerization of D-glucose to D-fructose which emerged from the highly developed skills of Japanese biochemists, was the result of a well-planned national economic policy which in turn was reflected in government support of the research effort.

Japan is sucrose poor and reluctant to spend hard currency on cane sugar imports; but sweet potatoes are a major starch crop and maize is a supplementary starch source which is imported from the United States and Thailand. Japan also has a highly developed starch hydrolysis technology including both enzyme and acid catalysis. Some years ago the Japanese government was ready to carry out alkali catalyzed isomerization of D-glucose to obtain D-fructose and thus an invert sugar replacement, but the development and large scale production of glucose isomerases has superseded the expected reliance on the older and less specific method.

In the United States, one company is offering a high dextrose equivalent starch hydrolyzate in which about ⅓ of the D-glucose is enzymatically isomerized to D-fructose. While details of the process are unavailable, it is probable that the basic starch hydrolyzate is produced by a combination of acid and enzyme catalyzed hydrolysis and the isomerase action is superimposed on this substrate. What emerges is a highly sophisticated process reflecting a new and important capacity for source interconvertibility.

Source variation in which the importance of function and national convenience outweigh tradition also poses nomenclature problems. About 40 yr ago when pure, crystalline dextrose was first introduced as a commercial product and at times referred to as "corn sugar," federal authorities objected to this nomenclature saying that corn sugar was the refined juice expressed from corn stalks. Technological and plant breeding advances have now reached the stage where for the past 2 or 3 yr a company in North Dakota has been experimenting with a maize variety which instead of laying down starch retains soluble carbohydrates in the stalk. This juice when expressed and refined contains about 85% sucrose and 15% invert sugar and falls into a sugar refining industry product category called liquid sugar. It is interesting to note that this product which might in the older sense be called "corn sugar" is having difficulty finding a name acceptable to all branches of the sweetener industry.

Bread

A physical interconvertibility problem is represented by bread. Slightly over a third (35.5%) of the world's population share wheat as a food staple—this is eaten often as a variety of bread. Better than one-half (53.7%) of the world's population eat rice. Yet bread is a status food and also a convenience food, and its use is increasing as industrialization increases per capita disposable income in developing countries. Many developing countries do not produce wheat and must purchase it out of limited hard currency funds to satisfy the increasing status demand for bread.

The need for processes, therefore, by which wheat flours can be extended or replaced by other indigenous cereal or root flours and starches to produce acceptable "bread" structures is great. New developments of the British Chorly Wood continuous bread process soon to be published promise important progress. In many ways this appears to be a carbohydrate problem, and the role of starches in bread structure needs further elaboration. The baking industry is in many ways tradition-bound and the introduction of continuous fermentation needs associated continuous loaf production. Perhaps foamed resin technology offers possibilities for adaptation.

WASTE TO FEED TO FOOD CYCLE

Another aspect of interconvertibility is provided by the affluence of our culture as well as our attention to convenience. Our consumption of cellulose grows ever larger until it almost exceeds our ability to discard the structural carbohydrate. Collection of waste paper is prohibitively costly. Already there are waste conversion processes, but it is interesting to note that a critically significant International Symposium on Cellulases will be held at the American Chemical Society Meeting in September, 1968. The conversion of cellulose to more easily biodegradable or biologically useful products, for example, animal feeds and ultimately human foods, requires more attention within the ecosystem context. The waste to feed to food cycle and its many interconversions require much more exploration if we are to make best use of our carbohydrate resources.

NUTRITION

The subject of nutrition in its broadest interpretation requires more attention from carbohydrate scientists—even as protein malnutrition occupies the major attention of clinical nutritionists. Marasmus complicates every case of kwashiorkor, particularly on a level where usual clinical observations are not evident. The need for calories in the developing world must not be neglected as well-deserved attention is devoted to proteins.

Carbohydrate foods are relatively cheap as well as abundant. Carbohydrate consumption plotted against per capita income in various

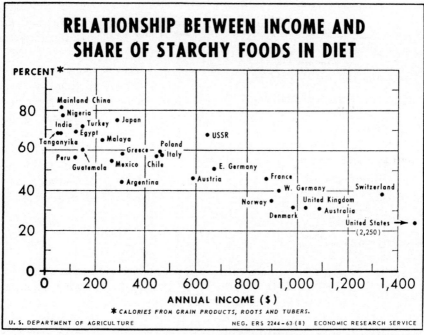

From Foreign Agr. Econ. Rept. No. 11 (1963)

FIG. 1. MAN, LAND, AND FOOD

Looking ahead at world food needs, by Lester R. Brown.

countries is shown in Fig. 1. A comparable curve for protein consumption is shown in Fig. 2.

There is no question but what adequate protein consumption is related to adequate economic status. Neither is there much question as to the unlikelihood of solving developing country problems with meat, milk, and eggs.

The greatest promise resides in the agriculture revolution already beginning to be apparent. This is significant with respect to both quantity and quality and affects all major nutritional elements: carbohydrates, fats, and proteins. The relationship of function in a food product to structure becomes increasingly important as our ability to modify structures genetically becomes more practical. As mass feeding becomes more necessary, functional efficiency will need to be much more efficient.

On a longer range basis, however, nonconventional resources will also be required to answer to growing nutrient requirements. Inter-

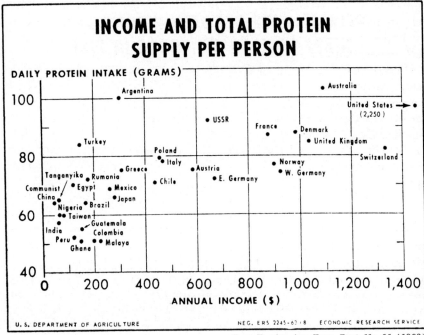

From *Foreign Agr. Econ. Rept. No. 11* (1963)

FIG. 2. MAN, LAND, AND FOOD
Looking ahead at world food needs, by Lester R. Brown.

convertibility of resources must therefore be a continuing area for study. Again the objective is efficiency of function (including cost) as well as economy of use.

The attention being paid to hydrocarbon to protein conversion needs more than parallel input for the study of carbohydrate to protein conversion despite the large experience of the subject of food and feed yeasts. Single cell protein from carbohydrates is being studied and shows a basic economic attractiveness because of the availability of carbohydrates in so many developing areas. But single cell protein converted to a bland, widely utilizable protein isolate is another matter, and this has held up food yeast utilization.

INTERCONVERTIBILITY

The interconversion of one material to another for taste, economic advantage, status, or other specific reason, dominates our development activities. Interconvertibility will join the other important terms—

convenience, marketing, function, formulation, which impinge on and serve to guide food scientists and technologists. The food and associated industries to an astonishing degree are involved in a vast culture transfer—another kind of conversion.

On a recent trip to Uganda to study its emerging food processing requirements, I was struck by the common denominator expressed as convenience and shared by a banana, an ear of corn, and a slice of bread. All three are portable, they vary in terms of shelf-life and most significantly in nutritional balance. Yet it appears that in the long run bread will win out and thus make the case for commodity processing and flexibility of the vehicle as a carrier of many kinds of consumer appeal.

This is culture transfer but it must be done in a way that is sensitive to local custom and understanding of local economic requirements.

It is apparent that more and more the food technologist must be guided by a knowledge of the market target in the broadest of terms. Moreover, the manner in which such knowledge is conveyed to the scientist who supplies the fundamental information is of critical importance. The development of a truly interdisciplinary communication system is essential, and the program presented in this symposium is a calculated effort which recognizes this need. The vast significance of food in today's world and the need for maximum clarity of understanding among the many disciplines needed to attack the problems might well be a subtitle to our discussions of the next few days.

DISCUSSION

T. J. Schoch.—I raise the question, which is very close to my heart for many years, of bread—bread as a world staple, something that can be produced in any part of the world, regardless whether it is derived from wheat or not. Bread is not a carbohydrate problem but a protein problem. It has a most unique type of protein in it, a protein which is elastic enough to maintain a structure before baking and which cures or denatures to a gel-like elastic structure. I visualize the role of starch here as being very minor, contributing somewhat to crumb structure, supplementing the gluten structure, but not in itself being really important. You can make bread without wheat starch, i.e. take wheat gluten and reconstitute it with any starch you wish. Some of these breads made this way are more acceptable than those made with wheat gluten and wheat flour. For example, the old Scottish practice of putting mashed potatoes in bread to get the moisture retention, the softness, and the mouth feel which are so much more acceptable to me than the properties contributed by wheat starch. Now if we could imagine a cheap available source of protein which could be modified to have this unique property of giving a crumb structure that wheat gluten has, then we could manufacture bread in any part of the world. In Japan, for example, which imports most of its flour, we could do the stunt of using available fish which might provide a cheaper source of protein and combining it with any starch that was available, e.g. the starch of sweet or white potatoes. Then can we get away from protein completely (not that this is desirable as the amino acid value of the protein is not to be despised in the least!) but could we make a bread in which by some other means—cellulose—we could get this type of crumb structure without the use of any protein gluten at all? I think that these are primarily protein problems and require the attention of protein chemists to see what the peculiar properties of wheat gluten are, how they relate to physical structure, what are the mechano-chemical properties of the crumb itself, and how this can be duplicated with other material.

S. M. Cantor.—Tom, I really don't have any comment, I think that you've said what there is to say on the subject. I would hesitate to contradict you on the function of the protein as opposed to the function of the starch, however, I've seen just enough of some of the products that are made at high carbohydrate content in which energy inputs are very carefully measured which provide some rather interesting structures at very low protein concentration. This is encouraging enough so I am told, that a great deal of work is going to go on in this area. This is in England and at Beltsville. There is a great deal of work incorporating oil seed flours into traditional bread structures. I think that there is at least a need for a look at some fundamental structures that you've indicated, in order to substitute the protein. The vital character of the gluten is one aspect which has really nothing to do with the nutrition. If that unique property can be substituted, the protein can come from other sources.

D. French.—If I could be permitted to tell a very brief personal story. My sister lives in Mexico and her children grew up on a Mexican diet. When one of them was about 5 yr old, she came to the States to visit us and we tried to treat her the best we could and during the visit we asked her if there was anything she thought she would like to have. She said yes she wanted some tortillas. Now for the average American the tortilla is nothing much but I can assure you that for people that have been brought up on them they will not accept what we consider is a highly superior type of bread. This illustrates one of Dr. Cantor's points that so much depends on one's upbringing and culture rather than what one considers the desirable properties in a bread material.

S. M. Cantor.—The tortilla is very close to Indian chapatties but in India our sliced bread (modern bread) is enjoying a tremendous popularity to the exclusion of the chapattie. This is sad in my view because there's nothing wrong with chapattie for Indians (I like it too). The whole idea of traditional culture and so on is something that I appreciate very much. I don't know why it's necessary to substitute sliced, white, fortified bread, but it's happening, particularly in India. It's so popular that separately wrapped slices are being sold on the black market.

R. Montgomery.—I was interested to learn that the biochemist was getting into this act with the enzyme conversion of glucose to fructose. But, recognizing that sucrose is still much sweeter than its invert, I'm wondering why the next step is not being taken and sucrose phosphory-lase brought in. Maybe the kinetics or the equilibrium is unfavorable. I see Dexter French shaking his head. My question is why not put the two together and make the sucrose?

S. M. Cantor.—Rex, I suppose it's a matter of cost.

J. L. Hickson.—Invert sugar is sweeter than sucrose under normal conditions by a significant factor, however, if the partial conversion is from dextrose then the totality of dextrose plus the invert may or may not be as sweet as the sucrose. The enzymic conversion to the sucrose molecule has been achieved quite some years ago, and I can assure you it is an issue of cost. However, in Denmark and Britain, over the past year there has been a proposal from a (may I say mad?) Dane to sell dextrose as a life sugar in place of sucrose because of the arguments for the differential metabolic value. This is a very live issue. The sucrose industry is looking at this enzymic conversion with great interest because if you do have a mixture of fructose and dextrose you do have a true invert sugar.

Physical and Chemical Structures of Carbohydrates

CHAPTER 2

Melville L. Wolfrom | Mono- and Oligosaccharides

INTRODUCTION

In the development of the general subject of the physical and chemical structure of carbohydrates, this section will be concerned with a group of naturally occurring compounds once designated the "simple sugars and glycosides" (Armstrong 1924). There are a host of substances known to the organic chemist as mono- and oligosaccharides (Staněk et al. 1963, 1965), but we will select from this large group only certain illustrative examples important in the chemistry of foods and food processing. Fortunately, nature has been highly selective in using only certain ones for the production of plants and animals useful in human and animal nutrition. The organic chemist, however, needs to consider the entire group and to do so there is required a nomenclature specifying exactly just which member is being dealt with. Modern exact carbohydrate nomenclature is rather new, in some respects, but will be adhered to in this treatment.

To this section belong the "Bausteine" (Fischer 1909) of the carbohydrates. Our basic conceptions are then very fundamental and this discussion will mainly be a review and extension of rather familiar principles. The substances to be considered are polyhydroxy aldehydes and ketones and their condensation products, by which we mean combinations between each other involving the elimination of water. This may result in a high polymer of the condensation type, known as a polysaccharide. If the number of units involved in this chemical condensation is relatively small and has been established with high precision, the substance is designated an oligosaccharide. The upper degree of polymerization for an oligosaccharide is generally about 12 but cannot be defined with exactitude. These condensations are effected

12

with great difficulty by the organic chemist, but are accomplished in nature with remarkable ease by enzyme systems (Neufeld and Hassid 1963). The chemical linkages between units are acetals and as such are hydrolyzable by aqueous acidity or by quite specific enzymes. The oligosaccharides are classified by the number of monosaccharides (glycoses) produced on hydrolysis while the monosaccharides are grouped according to the nature of the carbonyl group present and the number of carbon atoms in the chain. For the most part the carbon chains are normal and contain asymmetric carbon atoms which produce optical activity. The carbohydrates constitute by far the largest group of stereochemically related isomers known in organic chemistry.

The saccharides are crystalline solids (or should be) which readily tend to produce syrups because of their high degree of tautomeric isomerism. With a very few exceptions, the water-soluble saccharides have a sweet taste in varying degree. Some of the higher oligosaccharides are quite tasteless and at least one, gentiobiose, is bitter. With a few exceptions, such as sucrose, the saccharides (mono- or oligo-) have difficultly reproducible melting points. Their best constant is optical rotation. Modern chromatographic techniques have revolutionized the purification and analysis of these substances.

MONOSACCHARIDES

Aldoses (D-Glucose)

D-Glucose is the central saccharide in nature and as such is of wide occurrence, generally in the combined state. It is an aldohexose which has a mildly sweet taste and crystallizes with difficulty if at all impure. This property has found uses in food technology. It is only rather recently that economic methods for the large scale production of D-glucose (dextrose) have been established (Newkirk 1923). The present main commercial source is the enzymic hydrolysis of starches whereby the yield of high purity product is extremely high.

Figure three shows the various ways by which the chemist represents this substance. The original Fischer (Fischer 1891) formula shows terminal aldehyde and primary hydroxyl functions with 4 optically active secondary hydroxyl functions placed between, thus necessitating 2^4 (Van't Hoff 1874; LeBel 1874) or 16 isomers. In the principal ring structure, which is pyranose (Hirst 1926), the potential aldehydic carbon becomes asymmetric and is replaced by a monocyclic hemiacetal function, leading to two new isomers termed anomers. A saccharide will crystallize in 1 anomeric form and the 1 obtained with D-glucose when crystallized from water below 50°C, is the α-D-pyranose monohydrate, from which the anhydrous form may be obtained. Above ~100°C, the β-D-pyranose crystallizes and this form has a higher rate of solution than does its anomer but it is unstable. The Fischer formula is a projection while the Haworth (Drew and Haworth 1926) ring formula is a formalized pictorial representation. A formula now known to be closer to the actual shape of the α-D-glucopyranose ring is the conformational formula (Eliel *et al.* 1965), the principal forms encountered for the saccharides in general being those shown. These are termed *C1* D and *1C* D (Reeves 1949). The former is the principal type exhibited by the substance represented. It is of considerable interest that in the *C1* D

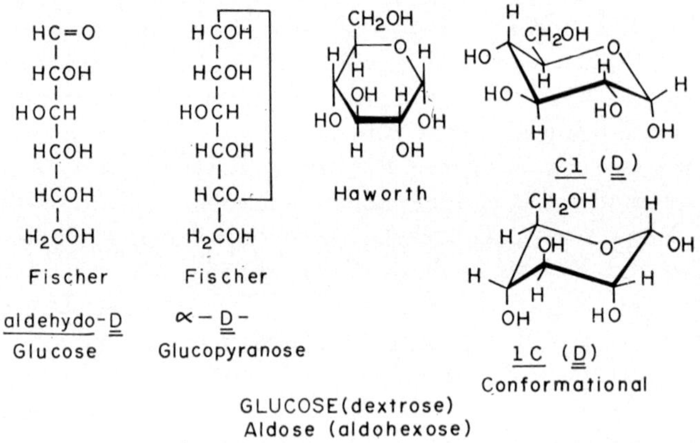

FIG. 3.   METHODS OF REPRESENTATION OF D-GLUCOSE

conformational formula for β-D-glucopyranose, all hydroxyl groups are equatorial. Since this is the only aldohexose possessing such a highly thermodynamically stable system, it may be speculated that this may be the cause of this aldohexose being selected by evolutionary processes to become the central sugar in nature. The anomers are in tautomeric equilibrium in solution so that such a solution contains several substances thus making a single one difficult to crystallize. Such tautomerism is the cause of the optical rotatory mutarotation exhibited by the reducing saccharides. An aqueous solution of D-glucose contains about ⅔ of the β-D and ⅓ of the α-D anomer at equilibrium at 20°C.

Ketoses (D-Fructose)

Figure four shows the various ways of depicting the formula of the principal natural ketose, the ketohexose D-fructose (2-ketohexose, 2-

hexulose, D-*arabino*-hexulose, levulose). This substance occurs mainly in the furanose form in the combined state. It is the sweetest saccharide known. It exhibits an unusual and rapid mutarotation and its solution contains both pyranose and furanose forms and probably also some of

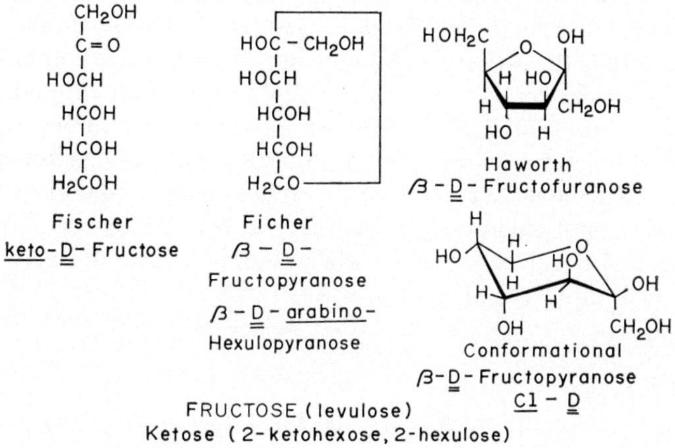

FIG. 4. METHODS OF REPRESENTATION OF D-FRUCTOSE

the acyclic keto form or its hydrate (ketrol). Because of this tautomeric behavior, D-fructose is extremely difficult to crystallize but under the proper conditions the β-D-pyranose form can be crystallized from water by slow cooling (Bates and Associates 1942). Unfortunately, this highly desirable crystalline saccharide is unstable to moist air.

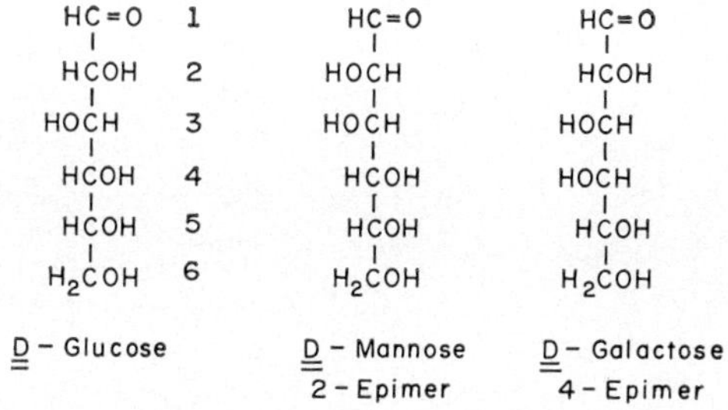

FIG. 5. ALDOHEXOSES OF WIDE DISTRIBUTION

In Fig. 5 are shown the few other natural saccharides of wide distribution and their relation to D-glucose. These are not to be construed as the only natural saccharides. Since the development of antibiotics, many rare sugars have been found in nature, even the L-form of glucose (in streptomycin). However, those shown constitute the main ones of concern in foods. D-Mannose is the 2-epimer of D-glucose. It is a very weakly sweet sugar which is also very difficult to crystallize. It is easily detectable as its water-insoluble phenylhydrazone (Fischer and Hirschberger 1888). It occurs mainly in the combined form, being, for example, the main component of the coffee bean where it occurs as the polysaccharide mannan.

D-Galactose is the 4-epimer of D-glucose. This sugar has a good sweetness and is easily crystallized. It is widely distributed in the combined form and occurs naturally in both enantiomorphous forms, sometimes together, as in the seaweed polysaccharide agar. In this connection it is of interest to note that reversal of the end functions will in this case lead to the enantiomorph and this may be the reason why both forms are of natural occurrence. It is, other than D-glucose, the principal saccharide in brain tissue.

Reactions

Being an aldehyde, D-glucose can be reduced to the corresponding primary alcohol [D-glucitol (sorbitol, not L-gulitol)] or oxidized to the carboxylic acid (D-gluconic acid, Fig. 6). D-Glucitol is again extremely difficult to crystallize although in recent times this substance has been made commercially available in crystalline form (Goepp 1943). It is interesting that DL-glucitol is a true racemic compound with differing

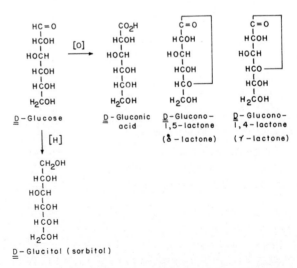

FIG. 6. OXIDATION AND REDUCTION OF D-GLUCOSE

physical properties from its components; it crystallizes readily (Wolfrom *et al.* 1946). The best method for D-glucitol synthesis is by reduction of D-glucose with hydrogen and a nickel catalyst (Ipatieff 1912). This alditol has a wide occurrence in fruits. It possesses a slightly sweet taste, as do all the alditols.

The aldehyde D-glucose is oxidizable to its carboxylic acid D-gluconic acid (Fig. 6). This can be effected by chemical methods (Kiliani and Kleemann 1884) or by fermentative procedures. This polyhydroxy acid readily forms an unstable 1,5-lactone or the more thermodynamically stable 1,4-lactone. All 3 of these forms are crystalline but by far the most readily crystallizable form is the 1,5-lactone, now commercially available. These lactones are neutral in the dry form but become acidic on solution in water. This property can be useful. The aldonic acids are fairly strong acids.

It has been stated that all the chemical changes producible in a compound fall under the headings: oxidation, reduction, action of acids, bases, and heat (pyrolysis). The latter three are of especial interest in food processing. Acids cause dehydrations, leading to the eventual formation of furan derivatives. Thus, hexoses produce 5-(hydroxymethyl)-2-furaldehyde and formic acid even on heating in water alone, since these polyhydroxy substances are mild acids to begin with. This furan substance can produce off-flavors. Bases cause enolization (Wohl and Neuberg 1900) and β-eliminations (Isbell 1944), causing very profound and complicated changes. Heat alone produces dehydrations and in the aldohexoses, 1,6-anhydride formation is especially favored. An interesting application of both heat and mild acidity is caramelization while a notable example of a favorable action of alkali on carbohydrate material is the formation of the pretzel glaze by strong alkali.

Other Types of Monosaccharide Occurrences Pertinent to Foods

Of the many higher (above 6 in carbon content) saccharides known, D-*manno*-heptulose (Fig. 7) occurs uncombined in the avocado (La-Forge 1916) together with some others (Sephton and Richtmyer 1966). The two trioses, D-glyceraldehyde and dihydroxyacetone (Fig. 7), are important intermediates, as phosphate esters, in carbohydrate metabolism controlled by enzymes. The nucleotides are the building stones of the nucleic acids found in both animals and vegetables. The nucleotide 9-β-D-ribofuranosylhypoxanthine 5'-phosphate (inosinic acid) was early isolated from muscle (Liebig 1847). The nucleoside obtained on phosphate removal belongs to the type of compound originally known as an "aldehyde-ammonia" derivative. Such substances should be designated by the general term glycosylamine. The purine nucleosides, of which this is a representative, are hydrolyzed by their specific enzymes and by both acids and bases. D-Ribose is the sugar component of one type of nucleic acid (RNA) while the closely related 2-deoxy-D-*erythro*-pentose is the characteristic sugar of the other type (DNA).

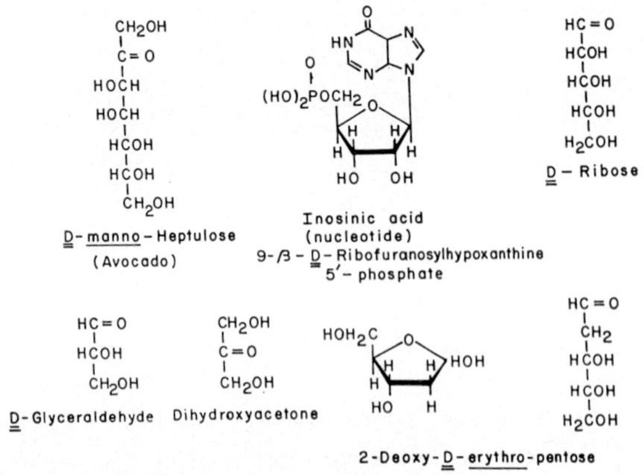

FIG. 7. OTHER TYPES OF MONOSACCHARIDE OCCURRENCES PERTINENT TO FOODS

GLYCOSIDES

A glycoside is a glycose (aldose or ketose) in which the hydrogen of the anomeric hydroxyl group has been replaced by an alkyl or aryl group. It is thus a mixed acetal and is hydrolyzed by acids or by enzymes. With a very few exceptions, glycosides are stable to bases and will not reduce alkaline oxidizing agents (as Fehling solution). There are a host of them present in nature, especially in plants. Those of natural occurrence are β-D or α-L anomeric forms. These anomers have the same order of groups on the anomeric carbon atoms (Hudson 1909). The glycosides are constituents of many natural flavors and pigments; they have a bitter taste. The red and blue pigments of fruits are mainly glycosides. Figure eight shows the formulas for the components of apiin (Vongerichten 1900), the principal glycoside of parsley. This substance contains as the aglycon the yellow pigment 5,7,4'-trihydroxyflavone, to the seventh position of which is attached a β-D-glucopyranosyl unit and to the C-2 position of the latter is found the branched-chain pentose apiose in a β-D-furanose form (Hulyalkar et al. 1965). The formula of apiose

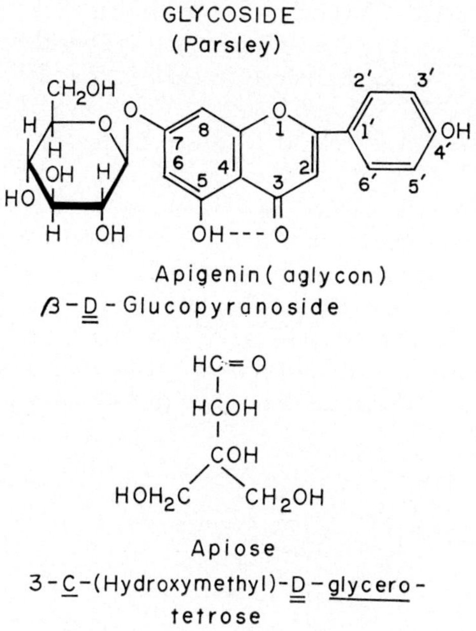

GLYCOSIDE
(Parsley)

Apigenin (aglycon)
β - $\underline{\underline{D}}$ - Glucopyranoside

HC = O
|
HCOH
|
COH
HOH₂C   CH₂OH

Apiose
3 - $\underline{C}$ -(Hydroxymethyl)- $\underline{\underline{D}}$ - glycero -
tetrose

FIG. 8. APIIN FROM PARSLEY

(Schmidt 1930) is of considerable interest in that the acyclic form has only one asymmetric center, but four isomers are possible for its furanoid ring.

OLIGOSACCHARIDES

Oligosaccharides can be viewed as glycosides involving an hydroxyl group of another saccharide but it is best to consider them separately.

Disaccharides

These substances are conveniently considered in two groups: reducing and nonreducing.

Nonreducing.—This is by far the smaller group but to it belongs sucrose, the circulating form of carbohydrate in all plants. This disaccharide is isolable from the saps and juices of any plant or tree. It is generally accompanied by its specific enzyme invertase so that some invert sugar (D-glucose + D-fructose) is generally present. Some sources contain the disaccharide in larger and more nearly pure amount than do others. The sugar cane and sugar beet are the commercial sources although in North America the sugar maple is valued for the flavor imparted by

characteristic impurities formed on concentration of the spring sap.   The
sugar cane grows wild in New Guinea and is considered to have origi-
nated there from whence it arrived in Louisiana and Florida by way of
India, Spain, and the Spanish colonies.   The sugar beet industry was
established in Germany (Achard 1809; see Lippmann 1929) in the early
part of the nineteenth century.   Achard and others succeeded in devising
a process which was carried out in numerous factories during the
years of the Napoleonic wars when the European continent was under
blockade.   These efforts had been stimulated by a prize offered by
Napoleon Bonaparte and won by Chaptal.

The structure of this commonest of the disaccharides was the most
difficult to unravel (Levi and Purves 1949).   The final solution is
depicted in Fig. 9.   The structure is β-D-fructofuranosyl α-D-glucopyrano-

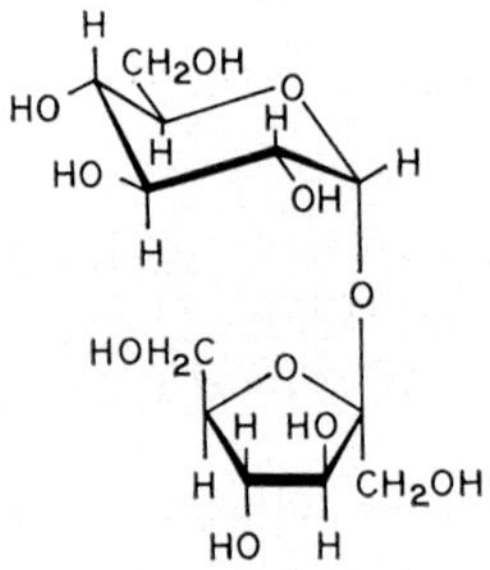

Sucrose

β−D−Fructofuranosyl α−D−glucopyranoside

FIG. 9. SUCROSE

side.   The substance is highly crystalline perhaps because it has no
tautomers and its seed crystals are universally at hand.   This ease of
crystallization must be circumvented in certain condiments.   In soft-
centered chocolates some invertase may be added to give D-fructose,
which will syrup most anything.

Another nonreducing disaccharide, trehalose or α-D-glucopyranosyl
α-D-glucopyranoside, is also an important disaccharide in some biological
areas.   It is the circulating form of carbohydrate in the blood of insects.

Reducing Disaccharides.—Of this large group not many occur naturally.   Most are found in nature as hydrolytic fragments formed by
enzymic action on polysaccharides, a process taking place especially in
animal digestive organs.   A notable exception is lactose (Fig. 10)
which is found in the mammary gland.   This substance is one of the

oldest known saccharides, the Italian pharmacists of the Renaissance being familiar with it. Lactose is an easily crystallizable, sweet sugar which is found in higher concentration in the human than in the bovine mammary gland. The structure of lactose or milk sugar is established as 4-O-β-D-galactopyranosyl-D-glucopyranose (Haworth and Long 1927). Both anomeric forms are known.

A reducing disaccharide of related structure is maltose (Fig. 10) or 4-O-α-D-glucopyranosyl-β-D-glucopyranose monohydrate. This crystalline sugar is obtainable in high purity only with difficulty and apparently no highly pure form is commercially available. The substance has a very characteristic and pleasant malt flavor. It is the principal end product of the enzymic breakdown of starches and glycogens.

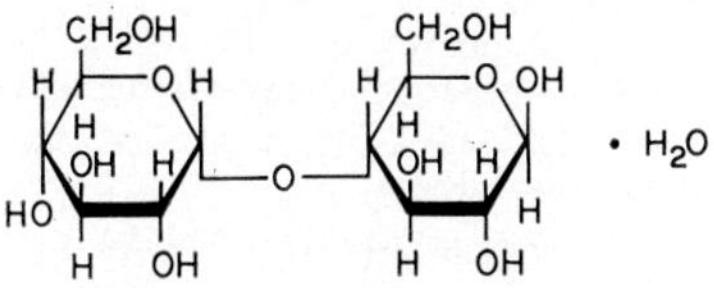

β − Maltose (monohydrate)
4 − O −α− D − Glucopyranosyl −β − D −glucopyranose
monohydrate

Lactose
4 − O −β − D −Galactopyranosyl − D − glucopyranose

FIG. 10. MALTOSE AND LACTOSE

Another disaccharide closely related to maltose is isomaltose (Fig. 11) which is 6-O-α-D-glucopyranosyl-D-glucose (Wolfrom et al. 1958). This saccharide has not been crystallized but is characterizable as its crystalline β-octaacetate (Wolfrom et al. 1949). Isomaltose is significant as being the disaccharide obtainable from the (1 → 6) branch point in starch (Thompson et al. 1953) and glycogen. The latter polysaccharide is a constituent of shellfish, liver, and muscle. Horse meat is known to have a higher glycogen content than beef.

Another (1 → 6)-linked disaccharide closely related to isomaltose is gentiobiose (Fig. 11) which is a component of the trisaccharide gentianose found in gentian root. This disaccharide is also a constitutent of commercial dextrins and of many glycosides, as amygdalin. As noted previously, gentiobiose has a characteristic bitter flavor and here the glycosidic union, known to produce bitterness in glycosides of mono-hydric aliphatic alcohols and phenols, apparently overcomes the sweetness contributed by the polyhydroxy nature of the remainder of the molecule.   This disaccharide is crystalline and both anomeric forms are known.

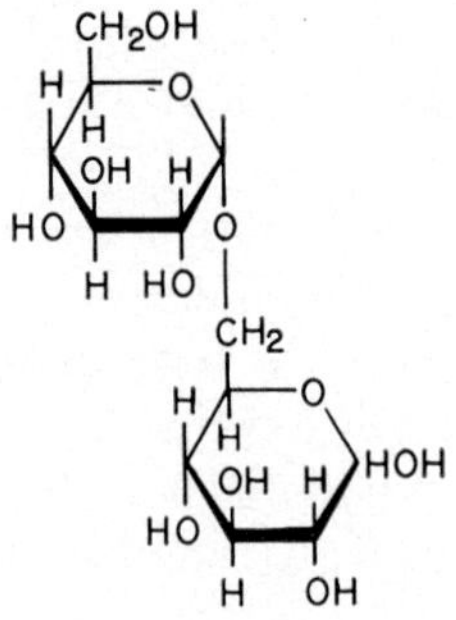

Isomaltose
6 – O – α – D – Glucopyranosyl – D – glucopyranose

Gentiobiose
6 – O – β – D – Glucopyranosyl – D – glucopyranose

FIG. 11.   ISOMALTOSE AND GENTIOBIOSE

Higher Oligosaccharides.—Many of these are now known as partial acid or enzymic hydrolytic products of natural polysaccharides.   The highest degree of polymerization characterized on a crystalline basis is about 7 (Wolfrom and Dacons 1952) although higher degrees of polymerization (about through 14) have been established by chromatographic methods (French *et al.* 1966).   A trisaccharide produced

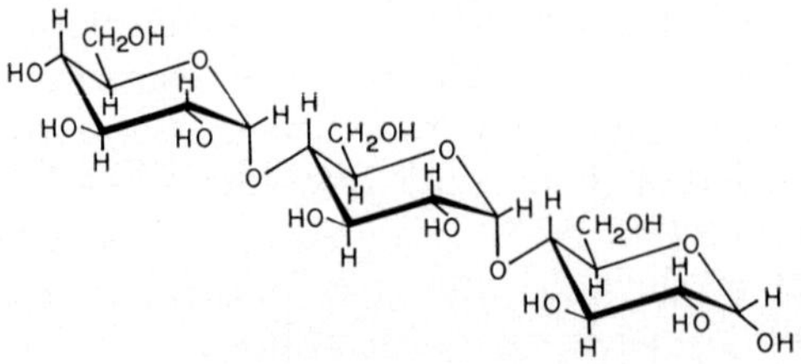

Maltotriose

O – α – D – Glucopyranosyl – (1→4) – O – α – D – glucopyranosyl –
(1→4) – D – glucopyranose

FIG. 12.   MALTOTRIOSE

by the action of enzymes or acid on starch is maltotriose (Fig. 12), a reducing trisaccharide which has not been crystallized but which can be characterized as a crystalline hendecaacetate (Wolfrom *et al.* 1949).

This substance is designated rationally as O-α-D-glucopyranosyl-($1 \rightarrow 4$)-O-α-D-glucopyranosyl-($1 \rightarrow 4$)-α-D-glucopyranose.

Another trisaccharide which is of natural occurrence is the highly crystalline, nonreducing raffinose, which is sucrose with a β-D-galactopyranosyl unit substituted on the C-6 hydroxyl group of the D-glucopyranose entity. This trisaccharide is obtainable as a by-product in the refining of beet sugar.

DISCUSSION

P. C. Markakis.—Is it the accepted convention to put two lines under the D and one line under the O?

M. L. Wolfrom.—Well, now, I do that because those are English printing instructions; a single underscore indicates that the letters should be set in italicized capitals; the doubled underscoring means small cap Roman type on the configurational symbols.

D. R. Lineback.—In your second slide you showed D-fructose in a $C1$ D conformation (Wolfrom "For the Pyranose"). Is this the preferred conformation with the 1-3 diaxial hydroxyl–hydroxymethyl interaction and the Reeves $\triangle 2$ effect?

M. L. Wolfrom.—Well, I am no authority on conformation. I'm not sure. Yes, I think it is because that's the one Lemieux put in when he wrote the sucrose formula conformation.

D. R. Lineback.—This has a 1-3 diaxial interaction between the hydroxyl on C-3 and the hydroxymethyl at C-1. It also has the Reeves $\triangle 2$ effect.

M. L. Wolfrom.—Well you know fructose is pretty unstable.

D. R. Lineback.—This is what I wondered concerning this particular conformation.

M. L. Wolfrom.—I think that's what it probably has. What you've got there is what fits into the crystal lattice. The crystal lattice doesn't particularly care about anything else if it fits. Also solubility, too, is concerned. That particular form is the least soluble of the many forms present in aqueous solution at the temperature at which fructose ordinarily crystallizes.

T. J. Schoch.—What is your preferred differentiation between the terms conformation and configuration? This seems to be quite smeared today.

M. L. Wolfrom.—It is very distinctly smeared by the general organic chemist who doesn't know anything about sugars. It was never in any confusion in the carbohydrate field. Configuration is the order of groups about each carbon center and conformation is the shape of the molecule.

W. Nagel.—Why didn't you discuss the uronic acids?

M. L. Wolfrom.—Well, perhaps I should have.

BIBLIOGRAPHY

ACHARD, F. C. 1809. The European Industrial Sugar Manufacture from Beet Root. Hinrichs, Leipzig. (German)

ARMSTRONG, E. F. 1924. The Simple Carbohydrates and the Glucosides. Longmans, Green and Co., London.

BATES, F. J., and ASSOCIATES. 1942. Polarimetry, Saccharimetry and the Sugars. Circ. Natl. Bur. Std. C440. U.S. Gov. Printing Office, Washington.

DREW, H. D. K., and HAWORTH, W. N. 1926. A critical study of ring structure in the sugar group. J. Chem. Soc., 2303–2310.

ELIEL, E. L., ALLINGER, N. L., ANGYAL, S. J., and MORRISON, G. A. 1965. Conformational Analysis. J. Wiley and Sons, New York.

FISCHER, E. 1891. On the configuration of grape sugar and its isomers. II. Ber. 24, 2683–2687. (German)

FISCHER, E. 1909. Investigations on Carbohydrates and Enzymes. Vol. 1. J. Springer, Berlin. (German)

FISCHER, E., and HIRSCHBERGER, J. 1888. On mannose. I. Ber. 21, 1805–1809. (German)

FRENCH, D., ROBYT, J. F., WEINTRAUB, M., and KNOCK, P. 1966. Separation of maltodextrins by charcoal chromatography. J. Chromatog. 24, 68–75.

GOEPP, R. M., Jr. 1943. Crystallization of sorbitol. U.S. Pat. 2,315,699.

HAWORTH, W. N., and LONG, C. W. 1927. The constitution of disaccharides. XII. Lactose. J. Chem. Soc., 544–548.

HIRST, E. L. 1926. The structure of the normal monosaccharides. IV. Glucose. J. Chem. Soc., 350–357.

HUDSON, C. S. 1909. The significance of certain numerical relations in the sugar group. J. Am. Chem. Soc. 31, 66–86.

HULYALKAR, R. K., JONES, J. K. N., and PERRY, M. B. 1965. The chemistry of D-apiose. II. The configuration of D-apiose in apiin. Can. J. Chem. 43, 2085–2091.

IPATIEFF, V. 1912. Catalytic reactions at high temperatures and pressures. XXV. Ber. 46, 3218–3226. (German)

ISBELL, H. S. 1944. Interpretation of some reactions in the carbohydrate field in terms of consecutive electron displacement. J. Res. Natl. Bur. Std. 32, 45–59.

KILIANI, H., and KLEEMANN, S. 1884. Transformation of gluconic acid into normal capronic acid or its lactone. Ber. 17, 1296–1303. (German)

LaFORGE, F. B. 1916. d-Mannoketoheptose, a new sugar from the avocado. J. Biol. Chem. 28, 511–522.

LeBEL, J.-A. 1874. On the relationship between the atomic formulas of organic molecules and their rotational powers in solution. Bull. Soc. Chim. France [2] 22, 337–347. (French)

LEVI, I., and PURVES, C. B. 1949. The structure and configuration of sucrose. Advan. Carbohydrate Chem. 4, 1–35.

LIEBIG, J. 1847. On the components of meat extracts. Ann. 62, 257–369. (German)

LIPPMANN, E. O. von. 1929. The History of Sugar. 2nd Edition. J. Springer, Berlin. (German)

NEUFELD, E. F., and HASSID, W. Z. 1963. Biosynthesis of saccharides from glycopyranosyl esters of nucleotides ("sugar nucleotides"). Advan. Carbohydrate Chem. 18, 309–356.

NEWKIRK, W. B. 1923. Crystallized grape sugar. U.S. Pat. 1,471,347.

REEVES, R. E. 1949. Cuprammonium glycoside complexes. III. The conformation of the D-glucopyranoside ring in solution. J. Am. Chem. Soc. 71, 215–217.

SCHMIDT, O. T. 1930. Structure and configuration of Apiose. On sugars with branched carbon chains. II. Ann. 483, 115–123. (German)

SEPHTON, H. H., and RICHTMYER, N. K. 1966. The isolation of D-erythro-L-galacto-nonulose from the avocado, together with its synthesis and proof of structure through reduction to D-arabino-D-manno-nonitol and D-arabino-D-gluco-nonitol. Carbohydrate Res. 2, 289–300.

STANĚK, J., CĚRNÝ, M., KOCOUREK, J., and PACÁK, J. 1963. The Monosaccharides. Academic Press, New York.

STANĚK, J., CĚRNÝ, M., and PACÁK, J. 1965. The Oligosaccharides. Publishing House of the Czechoslovak Academy of Sciences, Prague.

THOMPSON, A., WOLFROM, M. L., and QUINN, E. J. 1953. Acid reversion in relation to isomaltose as a starch hydrolytic product. J. Am. Chem. Soc. 75, 3003–3004.

VAN'T HOFF, J. H. 1874. On structural formulas in space. Arch. Neerl. Sci. (French)

VONGERICHTEN, E. 1900. On luteolin methyl ether as a hydrolytic product of a new glucoside of parsley. Ber. 3, 2334–2342. (German)

WOHL, A., and NEUBERG, C. 1900. On the characterization of glyceraldehyde. Ber. 33, 3095–3110. (German)

WOLFROM, M. L., and DACONS, J. C. 1952. The polymer-homologous series of oligosaccharides from cellulose. J. Am. Chem. Soc. 74, 5331–5333.

WOLFROM, M. L., GEORGES, L. W., and MILLER, I. L. 1949. Crystalline derivatives of isomaltose. J. Am. Chem. Soc. 71, 125–127.

WOLFROM, M. L., GEORGES, L. W., THOMPSON, A., and MILLER, I. L. 1949. Enzymic hydrolysis of amylopectin. Isolation of a crystalline trisaccharide hendecaacetate. J. Am. Chem. Soc. 71, 2873–2875.

WOLFROM, M. L., LEW, B. W., HALES, R. A., and GOEPP, R. M., JR. 1946. Sugar interconversion under reducing conditions. IV. D,L-glucitol. J. Am. Chem. Soc. 68, 2342–2343.

WOLFROM, M. L., THOMPSON, A., and BROWNSTEIN, A. M. 1958. Structures of isomaltose and gentiobiose. J. Am. Chem. Soc. 80, 2015–2018.

CHAPTER 3

Dexter French

Physical and Chemical Structure of Starch and Glycogen

INTRODUCTION

Starch is a principal constituent of many foods, and it constitutes not only a major energy source, but it is also essential to the gross structure, texture or consistency of many food preparations. Although food science has been limited in the past to the natural starches, there have now been developed, or are under development, many new starch types or starch modifications which will be of increasing importance in the years ahead.

In the limited space that we have available, it will be impossible to go into extensive details regarding all facets of starch and glycogen structure. Therefore, we will have to focus primarily on just a few details which are of principal significance in the utilization of starch as a food rather than, for example, its industrial applications. The same applies to glycogen. Traditionally starch and glycogen have been considered to be essentially the same from a *nutritional* standpoint. However, as we will see, there are major differences in the *physical* behavior of these substances which stem from seemingly minor differences in chemical structure.

Starch is not a single chemical substance, but a natural product which occurs as a reserve food in most green plants. Most starches consist of complex mixtures of *amylose* (straight chain) and *amylopectin* (branched chain or tree-like) molecules. While the notion of different fractions in starch goes back into the 19th century, it was not until the 1940's that practical methods for separating these fractions were worked out (Schoch 1941, 1942). Thus, much of the earlier work dealing with starch chemistry and structure was unreliable and had to be repeated.

Starch is one of nature's answers to the problem of storing substantial amounts of reserve food energy without large increases in osmotic pressure or bulk. Starch, being macromolecular and uncharged, has insignificant osmotic pressure. Furthermore, owing to its deposition in the form of dense granules, it is hydrated to only a trivial degree, in comparison with polysaccharides or sugars as they would exist in aqueous solutions in plant cells or sap. Starch granules are synthesized from sugars by the plants' metabolic enzymes and laid down more or less *in situ* in leaves, stems, fruits, seeds, roots, tubers, and even in pollen grains. When the plant has a need to mobilize its energy resources

26

(e.g., during germination of a seed) the starch granules are reconverted to sugars. This reconversion is again effected by a different set of metabolic enzymes. It is remarkable, and esthetically beautiful, that delicate control mechanisms in the plant are able to regulate the overall process

$$CO_2 + H_2O \xrightarrow{\text{sunlight}} \text{sugars} \rightleftharpoons \text{starch}$$
$$\longrightarrow \text{other plant constituents}$$

according to the plant's nutritional needs; at present very little is known of the details of the mechanism by which this regulation is achieved. It is as if silent messengers within the plant system inform the cells that certain enzymes must be produced, or that enzymes present must be activated or deactivated. We may confidently expect a far more sophisticated understanding of this regulation within a few years.

Inasmuch as these starch granules represent choice morsels of food energy, it is easy to understand that numerous predators (including man) have developed an appetite for starch, and digestive machinery to convert starch into sugars. As with digestion of many natural polymers, starch digestion usually involves action of two (or more) enzymes: the first a *solubilizing* or *liquefying* enzyme (usually an alpha amylase), to convert starch to water-soluble dextrins or oligosaccharides, and a *saccharifying* enzyme (usually an exoglucanase or endwise acting enzyme) to convert the intermediate products into glucose or maltose. The complete story of the enzymic digestion of starch is fascinating and studded with jewels of modern biochemistry; however, this story cannot be told here. Suffice it to say that a wide variety of amylases and glucosidases have evolved to handle the problem of starch digestion by various organisms. A few of these enzymes constitute the economically significant amylases; many are of a secret character known only to the organisms which elaborate them.

The polysaccharide *glycogen* is likewise a complicated mixture of closely-related, highly branched molecules, for which it is impossible to assign a definite structural formula except on a statistical basis. In the same way that starch constitutes a highly concentrated food reserve for plants, glycogen provides a food storage system for all forms of animal life (as well as for certain bacteria, yeasts, and even higher plants). The term *glycogen* was coined to connote the capacity for regenerating sugars, and it is this feature which makes glycogen not only a useful reserve food for the form of life which elaborates it, but also a readily utilizable food for predators. Like starch, the physical properties of glycogen appear to depend somewhat on its previous treatment. The old Pflueger method of isolating glycogen, involving heating with strong alkali, was originally regarded as producing "pure" glycogen, the major effect of the alkali simply being to digest away the protein and to dissociate the glycogen from other cell constituents. We now know that such treatment degrades glycogen rather seriously and destroys its native structure as judged by various physical techniques such as electron microscopy and the ultracentrifuge.

It is customarily thought that the digestion of starch and glycogen in foods parallels the degradation of the isolated polysaccharides when treated with various digestive enzymes. However, I am sure that you appreciate that this is a gross oversimplification which has been fostered by writers of textbooks simply to avoid the complications which we know occur when starch in foods is subjected to the normal digestion process.

The Simple, the Complex, and the Commonplace

At this point I wish to digress briefly for a discussion of the simple, the complex and the commonplace. Perhaps this can best be illustrated by an anecdote from several years ago when our undergraduate Foods and Nutrition majors were required to take a course in colloid chemistry. At that point in the course where the instructor was developing the concept of solution viscosity, he was talking about the classical system polystyrene–toluene, and the influence of molecular size and cross-linking on viscosity behavior. After about 20 min he noted the onset of apparent coma in most of the students, and asked the students what was wrong. One, less shy than the rest, said, "Well, this polystyrene in toluene is all right. But, why not talk about something *simple* like egg-white?" This student had failed to distinguish between the *simple* and the *commonplace*, and we are all guilty of this to a greater or lesser degree. A material such as starch is very commonplace, but it is by no means *simple*. It is my hope therefore that I may be pardoned if I gloss over some of the complexities of starch and glycogen in order to provide a more easily presentable story.

CHEMICAL STRUCTURE OF STARCH AND GLYCOGEN

The major outlines of starch and glycogen structure have been worked out 20 to 40 yr ago with the pioneering studies of such people as Haworth and Meyer. In fact, the Meyer picture of starch and glycogen is the one which is currently accepted, at least in its broad aspects. This picture shows starch and glycogen as continuing long chains of glucose units linked in a monotonous, repetitious pattern, in which each glucose unit is linked to the next by an α-1,4-glycosidic bond. These chains may range in length from just a few glucose units (perhaps 6) as in the A-chains of glycogen to many hundreds or even thousands of units in the long amylose molecules of starch (Fig. 13). In the amylopectin fraction of starch, and especially in glycogen, these "unit chains" are linked to one another in a tree-like pattern. The mode of linkage, originally established by methylation analysis, is one in which the reducing end of one unit chain is attached by an α-1,6-linkage to a glucose unit of another unit chain (Fig. 14). Such a "branching" as it is commonly called, could be either very uniform (as originally

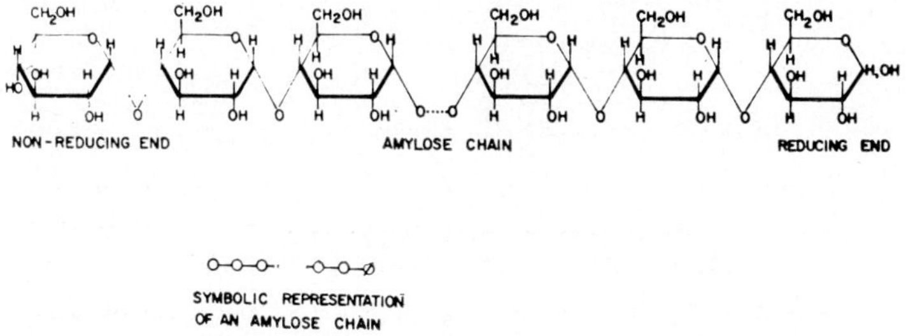

FIG. 13. STRUCTURE AND SYMBOLIC REPRESENTATION OF AMYLOSE

In the native amylose molecule there may be from a few hundred to 10,000 or more glucose units linked by α-1,4-glycosidic bonds.

depicted by Haworth, Staudinger, and others); it could be random as has been considered by Erlander and French (1956); or it could follow a pattern which is governed by (a) the rules of specificity of the enzymes which synthesize starch and glycogen and (b) steric factors. Most biochemists today accept the randomly branched or "tree" structures for glycogen (Fig. 15) and amylopectin (Fig. 16) as originally proposed by Meyer (Meyer and Bernfeld 1940; Meyer and Fuld 1941; Meyer 1943). The question of possible α-1 → 3-links in amylopectin or glycogen is currently unsettled. However, it is the author's opinion that such links either do not occur at all, or at least they are of extremely minor significance.

A major unknown factor in starch and glycogen structure has to deal with what we call its "fine structure." Up to now we know very little about the details of the spacing of branches in starch and glycogen, and in fact no one has devised adequate methods for getting at this

BRANCH POINT IN AMYLOPECTIN OR GLYCOGEN

SYMBOLIC REPRESENTATION OF A
BRANCH POINT IN AMYLOPECTIN OR GLYCOGEN

FIG. 14.   STRUCTURE AND SYMBOLIC REPRESENTATION OF AN
α-1,6-BRANCH POINT IN AMYLOPECTIN OR GLYCOGEN

The vertical arrow indicates an α-1,6-bond.

problem except in a few special cases as we will see later on.   Another
problem stems from the fact that these macromolecules are very easily

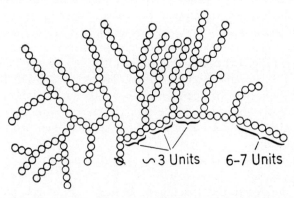

FIG. 15.   STRUCTURE OF GLYCOGEN, AS PROPOSED BY
MEYER (1943)

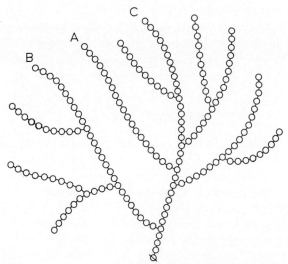

Adapted from Meyer and Bernfeld (1940)

FIG. 16. BRANCHING PATTERN OF AMYLOPECTIN, AS
PROPOSED BY MEYER

The A chains are those which are linked solely at the
reducing end by an α-1,6-link to another chain. Each
B chain is also linked at the reducing end by an α-1,6-
link to another chain, and in addition it is also linked
through one or more α-1,6-links to the reducing end or
ends of A or B chains. The C chain carries the reducing group of the molecule. This molecule contains
only 11 chains with 170 glucose units. Molecules of
native amylopectin range from several hundred to many
thousand glucose units in size. One should imagine 5
to 100 models such as the above joined together to
make a single amylopectin molecule.

degraded, so that we have little confidence that the "pure" polysaccharides which we work with in the laboratory are identical with these
substances as they occur in the original sources. A third factor is still
more difficult to resolve: this is the *distribution* of various structural
features (e.g., points of branching, chain lengths, molecular size, and
conformation) of the various molecules which constitute a sample under
study. Unless the substances involved have a *regular* structure, the
results of most types of experimental measurement give only average
values, with little or no evidence regarding statistical distribution of
these structural features. Therefore, it is essentially hopeless to obtain
a single structure formula for starch or glycogen, inasmuch as these

structural details can only be known on the basis of statistical distributions. Therefore, it is obvious that we need to develop techniques for getting at the *distributions* of structural elements in starch and glycogen. As we will see, there has already been a modest beginning in this direction; however, we still have a long way to go.

GLYCOGEN PARTICLES

The polysaccharide glycogen occurs in many cells as highly condensed "alpha particles" which are easily seen in the electron microscope. These particles may have dimensions of several hundred Å, and are easily recognized by their characteristic appearance. They may be isolated by simply grinding the tissue, centrifuging, and water washing. In this way, Orrell and Bueding (1958) have been able to obtain glycogen particles which are essentially protein-free, and which retain their characteristic appearance under the electron microscope. Molecular weights of such particles are in the range of 100,000,000-200,000,000 or more, and it is not certain whether these particles constitute single molecules or whether they represent some type of physical aggregation. The particles are easily broken down into much smaller particles by chemical, physical, or enzymatic treatment. Even such a mild procedure as heating a neutral aqueous suspension to boiling destroys the characteristic appearance. Therefore, it seems likely that these particles are composed of many molecules held together by hydrogen bonding, entangling (intergrowth), or linking through specific protein-polysaccharide interactions.

Glycogen isolated by homogenizing animal tissue and centrifuging in a density gradient contains bound enzymes: specifically phosphorylase and glycogen synthetase (Barber *et al.* 1967). There is no evidence for a membrane as such surrounding the glycogen particle. However, in view of the high affinity of glycogen for the enzymes of glycogen metabolism, it is likely that *in vivo* every spot on the particle surface is normally covered with an enzyme molecule. This association of glycogen with enzymes undoubtedly plays a determining role in the chemical and physical structure of glycogen. However, at present, we have very little understanding of this role (Fig. 17).

On breaking down the highly aggregated alpha particles, one obtains beta particles of somewhat lower molecular size. These particles still retain a characteristic appearance in the electron microscope and may well consist of individual molecules. In nature there appears to be a dynamic equilibrium between alpha and beta particles, especially in

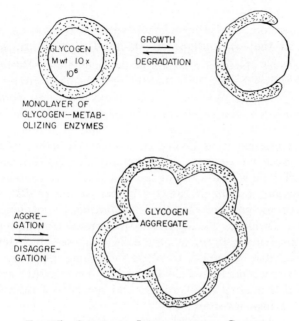

FIG. 17.   SCHEMATIC ILLUSTRATION OF GLYCOGEN
ALPHA AND BETA PARTICLES

The small "subunits" or beta particles during growth
may become fused together to give the larger alpha
particles. *In vivo* it is likely that each glycogen particle is completely covered by a layer of the enzymes
of glycogen metabolism, specifically phosphorylase, glycogen synthetase, branching, and debranching enzymes.

the liver where the alpha particles preponderate under conditions of
high nutritional level.   On the other hand, muscle tissue appears to
contain primarily the smaller beta particles; this may be related to the
well-known capacity of liver to store large quantities of glycogen in
comparison to muscle.   It might also be related to the very high level
of glycogen-metabolizing enzymes in muscle tissue.

Under the electron microscope one can see still smaller particles, the
gamma particles.   These gamma particles do not correspond to any
known chemical subunit of glycogen.   Judging from the size of these
particles as seen in the electron microscope (about 30-50 Å in diameter),
they would have molecular sizes in the range of perhaps 20,000 to
100,000.   There is no means of breaking down glycogen to particles of
this size which does not involve extensive chemical (or enzymatic)
degradation (Barber *et al.* 1965).

GLYCOGEN FINE STRUCTURE

Most of the work on the finer details of glycogen structure has been carried out using chemical and enzymatic techniques. By methylation analysis it has long been known that the unit chains of glycogen contain approximately 12 glucose units. This result has been confirmed by periodate oxidation; the exact value differs from one source or sample to another. In aqueous solutions, starch chains of this length are not capable of crystallizing or giving an appreciable iodine stain. Some samples of glycogen are capable of staining deeply; this was perhaps one of the first clues that some of the unit chains in glycogen must be substantially larger than 12 units. Larner *et al.* (1952) developed an enzymatic method for the stepwise breakdown of glycogen into "tiers." The results of this study clearly showed that glycogen has a tree-like branching pattern, as originally proposed by Meyer. One might expect that in these tree-type branching patterns, the main branches (B & C chains) would be longer than the side branches (A chains). Unfortunately these workers did not have a suitable method for getting at this question.

The most significant breakthrough in recent years has been the discovery and application of the enzyme *pullulanase* (Bender and Wallenfels 1961). This enzyme is capable of cleaving the branching linkages in starch and glycogen and at least in principle it should be capable of completely breaking down glycogen into its unit chains. As yet this enzyme has not been fully characterized, and we are not certain whether it is capable of cleaving all the branching linkages in complex polysaccharides. From various studies it is well-known that pullulanase can cleave low molecular weight oligosaccharides, when these contain a minimum of two maltose units linked α-1,6 (Fig. 18) (Abdullah *et al.* 1966; French and Abdullah 1966). With increasing size of the oligosaccharide chains, it appears that one goes through a maximum rate in the action of this enzyme. Virtually nothing is known about the action of pullulanase on very long chains. Thus it is possible that some of the longest branches in starch or glycogen may be immune to debranching by this enzyme. From studies on high molecular weight polysaccharides, it is clear that these macromolecules are at least partially resistant to pullulanase action; see Table 1 (Abdullah *et al.* 1966). Whether this resistance is the result of steric hindrance or the result of enzyme specificity is not known. Even with this limitation, pullulanase has an enormous potential in the further elucidation of the branching patterns of starch and glycogen. Some of the

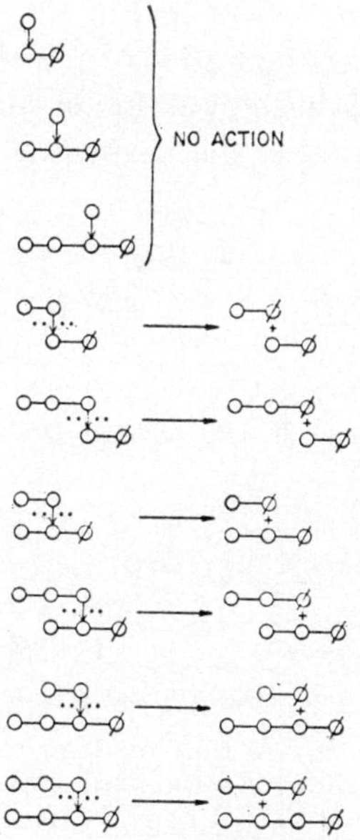

NO ACTION

From Abdullah and French (1968)

FIG. 18. ACTION OF PULLULANASE ON OLIGO-SACCHARIDES CONTAINING A SINGLE α-1,6-LINK

TABLE 1

EXTENTS OF DEBRANCHING OF AMYLOPECTIN AND GLYCOGEN BY R-ENZYME AND PULLULANASE AS JUDGED BY THE INCREASE IN DEGREE OF BETA-AMYLOLYSIS

Substrate	Beta-Amylase Alone	Conversion into Maltose			
		Successive Actions of Debranching Enzyme and Beta-Amylase		Simultaneous Actions of Debranching Enzyme and Beta-Amylase	
		R-Enzyme	Pullulanase	R-Enzyme	Pullulanase
	%	%	%	%	%
Waxy-maize amylopectin	52	64	92	101	99
Waxy-maize amylopectin } Beta-limit dextrin	0	73	99	...	103
Rabbit-liver glycogen	48	47	56	...	97
	40	..	..	40	...
Rabbit-liver glycogen } Beta-limit dextrin	0	0	39	...	...

Source: Abdullah *et al.* (1966).

results of recent studies are shown in Fig. 19. It is seen that glycogen contains some branching configurations in which branches are located on glucose units separated by only a single glucose unit. Such closely spaced branches can have either the "Haworth" or "Staudinger" arrange-

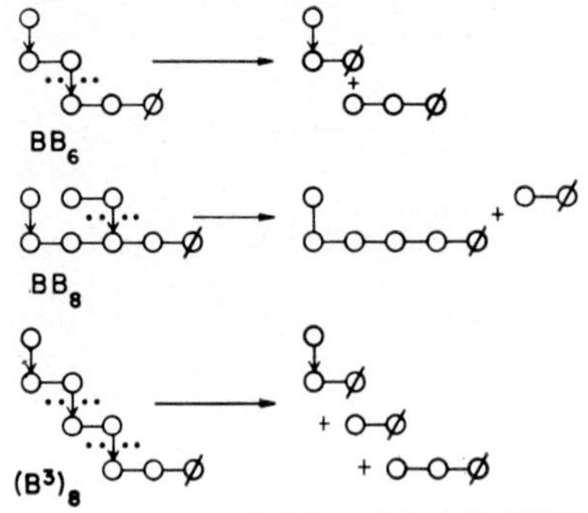

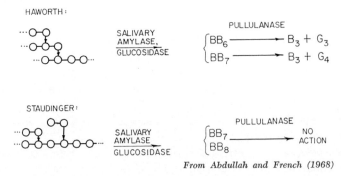

From Abdullah and French (1968)

FIG. 19. ACTION OF PULLULANASE
ON MULTIPLY-BRANCHED ALPHA
AMYLASE LIMIT DEXTRINS

ment as shown in Fig. 20. If some branches in glycogen are this closely spaced, it is obvious that others must be somewhat more widely spaced than the average values obtained in previous analyses. The occurrence of closely spaced branches also suggests that there may be

HAWORTH :

SALIVARY
AMYLASE,
GLUCOSIDASE

PULLULANASE

$BB_6 \longrightarrow B_3 + G_3$
$BB_7 \longrightarrow B_3 + G_4$

STAUDINGER :

SALIVARY
AMYLASE
GLUCOSIDASE

PULLULANASE

$BB_7 \longrightarrow$ NO ACTION
BB_8

From Abdullah and French (1968)

FIG. 20. TIGHTEST DOUBLY-BRANCHED STRUCTURES WHICH
OCCUR IN GLYCOGEN

statistical fluctuations in the density of branching which could give rise to pockets of dense branching. In fact such densely branched regions were identified as such by Heller and Schramm (1964), who showed that "macrodextrins" resulted from action of alpha-amylase (salivary or pancreatic) on amylopectin or glycogen (Fig. 21). Depending on the polysaccharide used, and the extent of the enzymatic treatment, the macrodextrins obtained vary in amount, molecular size, and resistance to further enzymatic hydrolysis. With shellfish glycogen, we have found that about 15% of the polysaccharide appears in a macrodextrin fraction which is essentially totally resistant to further degradation by salivary or pancreatic amylase. This material has an average molecular size of about 30 to 40 glucose units and an average

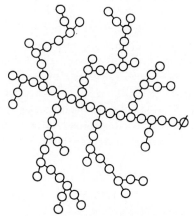

From Schramm (1968)

FIG. 21. GLYCOGEN "MACRODEXTRIN"
RESULTING FROM ACTION OF ALPHA
AMYLASE ON GLYCOGEN

chain length of only 4 glucose units. Verhue and French (1968) have carried out a very thorough structural study of this material. There is no doubt but that the branching pattern is very dense and irregular, as is suggested by Schramm (1968). It also appears the macrodextrins have a preponderance of Haworth branching, inasmuch as the A chains constitute less than half the total number of chains in the molecule. There is no evidence for branches separated by more than 2 glucose units; in fact, such a separation would hardly be permitted with an average chain length of only 4 units. Moreover we have shown that salivary and pancreatic amylases can readily attack the inner branch regions of starch chains where these branches are separated by three or more glucose units. Thus these macrodextrin structures must represent pockets of dense branching in the glycogen molecule. From the molecular size and yield of macrodextrin, one may readily calculate that there is one macrodextrin unit to each 50,000–100,000 molecular weight units of the original glycogen. It is tempting therefore to suggest that there may be some relationship of the macrodextrins to the gamma particles mentioned earlier. However, at the moment, this is pure speculation.

Although we know relatively few additional details of the inner branch structure of glycogen, it has been pretty well established that the outer branches average perhaps 6 to 8 glucose units in length. In a study of the specificity of "branching enzyme," Verhue and Hers (1966) showed that the minimum chain length which could be transferred was 6 glucose units. The minimum chain length of A chains obtained in phosphorylase degradation is four glucose units. Therefore, it is likely that the outer chains of glycogen are fairly uniform in length; for example, it is hardly likely that there would be many chains substantially longer than those required for branching enzyme action. Degradation of glycogen by bacterial amylase converts the outer chains primarily into maltohexaose and maltoheptaose (Robyt and French 1963). This again is evidence that these outer chains are more or less uniform in length and that they are accessible to the action of amylases.

AMYLOPECTIN FINE STRUCTURE

In spite of the great commercial interest in starch, there is relatively little evidence for the details of its fine structure. The Meyer structure depicted in Fig. 15 is generally accepted. In view of the relatively open structure, in comparison with glycogen, alpha amylases give only very small amounts of macrodextrins. Limited studies on doubly-branched oligosaccharides from waxy maize starch (Fig. 22) (Wild 1953; French 1960) show that these oligosaccharides are similar or identical to those produced from glycogen (Brammer and French 1968). Therefore, the branching enzymes of starch synthesis are capable of creating new branch points separated by no more than one glucose unit from previously formed branches.

With the availability of the enzyme *pullulanase*, it is inevitable that new researches will show further details of amylopectin structure. It is also likely that further studies on the specificity of the enzymes of starch biosynthesis will give important clues of the details of the branching patterns.

PHYSICAL STATES OF STARCH

With low molecular weight, nonaggregated organic substances, the physical behavior is almost completely independent of the source from

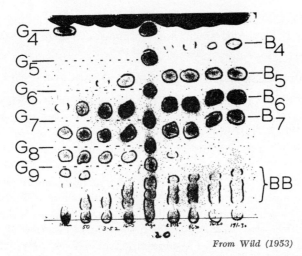

From Wild (1953)

FIG. 22. ACTION OF ALPHA AMYLASE ON AMYLOPECTIN
(WAXY MAIZE STARCH)

This is a 20-ascent chromatogram to show the formation
of branched oligosaccharides (central area of chromatogram). The center vertical row of spots is a control
series of linear starch oligosaccharides ("oligo"). G_4,
G_5, etc., represent maltotetraose, maltopentaose, etc. B_4,
B_5, etc., represent branched tetrasaccharide, branched
pentasaccharide, etc. Doubly-branched oligosaccharides
of low R_F form a band in the lower part of the chromatogram (BB). Samples were withdrawn from the
digest after 10, 50, 232, 965, 1965, 2810, 4580, and
10,290 min.

which the material was isolated, or the previous treatment and processing of the material. Thus the differences which may exist between
samples of ordinary sugar (sucrose) reflect minor amounts of impurities
which along with processing variations may influence crystal size and
form. However it is not difficult for the sugar molecules to "forget"
their past. By simply dissolving pure sucrose, from any source, and
of any crystal size or form, one will obtain solutions of identical
behavior.

With high molecular weight polymers, such as starch, the situation
is far different. In the first place, it is difficult if not impossible to
obtain a true solution of the molecules of starch, without degrading
these molecules. Secondly, for many food or industrial applications,
starch is much more useful as a high-viscosity paste or semigel than it
would be as a solution. Finally, neither the true solutions nor the
pastes or sols are stable, but tend to revert to an insoluble form—a

phenomenon called "retrogradation." Our understanding of these phenomena is very incomplete and unsophisticated, to say the least.

CONFORMATIONS OF STARCH CHAINS IN AQUEOUS SOLUTIONS

Interplay of several factors will influence the actual physical arrangement of starch chains in solution. Actually, these chains are not at all static, but under the influence of tireless Brownian motion they constantly twist, bend, vibrate and undergo every conceivable type of contortion.

In the simplest case, where we have very short starch chains (e.g., 4–6 glucose units in length), we can make (or compute) models which illustrate the behavior of molecules. To start with, we know fairly accurately the *most stable* arrangement ("conformation") of the D-glucose unit: this is the C-1 conformation in Reeves' designation. The bond lengths and angles have been very accurately determined by X-ray methods (Fig. 23) (Chu and Jeffrey 1967; Hybl *et al.* 1965). It might be thought that this information would suffice to determine the arrangements of starch chains in solution (Storey and Merrill 1958). Furthermore, the interactions of glucose units around glycosidic bonds (Fig. 24) have been subjected to computer analysis (Rao *et al.* 1967) as well as X-ray diffraction analysis (Marchessault and Sarko 1967). However, even if we know the most stable arrangement for a given pair of adjacent glucose units, the laws of chance (which are bound up with the laws of thermodynamics) tell us that the most stable arrangement is not the only one which occurs. This means that starch chains could not have a perfectly regular structure, but that there will be a certain amount of randomness (coiling). The same statistical arguments require that these irregularities are constantly changing in *position* as well. This gives a *dynamic* molecule.

A second factor influencing molecular conformation is the fact that in a nonrigid chain molecule, we can have *intramolecular* interactions, since the molecule can be twisted and looped, so that seemingly distant parts of the same molecule are brought into contact. These intramolecular contacts can be either regular (for example, as in helix formation) or very irregular (as in entangling and knotting). The longer the molecule, and the more flexible, the more likely that such interactions may occur.

A third factor of substantial significance influencing molecular conformation is the possible interaction with additives. With starch, it has long been realized that iodine and many other substances are capable

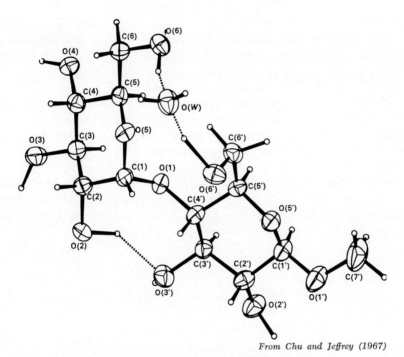

From Chu and Jeffrey (1967)

FIG. 23. STRUCTURE OF β-METHYL MALTOPYRANOSIDE

of stabilizing a regular helical structure. The degree to which such helices are present *before* addition of a complexing agent is uncertain (Rao and Foster 1963; Banks and Greenwood 1967). Many lines of evidence suggest that starch chains in solution have at most only a loose and imperfect helical arrangement, whereas in the iodine complex the helices are compact and may extend for hundreds of turns. In

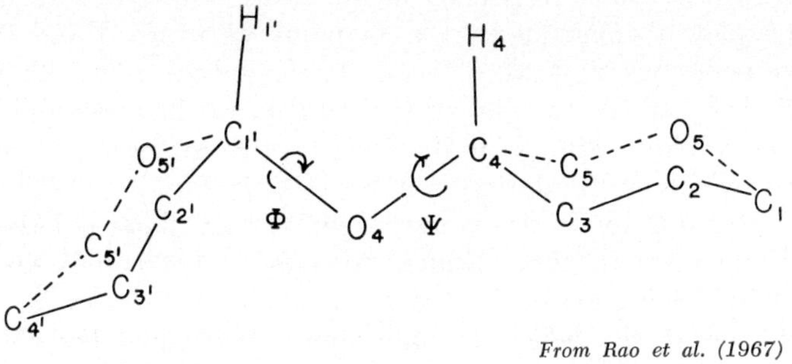

From Rao et al. (1967)

FIG. 24. ROTATIONAL ANGLES ABOUT GLYCOSIDIC BOND IN MALTOSE

principle, measurements of solution viscosity can give clues to the conformations of macromolecular solutes. In the case of starch chains, such measurements have been somewhat ambiguous. Holló and Szejtli (1958) have suggested that solutions of starch contain *segmented* helices (Fig. 25). In other work, they showed that the optimum chain length for iodine complex formation is about 120 glucose units. Beyond this length, amylose helices are less perfect and presumably bent or folded. The concept of an interrupted or segmented helix is an appealing one, either for the amylose complex or for amylose alone.

One type of arrangement, for which there is at present little evidence, is the "double helix." Such an arrangement was brought into prominence with nucleic acids; however it is equally possible for many other substances and has been shown in a β-1,3-linked xylan (Frei and Pres-

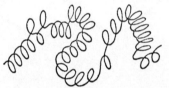

From Holló and Szejtli (1958)

FIG. 25. DEFORMED OR SEGMENTED
HELICES IN SOLUTIONS OF STARCH
CHAINS

ton 1964). Models of starch chains easily give such double helices, and it seems likely to the author that this arrangement will turn out to be significant in starch behavior.

STARCH PASTES, DISPERSION AND GELS

Because so many of the applications of starch require a prior pasting, gelatinization or other dispersion, and because the colloidal properties of starch play a dominant role in so many of its food and industrial uses, an enormous amount of work has revolved about the pasting properties of starch. In spite of the number of man-years of effort in this field, we are still painfully ignorant of the many complicated processes which occur during starch gelatinization.

When starch granules are warmed with water, they undergo several types of more-or-less well defined changes. At first, the dry starch granules *imbibe* a good deal of water—up to 40 or 50% of their weight —without loss of the typical granule appearance under the microscope. As the granules are warmed, they reach a critical temperature at which they suddenly swell to many times their original volume. This gelatinization is accompanied by a large and rapid increase in the suspension viscosity and increase in the transparency of the suspension. Although the starch chains become much more hydrated, they are still thoroughly entangled. The original granule retains its identity. Further heating to autoclave temperatures in excess water leads to a gradual solution of the macromolecular components and eventual complete breakdown of the granule structure. At lower temperatures, the granule breaks down much less readily. It is very subject to mechanical shear. The smaller amylose molecules may be leached out of the granule network, and this forms the basis for one of the original methods of starch fractionation. If swelling occurs in a limited amount of water (as is always the case in any industrial process or food material), swelling is limited by the amount of water available. Starch granules swelling in contact with each other will tend to adhere to each other, leading to an enormous structural viscosity, even rigidity (pastes or gels). As the amylose diffuses from one granule to the intergranule region, it too acts as a cement giving added structure to such pastes. On cooling, further starch-starch associations occur, primarily involving the "crystallization" of the longer starch chains. Storage of aqueous starch pastes or dispersions eventually leads to "retrogradation"—a phenomenon in which much of the hydration is lost, and the starch chains become even more difficult to hydrate than in the original granules.

During gelatinization of starch, there is a manyfold increase in the ease of digestion of the starch by enzymes. It is usually thought that starch digestion is confined to *cooked* starches; this is not strictly true inasmuch as the amylases are capable of acting at a very slow rate on raw starch. Breaking up the granule structure through swelling makes the starch chains available to *rapid* attack by amylases. As starch dispersions age and retrograde, they again become less readily attacked by enzymes. As Walker and Hope (1963) have shown, various raw starches differ greatly in their ease of attack by amylases. Potato starch is hardly attacked at all, while rice and corn starch are fairly readily attacked, especially by salivary and pancreatic amylases. Thus it is important to realize that there are many factors which influence the digestibility of starch. By the same token, these variations in response to enzymes or other agents provide a delicate means for elucidating the nature of starch in the various types of colloidal structure. As yet, this potential has scarcely been touched.

STARCH GRANULE STRUCTURE

The structure and behavior of starch granules represents the number one enigma in starch science. It is universally appreciated that granule behavior is of supreme importance in the reactions and applications of starch, and a great many man-years of effort have been spent trying to understand it. It is likely that the following factors play a role in dictating the structure of the starch granules from a given plant source. (1) Chemical constitution of the starch molecules: molecular size, degree of branching, branching patterns. (2) Specificity of the enzymes involved during starch synthesis. (3) The rate at which starch is laid down—availability of starch precursors. (4) Temperature and moisture relationships during granule development. (5) The number of granule nuclei per unit volume or mass of starch-synthesizing tissue ("crowding"). (6) Presence of noncarbohydrate "impurities," e.g., fats, proteins. (7) Physiological fluctuations in the plant such as day and night, hormone and other control mechanisms, plant nutrition. (8) "Age" of the granule. (9) Possible degradative effects of starch-metabolizing enzymes within the plant tissue. (10) Changes in granule structure during harvesting, storage, and processing.

This is undoubtedly not a complete list, so it is no wonder that the starch granule is not better understood. However, the following points are generally recognized features of starch which must be taken into account in any overall picture.

Crystal Structure

All common native starches (except "high amylose") give a well-defined X-ray diffraction pattern. Most cereal starches give an A-pattern; most tuber starches give a B-pattern, and there are intermediate forms called C-patterns. Possibly these crystallization types relate to temperature and moisture levels while the starch is being laid down; the B type of crystallization is generally favored by low temperature and abundant moisture. It is generally thought that the amylopectin fraction contributes primarily to the X-ray pattern inasmuch as waxy maize starch (which contains no amylose) gives an excellent pattern, while some high-amylose starches give only a very poor or amorphous type of pattern. Starch chains as short as 9 glucose units can give A patterns (French and Youngquist 1960). On extensive treatment of starch granules with cold aqueous acid, about half of the weight of the starch is degraded and dissolved out leaving a highly-crystalline, short-chain residue ("Nageli amylodextrin") which may retain the microscopic appearance, birefringence, and X-ray pattern of the original granules. On further examination, this residual material has been found to consist of relatively unbranched chains of average length of about 25 glucose units. It can be readily dissolved in hot water and recrystallized to give either A or B patterns. Thus the crystalline part of starch granules consists mainly of the more linear parts of amylopectin molecules—possibly the outer branches. The more highly branched parts of amylopectin are incapable of participating in a crystal structure; they are thus more amorphous and probably much more susceptible to the action of chemical reagents and enzymes.

Birefringence and Dichroism

Starch granules have a unique appearance under the polarizing microscope, resembling "spherocrystals" formed by crystals of many high polymers. From the sign of birefringence (*positive* with respect to the spherulite radius) one may infer that the orientation of chains within the crystalline birefringent areas is parallel to the radius, and thus essentially perpendicular to the surface of the growing granule. The birefringence remains undiminished during cold acid treatment, but it quickly disappears during gelatinization. Indeed, loss of birefringence is sometimes used as one index of starch swelling. Physical distortion or damage of starch granules is often accompanied by loss of crystallinity and increase in cold water solubility.

When starch granules are lightly stained with iodine, they show little or no dichroism—although dichroism can be easily observed with thin sections (e.g., 1 μ) (French and Outka 1968), with certain preparations of acid-treated granules, or with granules swollen by hot water or swelling agents (MacMasters 1953). This dichroism is obviously indicative of the orientation of the starch units which stain with iodine, primarily the amylose fraction. Clearly, the optical properties of starch have great potential for learning more about granule structure and behavior.

Electron Microscopy

The electron microscope represents the most powerful tool we have for examining structure and organization in the range below that of the optical microscope (less than 1 μ) and above that of X-ray, spectroscopic, and chemical methods (greater than 20 Å). Although resolution of electron microscopes has been vastly improved, so that now with suitable specimens 2 Å resolution can be attained, most biological materials including starch have too low a density of electrons to permit such fine details to be seen. Although most reported electron micrographs of starch granules show little or no structural detail, Buttrose (1960) and Mussulman and Wagoner (1968) have shown that by first acid treating the starch and then fixing and embedding it in a resin, one can prepare thin sections showing a good deal of structure. With corn starch, there are alternating regions of high and low density, presumably reflecting regions of greater and lesser crystallinity in the original granule (Fig. 26). These layers are about 1 μ apart, and thus at the limit of visibility in the optical microscope. Presumably the layers represent "growth rings"; such rings are easily visible with potato starch granules and can only be interpreted as successive layers of higher and lower density (refractivity).

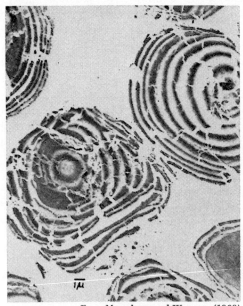

From Mussulman and Wagoner (1968)

FIG. 26. ELECTRON MICROGRAPH OF A 1000-Å SECTION OF ACID TREATED CORN STARCH GRANULE

HELICAL COMPLEXES

Starch in contact with many organic substances, particularly alcohols such as butyl or amyl alcohol, organic acids, and even hydrocarbons and halogenated hydrocarbons, tends to form insoluble complexes with the amylose constituent. This phenomenon was first developed by Schoch (1941, 1942) in his now-classical "butanol fractionation" method. In this, an autoclaved dilute starch sol is treated hot with butyl or amyl alcohol and allowed to cool slowly to room temperature. During cooling, the amylose constituent of the starch forms a crystalline complex which can readily be recovered and recrystallized. This was the first and is still the best method of obtaining good starch fractions.

Studies of the amylose complex by optical, X-ray diffraction and electron microscopy has shown convincingly that the insoluble complex consists of bundles of helices, packed in a more-or-less hexagonal fashion. With butanol and other linear aliphatic alcohols, the helices contain 6 glucose units per turn, while with the more bulky tertiary alcohols and many other complexing agents, there are 7 units per turn. Optical studies of the crystalline complexes has shown that the helix axis is directed along the thin direction of the crystal. In a recent paper, Manley (1964) has also shown that a single amylose molecule is folded back and forth many times in a given crystal. The thickness of these crystals is about 75 Å, or enough to accommodate about 60 glucose units.

For a variety of reasons, the author feels that starch helices in aqueous solution are relatively loose and expanded, with perhaps 7 or 8 glucose units per turn, and the turns not bonded very tightly to each other. With strong complexing agents, there is a strong interaction with the complexing agent leading to a tightening of the helix—down to a 6-unit helix if the size of the complexing agent will permit it, and formation of interturn hydrogen bonds. Such rigid, compact helices crystallize readily. Formation of a tight complex requires a good deal of dehydration of the amylose chains, which are therefore less soluble. Incidentally, the formation of compact helices would be expected to retard greatly action of enzymes or chemical reagents.

It is generally believed that the fatty materials which occur in cereal starches (e.g., about 0.6% in corn starch) are involved in helical complex formation with a part of the amylose fraction of the starch. Removal of these lipids by alcohol extraction has a very significant effect on starch pasting and other properties. It is also likely that when starch is cooked in the presence of food lipids, there may be strong interactions of helical starch with some of the lipid constituents. This might well influence the emulsification of the lipids, the digestion of the starch and many other factors of interest in food science.

STARCH FILMS AND FIBERS

Ability of starch to form strong films or fibers depends mainly on the amylose fraction. At present films of starch or starch modifications

are being investigated as possible food ingredients. The type of organization of amylose in the films and fibers depends on the conditions
during the preparation of the film and its subsequent handling. The
temperature at which the film is dried, the humidity, the rate of drying,
the concentration of the starch solution and the thickness which is
dried down, as well as the presence of additives or plasticizers, all
have an important effect on the type of crystallization and other organization in the films. Films range from highly crystalline to amorphous,
from translucent to clear, from strong and tough to brittle, depending
on these factors.

Optical and X-ray studies on amylose films and fibers has given us
much information about the arrangement and organization of starch
chains. Starch fibers have been prepared which show all the common
types of crystallization—A, B, and V—as well as complexes with various
salts and other agents. Since the interpretation of the X-ray patterns
of fibers is much simpler than that of starch powders, recent X-ray work
has focused strongly on films or fibers, and we can expect substantial
results in this area. The study by Sarko and Marchessault (1967) on
amylose acetate is a fine example of the type of information which can
be obtained from fibers. By stretching thin strips of amylose acetate
film at a high temperature, fibers were obtained which gave remarkably
detailed X-ray patterns. Although these patterns were still by no
means as perfect as those obtained from simple crystalline materials,
by an ingenious use of computer methods Sarko and Marchessault were
able to work out the detailed structure of amylose acetate (Fig. 28).
They found that the amylose chains were in the form of a left-handed
extended helix, with 4⅔ glucose units per turn. The resulting amylose
chain conformation is qualitatively similar to that obtained by Senti
and Witnauer (1952) and Jackobs et al. (1968) who found four units per
turn for the amylose-KBr complex (Fig. 27).

Another type of crystallization in films and fibers is the unique near-
tetragonal compact helix found in the dimethyl sulfoxide (DMSO)
complex. By stretching DMSO-containing films, French and Zobel

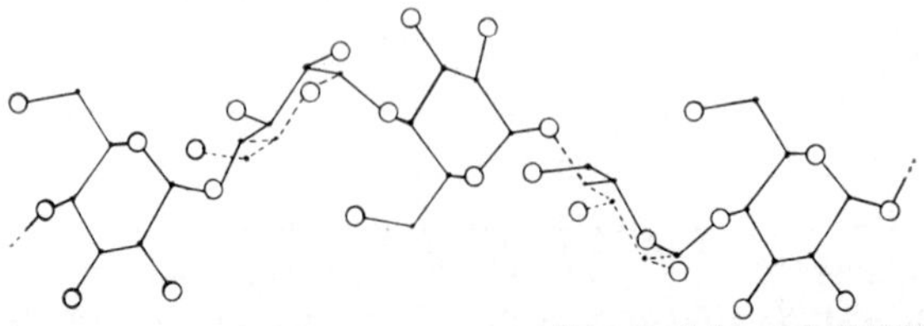

From Jackobs et al. (1968)

FIG. 27.   EXTENDED AMYLOSE CHAIN AS IN THE POTAS-
SIUM BROMIDE COMPLEX

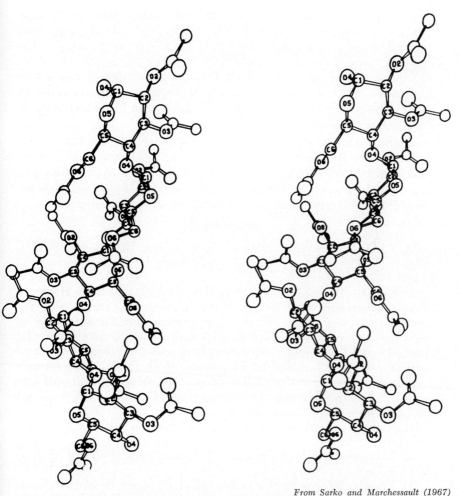

From Sarko and Marchessault (1967)

FIG. 28. STEREO DRAWING OF AMYLOSE ACETATE STRUCTURE

The figure may be viewed using a stereo viewer or by placing
a card between the left and right halves of the picture.

(1967) were able to obtain excellent fibers giving good X-ray patterns.
These indicated that the amylose is in a compact, 6-unit helix. In
contrast to the alcohol complexes, these helices pack in a tetragonal
pattern, apparently with the DMSO in the interstices *between* the
helices, rather than along the helix axes. Hinkle and Zobel (1968) and
Zobel *et al.* (1967) were also able to obtain fibers of V-amylose. Analysis of these fibers has not been completed, but presumably may show

whether the V-helix is right or left handed (Marchessault and Sarko 1967).

EPILOG

From the foregoing one may see that our knowledge of starch and glycogen is far from complete. Nevertheless, great progress is being made—new tools are beng invented, and there is an increasing body of basic information which is badly needed.

At the present time many persons are interested in the further details of the structures of starch and glycogen. I wish that I could look into the crystal ball and tell you how these studies will turn out. However, I believe that there are many surprises ahead and that we all will find this story to be a fascinating one as it is developed.

DISCUSSION

M. L. Wolfrom.—What evidence do you have that these beautiful oligosaccharide structures have not been made by glycosyl transfer in part?

D. French.—The evidence is not as clean as one might like. However, these oligosaccharides occur only if one treats amylopectin or glycogen with this enzyme system. If one treats amylose with this same type of enzyme, these compounds are not produced. There is no evidence for any α-1,4 $\rightarrow$ 1,6 transferase action occurring with the α amylases. Pazur and Okada (1968) as well as Greenwood et al. (1968) have recently suggested that some 1,4 $\rightarrow$ 1,4 transferase does occur under special conditions. With pullulanase, the enzyme which is used to cleave the 1,6 linkage, there would be no significant transfer under these conditions. The evidence is primarily on the basis of controls with amylose as a substrate.

K. J. Goering.—Did you indicate that pullulanase does not split all α-1,6-linkages?

D. French.—That is right. Pullulanase fails to act on single glucose "stubs," on extensively 1,6-linked polysaccharides such as dextran, and it is at least reduced in activity on densely branched glycogen.

K. J. Goering.—If you got a higher value of beta limit on the amylopectin using pullulanase, would this indicate longer branches?

D. French.—In the work by Abdullah et al. (1966) (Table 1), it is true that the beta limits were much higher for the pullulanase-treated amylopectin and its beta limit dextrin than for glycogen or its beta limit dextrin. In this work, digestion by pullulanase was not as extensive as it might have been. However, it was more than adequate to hydrolyze the normal substrate, namely pullulan, or the various branched oligosaccharides. We are not sure whether the failure of pullulanase is something inherent in the branching structure or whether we simply have a slight degree of resistance owing to steric hindrance.

K. J. Goering.—In our canarygrass starch we have suggestions of very long branches, and we're getting a higher limit than this with pullulanase and beta amylase.

D. French.—Is this joint action of beta amylase with pullulanase?

K. J. Goering.—Yes.

D. French.—Well, I wasn't trying to emphasize the joint action in Table 1, although it was on the slide. What I was referring to was the action of pullulanase itself followed by the action of β-amylase simply as an analytical device for measuring the degree of debranching.

K. J. Goering.—Has anyone had any success in making a starch granule in glass?

D. French.—Professor Sandstedt is the world's expert on this, and he has published his results in the monumental paper in *Cereal Science Today* called "Fifty Years of Progress in Starch Chemistry" (Sandstedt 1965). There are some beautiful pictures of what he calls starch spherocrystals. For further details you'll have to talk to Dr. Sandstedt.

R. W. Youngquist.—What is the present status of the so-called α-1,3 linkages that some people say are in starch and/or glycogen? Are they there, or aren't they there, or are they something else?

D. French.—We have no evidence for them. If they are present we don't think they are present to any high degree. Wolfrom and Thompson (1956, 1957) have isolated the α-1,3-linked disaccharide in extremely small amounts from both amylopectin and glycogen. However, Manners *et al.* (1965) have interpreted Wolfrom's data to indicate that some glycosyl transfer could have occurred through the acid catalyzed cleavage leading to these 1,3 links as artifacts.

R. W. Youngquist.—Could you estimate, is this one abnormality per 1,000 glucose units or one per 10,000?

D. French.—I refer this question to Professor Wolfrom.

M. L. Wolfrom.—Alva Thompson did this work, and he has also run a glucose control. He thought that it was definitely more than his control on glucose reversion. If there is any there it is very little.

D. French.—I would just like to add, if I might, that maybe the final answers will come if the biochemists are successful in showing that there is an enzyme that is capable of producing such linkages. If and when that time comes then I am going to be very willing to believe it. Up to now I am exceedingly skeptical.

W. A. Rock.—Is it known what happens to the structure in this gelatinous disarrangement of the starch granule, what structural changes seem to take place?

D. French.—My best guess is that the starch chains go from a relatively linear conformation to a relatively helical and hydrated conformation; see Maywald *et al.* (1968).

A. Neuberger.—May I ask just one question? There are obvious physical differences between starch and glycogen, and they must be based ultimately on certain definable chemical differences, i.e., differences in covalent bonding. Could Dr. French suggest what these differences might be?

D. French.—One obvious difference is that of chain length. The average chain length for glycogen is approximately 12 glucose units whereas that of starch is more like 20 to 25. The ability of starch chains to associate in a crystalline form depends very much on the chain length. The minimum chain length for crystallization of higher starch oligosaccharides is 9 glucose units (French and Youngquist 1960). This implies that the starch chain must be at least 9 glucose units in length in order to get a crystalline type of association. The unit chains in glycogen are longer than this, but most of them are branched so that the lengths of *straight* chains which are available for side by side association are presumably less than 9 glucose units. This accounts for the failure of glycogen to give the starch-type crystalline or semicrystalline aggregates.

BIBLIOGRAPHY

ABDULLAH, M., and FRENCH, D.   1968.   Structures of doubly-branched oligosaccharides from glycogen alpha amylase limit dextrins.   Unpublished studies.

ABDULLAH, M. *et al.*   1966.   The mechanism of carbohydrase action.   11. Pullulanase, an enzyme specific for the hydrolysis of alpha-1 → 6-bonds in amylaceous oligo- and polysaccharides.   Cereal Chem. *43*, 111–118.

BANKS, W., and GREENWOOD, C. T.   1967.   The hydrodynamic behavior of native amylose.   Conformation of Biopolymers, Vol. 2, 739–749.   Academic Press, London.

BARBER, A. A., HARRIS, W. W., and PADILLA, G. M.   1965.   Studies of native glycogen isolated from synchronized *Tetrahymena pyriformis* (HSM). J. Cell Biol. 27, 281–292.

BARBER, A. A., ORRELL, S. A., JR., and BUEDING, E.   1967.   Association of enzymes with rat liver glycogen isolated by rate-zonal centrifugation.   J. Biol. Chem. *242*, 4040–4044.

BENDER, H., and WALLENFELS, K.   1961.   Investigations on pullulanase. II.   Specific breakdown through a bacterial enzyme.   Biochem. Z. *334*, 79–95.   (German).

BRAMMER, G. L., and FRENCH, D.   1968.   Fine structure of amylopectin. Structure of doubly-branched oligosaccharides formed by action of porcine pancreatic amylase on waxy maize starch.   Unpublished research.

BUTTROSE, M. S.   1960.   Submicroscopic development and structure of starch granules in cereal endosperms.   J. Ultrastructure Res. *4*, 231–257.

CHU, S. S. C., and JEFFREY, G. A.   1967.   The crystal structure of methyl β-maltopyranoside.   Acta Cryst. *23*, 1038–1049.

ERLANDER, S. R., and FRENCH, D.   1956.   A statistical model for amylopectin and glycogen.   The condensation of A-R-B$_{f-1}$ units.   J. Polymer Sci. *20*, 7–28.

FREI, EVA, and PRESTON, R. D.   1964.   Non-cellulosic structural polysaccharides in algal cell walls. I. Xylan in siphoneous green algae. Proc. Roy. Soc. (London) B 160, 293–313.

FRENCH, D. 1960. Determination of starch structure by enzymes. Bull. Soc. Chim. Biol. 42, 1677–1689.

FRENCH, D., and ABDULLAH, M. 1966. Specificity of pullulanase. Biochem. J. 100, 6–7P.

FRENCH, D., and OUTKA, D. E. 1968. Dichroism of iodine-stained thin sections of starch granules. Unpublished research.

FRENCH, D., and YOUNGQUIST, R. W. 1960. Crystallization of starch oligosaccharides and oligosaccharide-iodine complexes. Abstracts Papers Am. Assoc. Cereal Chem. 45, 42–43.

FRENCH, A. D., and ZOBEL, H. F. 1967. X-ray diffraction of oriented amylose fibers. I. Amylose dimethyl sulfoxide complex. Biopolymers 5, 457–464.

GREENWOOD, C. T., MILNE, E. A., and ROSS, G. R. 1968. Observations on the possible occurrence of transferase activity in α-amylase preparations. Arch. Bioch. Bioph. 126, 244–248.

HELLER, J., and SCHRAMM, M. 1964. α-Amylase limit dextrins of high molecular weight obtained from glycogen. Biochim. Biophys. Acta. 81, 96–100.

HINKLE, M. E., and ZOBEL, H. F. 1968. X-ray diffraction of oriented amylose fibers. III. The structure of amylose-n-butanol complexes. Biopolymers, 6, 1119–1128.

HOLLÓ, J., and SZEJTLI, J. 1958. The structure of the amylose molecule in aqueous solution. Die Stärke 10, 49–52. (German).

HYBL, A., RUNDLE, R. E., and WILLIAMS, D. E. 1965. The crystal and molecular structure of the cyclohexaamylose-potassium acetate complex. J. Am. Chem. Soc. 87, 2779–2788.

JACKOBS, J. J., BUMB, R. R., and ZASLOW, B. 1968. Crystalline structure in oriented fibers of KBr-amylose. Biopolymers 6, 1659–1670.

LARNER, J., ILLINGWORTH, B., CORI, G. T., and CORI, C. F. 1952. Structure of glycogens and amylopectins. II. Analysis by stepwise enzymatic degradation. J. Biol. Chem. 199, 641–651.

MACMASTERS, M. M. 1953. The return of birefringence to gelatinized starch granules. Cereal Chem. 30, 63–65.

MANLEY, R. ST. J. 1964. Chain folding in amylose crystals. J. Polymer Sci. Part A 2, 4503–4515.

MANNERS, D. J., MERCER, G. A., and ROWE, J. J. M. 1965. α-1,4-glucosans. Part XIX. The action of acid on maltose and starch-type polysaccharides. J. Chem. Soc., 2150–2156.

MARCHESSAULT, R. H., and SARKO, A. 1967. X-ray structure of polysaccharides. Advan. Carbohydrate Chem. 22, 421–482.

MAYWALD, E. C., LEACH, H. W., and SCHOCH, T. J. 1968. Expansion and contraction of starch molecules in solution. I. Effects of temperature, pH, and alkali. Die Stärke 20, 189–197.

MEYER, K. H. 1943. The chemistry of glycogen. Advan. Enzymol., 3, 109–135.

MEYER, K. H., and BERNFELD, P. 1940. Research on starch. V. Amylopectin. Helv. Chim. Acta. 23, 875–885. (French).

MEYER, K. H., and FULD, M. 1941. Research on starch. XII. The

54 CARBOHYDRATES AND THEIR ROLES

arrangement of the glucose residues in glycogen. Helv. Chim. Acta. *24*, 375–378. (French).

MUSSULMAN, W. C., and WAGONER, J. A. 1968. Electron microscopy of unmodified and acid-modified corn starches. Cereal Chem. *45*, 162–171.

ORRELL, S. A., JR., and BUEDING, E. 1958. Sedimentation characteristics of glycogen. J. Am. Chem. Soc. *80*, 3800.

PAZUR, J. H., and OKADA, S. 1968. The isolation and mode of action of a bacterial glucanosyltransferase. J. Biol. Chem. *243*, 4732–4738.

RAO, V. S. R., and FOSTER, J. F. 1963. Studies of the conformation of amylose in solution. Biopolymers *1*, 527–544.

RAO, V. S. R., SUNDARARAJAN, P. R., RAMAKRISHNAN, C., and RAMACHANDRAN, G. N. 1967. Conformational studies of amylose. Conformation of Biopolymers, Vol. 2, 721–737. Academic Press, London.

ROBYT, J., and FRENCH, D. 1963. Action pattern and specificity of an amylase from *Bacillus subtilis*. Arch. Biochem. Biophys. *100*, 451–467.

SANDSTEDT, R. M. 1965. Fifty years of progress in starch chemistry. Cereal Sci. Today *10*, 305–315.

SARKO, A., and MARCHESSAULT, R. H. 1967. The crystalline structure of amylose triacetate. I. A stereochemical approach. J. Am. Chem. Soc. *89*, 6454–6462.

SCHOCH, T. J. 1941. Physical aspects of starch behavior. Cereal Chem. *18*, 121–128.

SCHOCH, T. J. 1942. Fractionation of starch by selective precipitation with butanol. J. Am. Chem. Soc. *64*, 2957–2961.

SCHRAMM, M. 1968. Interaction of α-amylase with glycogen and its hydrolysis products. Control of Glycogen Metabolism, 179–186. Universitetsforlaget, Oslo.

SENTI, F. R., and WITNAUER, L. P. 1952. X-ray diffraction studies of addition compounds of amylose with inorganic salts. J. Polymer Sci. *9*, 115–132.

STOREY, B. T., and MERRILL, E. W. 1958. The rheology of aqueous solutions of amylose and amylopectin with reference to molecular configuration and intermolecular association. J. Polymer Sci. *33*, 361–375.

VERHUE, W., and FRENCH, D. 1968. Structure of glycogen macrodextrin. Unpublished studies.

VERHUE, W., and HERS, H. G. 1966. A study of the reaction catalyzed by the liver branching enzyme. Biochem. J. *99*, 222–227.

WALKER, G. J., and HOPE, P. M. 1963. The action of some α-amylases on starch granules. Biochem. J. *86*, 452–462.

WILD, G. M. 1953. Action patterns of starch enzymes. Ph.D. Thesis, Iowa State Univ., Ames, Iowa.

WOLFROM, M. L., and THOMPSON, A. 1956. Occurrence of the $(1 \rightarrow 3)$ linkage in starches. J. Am. Chem. Soc. *78*, 4116–4117.

WOLFROM, M. L., and THOMPSON, A. 1957. Degradation of glycogen to isomaltotriose and nigerose. J. Am. Chem. Soc. *79*, 4212–4215.

WOLFROM, M. L., THOMPSON, A., and MOORE, R. H. 1963. Effect of maltose on acid reversion mixtures from D-glucose in relation to the fine structures of amylopectin and glycogen. Cereal Chem. *40*, 182–186.

ZOBEL, H. F., FRENCH, A. D., and HINKLE, M. E. 1967. X-ray diffraction of oriented amylose fibers. II. Structure of V amyloses. Biopolymers *5*, 837–845.

CHAPTER 4

Kyle Ward, Jr. | ## Cellulose

INTRODUCTION

Although cellulose is a constituent of almost every food from the plant kingdom, and in many cases a major constituent, it differs from most of the other carbohydrate materials in that it is normally not digested in the human organism and contributes little to nutrition, but merely serves as roughage in the elimination process. The reasons for its inertness in this respect can be directly related to its physical and chemical structure. This structure has been studied in great detail by a multitude of scientists all over the world and there is more information on the structure of cellulose than on many simpler carbohydrates. This is due to its prevalence and its utility; it is probably the most abundant and widely used organic compound on earth.

Widespread as it is, it usually occurs as part of a complex system of related polymers and its isolation unchanged is difficult, but when this has been accomplished the cellulose obtained seems to be chemically the same regardless of source. The molecular structure of such cellulose (the chemical structure) is a determining factor in the molecular arrangement (the physical structure) and therefore it will be discussed first.

CHEMICAL STRUCTURE

The first points to be made are that cellulose is a polysaccharide and that it is composed exclusively of glucose units. Neither of these statements has gone unchallenged and 40 yr ago they were hotly disputed, but today they are firmly established. It will be well to discuss the matter of the glucose composition first, for it has become a matter of definition. It was recommended by Purves (1954) that the word "cellulose" be restricted to that portion of the cell wall derived exclusively from glucose, which is the usage in this chapter. Wise (1958) has discussed the semantic problems that arise in referring to wood cellulose, for wood pulps contain varying amounts of mannose and xylans, but they can be removed until only trace amounts remain (Ward and Murray 1959; Rapson and Morbey 1959). The amount of non-glucose sugars drops with increasing purification, eventually to less than one unit per chain, which is very good evidence that the occurrence is traceable either to the extreme difficulty of purification of cellulose or to the possibility of isomerization during hydrolysis and identification of the glucose (Adams and Bishop 1955; Matsuzaki *et al.* 1959).

A somewhat similar state of confusion occurs in the matter of foods. The word "cellulose" is frequently used for what might more properly be called "crude fiber." In the talk today I am using the word "cellulose" in its narrow sense.

The question of the nature of the linkage naturally arises and the first evidence here is also obtained by breaking down the molecule. It is possible by partial hydrolysis to obtain a whole series of oligosaccharides, the simplest of which is cellobiose with two glucose units, while the others have the same linkage and differ only in the number of glucose units (Zechmeister and Toth 1931). Good yields of the acetates of the oligosaccharides can be obtained by acetolysis of cellulose (Dickey and Wolfrom 1949; Wolfrom *et al.* 1956). The structure of cellobiose is known (Haworth *et al.* 1927). It is a beta-anomer; hence we assume that cellulose is also beta-linked. This assumption is strengthened by the discovery that cellulose and its derivatives are split by β-enzymes and not by α-glucosidases; Whitaker (1963) has reviewed the criteria for characterizing cellulases. The β-structure is finally clinched by the x-ray evidence to be discussed later (Meyer and Misch 1937).

The type of ring within the sugar units in cellulose is also predicted by analogy with cellobiose. This, too, is confirmed by the x-ray pattern. The sugar units consist of D-glucopyranose rings.

One more major point can be seen in our scrutiny of cellobiose. The linkage joins the 4-carbon of one glucose to the 1-carbon of the next, which gives us a complete structure for cellulose. The 1–4 linkage has been amply confirmed by studying the cleavage products of completely methylated cellulose (see Purves's review in 1954), and it agrees with the x-ray diagrams.

Up to this point it has been tacitly assumed that all the linkages in the cellulose molecule are alike. Since the early 1930's, it has been clear, especially from the investigations of Freudenberg and Blomqvist (1935), that 99% of the linkages are identical. The kinetics of hydrolysis indicate, however, that an initial high rate of hydrolysis is followed by a lower rate and a number of chemists have interpreted this to mean that there are a few easily broken bonds representing a different and less stable linkage. It is, of course, possible that these occur but it is significant that, whenever a specific type of chemical weak link has been proposed, evidence showing its absence within the limits of our analytical methods has always been adduced and, as our analytical methods have grown more and more sensitive, the possible number of weak links has decreased. The idea of a physical rather than a chemical difference producing weak links seems more tenable; at least, it is more difficult to disprove, perhaps because it is less easy to formulate its exact nature. A brief review of cellulose hydrolysis with literature references to some individual hypotheses has been given by Ward and Morak (1964).

The kinetics of hydrolysis is very similar to that of many other cellulose reactions—a high rate at first and a slower one later. The most usual explanation for this is physical rather than chemical and is based on a greater accessibility of some molecules to the reagent than of others. This will be discussed in more detail later.

We can indicate the structure of cellulose as in Fig. 29 and if it were an "ordinary" chemical compound of low molecular weight, there would

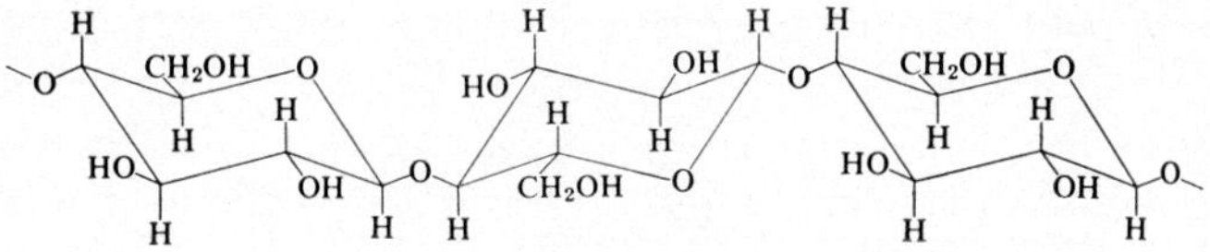

FIG. 29. SEGMENT OF A CELLULOSE MOLECULE

be little more to say of its chemical structure, but cellulose is a polymer; indeed, cellulose is one of the two natural polymers (the other is rubber, of course) which inspired such giants of the thirties and forties as Staudinger, Meyer and Mark to develop the science of polymer chemistry. Because cellulose is a polymer, there remain two unanswered questions—the length of the molecule and the nature of the end-groups.

As a matter of history, a great many of our methods of determining molecular weights of polymers were developed in order to know the molecular weight of cellulose, but it is unnecessary for us to go into this in great detail. It can be summarized, and perhaps oversimplified, by saying that, as our methods of molecular weight determination have improved and, particularly, as our methods of cellulose isolation and purification have become milder and more efficient, our estimates of the molecular weight have increased. In one man's lifetime, they have increased by four orders of magnitude, if we go back to the association theories (see the review by Purves 1954). Perhaps cellulose in its native state may sometimes be larger, but Goring and Timell (1962) and Marx-Figini and Penzel (1965) found degrees of polymerization above 10,000, which means a molecular weight above 1,620,000 since the weight of an anhydroglucose unit is 162.

One must be careful not to give the impression that cellulose has a definite and specific molecular weight. Cellulose is a polymer of high molecular weight, a long linear chain of glucose units; it represents, not a single molecular species, but an entire homologous series. No one knows how uniform the molecular weight distribution of native cellulose may be, but the distribution becomes broader with some very mild treatments; it is possible that native cellulose may have a fairly narrow distribution. Recent figures by Marx-Figini and Schulz (1963) show 3 maxima in the distribution curve at 11,500, 5,500 and 1,500 DP (degree of polymerization) for samples of very mildly treated cotton cellulose. They believe that the low DP material is morphologically different and that most cotton cellulose is contained in the very large, fairly uniform fraction of 11,500 DP.

In a chain of over 10,000 glucose units, the exact nature of the terminal units may seem to be relatively unimportant, and indeed it is,

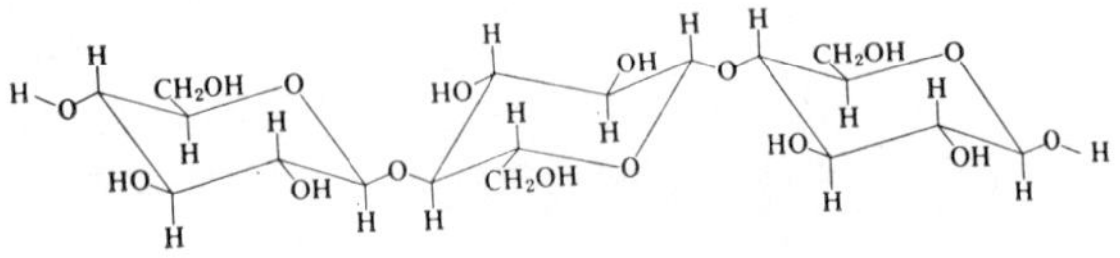

FIG. 30. CELLOTRIOSE

but it should be considered if one wishes to have a complete picture. There is no evidence that the terminal units in cellulose are necessarily much different from those in the middle of the chain. Cellotriose is a valid model compound for cellulose, then, especially as regards the terminal units (Fig. 30). If we look at cellotriose, we see that the two groups are not identical; one has the 4-hydroxyl free and the other the 1-hydroxyl.

The former end-group is nonreducing and has been used to determine the molecular weight of cellulose. If cellulose is completely methylated and then completely methanolyzed with methanolic hydrogen chloride, one obtains a mixture of methylated methyl glucosides. Every unit produces a trimethyl methyl glucoside but one; the nonreducing end-group produces a tetramethyl methyl glucoside and from the content of this substance one can calculate the average length of the chain. The method has only historical significance today (1954 review by Sookne and Harris).

The reducing end-group has also been used to measure molecular weights in several ways, but it does not seem to be unambiguous. One reason for this is that this group can easily be oxidized to a gluconic acid, at least in part. Since the isolation and purification of cellulose is frequently carried out in the presence of oxidants, most celluloses contain hemiacetals on some molecules and gluconic acids on others. Oxidation of the nonreducing end-group may also occur, but is much less common (1954 review by Sookne and Harris).

PHYSICAL STRUCTURE

In the preceding exposition, practically nothing has been said about the physical structure. The gross features are well-known; the details of fine structure are controversial, but it would be well to discuss those points that bear on the properties of the material.

Reviewing the information accumulated over the last century, one can see that the physical structure, i.e., the arrangement of the molecules, is a direct result of the structure of these molecules. As we have seen, the cellulose molecule is thread-like, several thousand times as long as it is broad, and studded with hydroxyl groups. Long linear polymers tend to align themselves to form bundles, especially where strong lateral forces can exist to stabilize such groups of molecules; in cellulose, the hydroxyls create such forces as hydrogen bonds between adjacent molecules which effectively "zipper" them together. Such bundles of molecules are called fibrils.

The microscope and the electron microscope conclusively demonstrate the existence of such a fibrillar structure. The names for the fibrillar subdivisions of the cellulosic fiber differ from one investigator to another and the dimensions which have been reported do not agree either. In this situation, some investigators have concluded that the subdivisions are simply irregular fragments of a more or less homogeneous structure. As the limit of resolution for the electron microscope has decreased, so has the size of the "ultimate" structure. However, studies on plant cells in the early stages of growth (that is, cellulose in the process of genesis) and the use of milder preparative techniques have made the evidence for the existence of a regular, preformed unit more convincing. A few years ago this unit was considered to be about 70 Å thick; a number of recent measurements indicate values of about 30 Å. Many investigators have found these units to be somewhat flattened. Since aggregation to larger bundles is very likely and further subdivision cannot be completely ruled out, the lack of agreement as to shape and size of these units (micellar strings, elementary fibrils, microfibrils) is not surprising. Since the cellulose molecule is about 5 Å across, however, there is not room for much more subdivision. Good reviews of the subject are available by Frey-Wyssling (1959) and Warwicker *et al.* (1966).

The molecules in these fibrillar bundles are laid down with sufficient regularity to diffract x-rays as crystals do. There are polymorphic forms but the pattern for all native cellulose from any source, with a few possible exceptions among marine plants, is the same (Cellulose I). The arrangement in this crystal lattice is such that the molecules run longitudinally and parallel along the fibrillar structure, although the fibrils are usually at an angle to the fiber axis. There seem to be several conformations of chains that will satisfy most of the details of the x-ray and infrared spectra of cellulose and none that is completely acceptable. There is a good review by Ellefsen *et al.* (1964). Chain folding, like that observed in some synthetic polymers, has been proposed by Tønnesen and Ellefsen (1960), Battista and Smith (1962), Dolmetsch and Dolmetsch (1962), and Manley (1964), but there is no direct evidence for it in native cellulose. The very high degree of order in the microfibril can be adequately explained by the Meyer-Misch structure (1937), although it does not satisfactorily account for *all* the x-ray and infrared spectra. The model has been modified by Liang and Marchessault (1959) and shows the proposed hydrogen bonds present in native cellulose (Fig. 31).

The order just mentioned is not perfect, however; the x-ray pattern shows a good deal of scatter, presumably from disordered regions. When this fact is considered along with the phenomenon of accessibility, a great many detailed explanations of the nature of order and disorder in the structure of fibrous cellulose have been advanced. The fringe micellar theory was in agreement with most of the observed phenomena (Fig. 32); according to most modifications of this theory, very small ordered regions alternated with even smaller disordered regions along the microfibrils. Individual molecules extended through several such regions. More recently, as one comes to accept a smaller microfibril, the correspondingly high surface will account adequately for both the x-ray scatter and the readily accessible, rapidly reacting portion of the material. In this case, all cellulose chains are in high order and the microfibril itself is essentially crystalline with a few crystalline defects. The early model by Meyer and van der Wyk (Fig. 33) was not very different from this. This subject has been admirably reviewed a few months ago by Warwicker *et al.* (1966), a review which covers a great deal more than the rather limited title.

The question of order and disorder is not merely academic and I

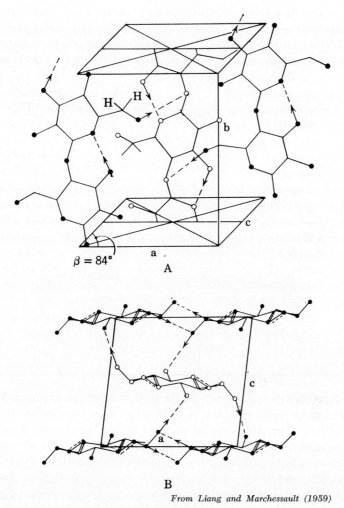

B

From Liang and Marchessault (1959)

Fig. 31.   Arrangement of Cellulose Molecules in Native
Cellulose

should like to say a little more about it before I close.   The sharpness
of the peaks in the x-ray diffraction diagrams is highest for fibers like
cotton and ramie, less for wood pulps, and lowest for regenerated celluloses like the rayons.   Moreover, it varies from one sample to another.
A similar state exists in regard to density and, indeed, samples are
usually ranked in the same order by x-ray methods and by density
methods.   Even more important to chemists interested in cellulose re-

actions the rank is generally the same as that obtained by studying the accessibility of the celluloses in question to small molecules, either in sorption or reaction. This correspondence between chemical behavior and physical structure is strong enough that we are forced to

From Mark (1940)

FIG. 32. FRINGED MICELLES AFTER MARK

devise an explanation in terms of the molecular arrangement of the molecules. Either the fringe micellar or the disordered surface accounts for most accessibility phenomena, however. The individual still faces a choice in regard to physical structure. There is little argument about chemical structure.

From Meyer and van der Wyk (1941)

FIG. 33. MEYER'S CONCEPT OF MOLECULAR ARRANGEMENT IN CELLULOSE

INERTNESS IN THE DIGESTIVE TRACT

The inertness of cellulose in foods to the digestive enzymes is probably related to both chemical and physical structure but particularly to the latter. The high degree of order, the crystallinity, if you will, of native cellulose is associated with a level of hydrogen-bonding so high that even water does not penetrate the crystal lattice. The accessibility to larger molecules like enzymes is simply too low for action to proceed without some means of opening up the structure. Cellulose preparations or derivatives without such structure (swollen cellulose, water-soluble cellulose ethers like carboxymethylcellulose, never-dried cotton fibers from unopened bolls) are split easily by cellulases.

More detail about this important subject should be given to a group interested in carbohydrates in food. Let us first look more closely at the acid hydrolysis of cellulose. If cellulose is heated with moderately strong acid, it loses its cohesiveness and becomes a friable mass which is shown by the electron microscope to consist of bundles of very tiny needles (Rånby and Ribi 1950; Morehead 1950). They are not completely uniform in size, but nearly so. Chemically they are still cellulose, but of lower molecular length. This chain length corresponds closely to the length of the microcrystallite and, like it, is rather uniform; after it once reaches the so-called "leveling-off degree of polymerization," it is affected very little by further hydrolysis (Battista 1950). How one pictures the course of this hydrolysis will depend upon how he pictures the fine structure discussed earlier. If he believes in the fringed micelle, he will envision the acid eating away the fringes. If he believes in folded chains, he will envision scission at the folds. If he believes in a crystalline microfibril, he will envision the development of transverse cleavage at crystal defects. In all three cases, however, uniform subunits are formed. This particular form of hydrolyzed cellulose is called microcrystalline cellulose (Battista and Smith 1962) and consists of particles of colloidal dimensions.

Microcrystalline cellulose, like native cotton or wood fiber, is not digested by the human organism (Tusing et al. 1964). In fact, it is marketed as "Avicel" and has been recommended for use as a nonnutritive additive for foods where too many calories are to be avoided (Trauberman 1961). When ingested, it passes through the human digestive tract and, like the cellulose from which it was made, appears in the stool apparently unchanged. The polarizing microscope does not show any deterioration of the particle. This does not mean that cellulose is not subject to enzymolysis. It simply means that the proper enzymes are not present. As a matter of fact, in most cases where cellulose is utilized by animals, like cows or termites, the breakdown is usually brought about by microbial flora which produce cellulolytic enzymes.

Even if we have the proper cellulase under the proper conditions, accessibility is of great importance in the process of cellulose breakdown (Walseth 1952). As in the case of acid hydrolysis, the susceptibility to enzyme hydrolysis is inversely related to the degree of order (Reese et al. 1957) (Table 2). Enzymolysis of cellulose, again like

TABLE 2
ENZYMIC HYDROLYSIS OF VARIOUS CELLULOSES

| Sample[1] | Crystallinity, % | Hydrolysis Loss, % | Relative Rates of Enzymolysis | |
			Expt. 1	Expt. 2
A	90	4	1.0	1.0
B	70	6.5	1.5	1.3
C	40–50	7	2.1	1.9
D	–	8.9	4.4	5.0

[1] Sample A is kiered cotton and sample B slack mercerized cotton. Samples C and D are decrystallized with ethylamine; C is cellulose I, and D a mixture of I and III.

TABLE 3

TABLE 3

EFFECT OF ENZYMOLYSIS ON CHAIN LENGTH

| | Degree of Polymerization | | |
Samples	Before Enzymolysis	After Enzymolysis	% of D. P. Retained
A Cotton	4970	4200	85
B Mercerized	5040	3040	60
C Decrystallized	4670	3100	66
D Decrystallized	3920	1630	42
AH Hydrocellulose from A	225	227	100
BH Hydrocellulose from B	138	145	104
CH Hydrocellulose from C	133	128	96
DH Hydrocellulose from D	112	104	93

acid hydrolysis, produces a residue with a "leveling-off degree of poly-merization" which is however larger than that produced by acid. If the enzyme reacts on hydrocelluloses produced by acid hydrolysis, the D.P. is not altered (Table 3). In fact, the residue looks very much like the original material before enzymolysis (Fig. 34).

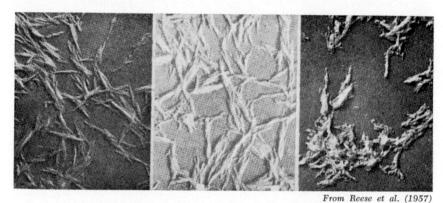

From Reese et al. (1957)

FIG. 34. HYDROCELLULOSE BEFORE AND AFTER ENZYMOLYSIS

Central figure is after enzymolysis.

CELLULOLYTIC ENZYMES

It is rather daring for someone so little experienced in the field as I to discuss enzymes, even if he is stressing the substrate rather than the enzyme. There are marked differences between enzymes from different sources; I shall neglect this aspect and refer those interested to the proceedings of a symposium sponsored jointly by the American Chemical Society and the Army Research Office (Reese 1963). There are probably several different enzymes in most cellulase systems, although systems consisting of single enzymes have been reported by some investigators (Whitaker *et al.* 1963; Pal and Ghosh 1965).

Two experimental facts stand out. First, many organisms flourish on soluble or swollen forms of cellulose, but will not grow on fibers of native cellulose. Second, many other organisms attack cellulose, but relatively few isolated enzymes do this: the breakdown of cellulose, moreover, usually occurs in the immediate neighborhood of the organism. Gascoigne and Gascoigne (1960) have reviewed the subject, as has Cowling (1963), the latter also pointing out some evidence supporting enzymatic action at some distance from the fungal hyphae in the attack on wood. For a long time it was believed that cellulolytic action could only take place in the presence of the microorganism; however, breakdown by cell-free enzymes are now well-known (Selby *et al.* 1963; Mandels and Reese 1964). The localized attack and the resistance of native cellulose to enzymolysis are probably due to the morphology and hydrogen-bonding discussed earlier. This also accounts for the fact that relatively few cellulolytic organisms attack the particularly well-ordered cotton fibers.

In 1950, Reese, Siu, and Levinson postulated an enzyme which is able to separate linear chains from a substrate of crystalline cellulose. This enzyme, which they designated as "C_1" is not produced by most organisms. The isolated molecules of cellulose are then cleaved by a system of "Cx" enzymes. Many such "Cx" enzymes have since been described. Figure thirty-five is taken from a more recent paper by Mandels and Reese (1964) and it indicates that the last stage in the enzymatic pro-

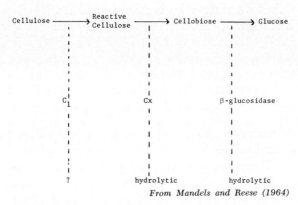

From Mandels and Reese (1964)

FIG. 35. REESE'S SCHEME FOR ENZYMOLYSIS OF NATIVE CELLULOSE

duction of glucose from cellulose is a cellobiase, a β-glucosidase. This is a hydrolytic enzyme like the "Cx" enzymes, but can easily be distinguished from them.

It is the "C_1" enzyme whch makes it possible for the "Cx" enzymes to hydrolyze the individual cellulose molecules (Mandels and Reese 1964), and in its absence neither cotton nor the highly ordered microcrystalline cellulose are attacked appreciably by enzymes. Fractionation of the extracellular enzymes of *Trichoderma viride* gives several enzymes, one of which consists essentially only of "C_1"; it attacks cotton but has only a trace of activity for carboxymethylcellulose, which is readily hydrolyzed by "Cx". The activity of "C_1" on cotton decreases greatly with increasing purity and a marked synergism occurs when "C_1" and "Cx" act together (Gilligan and Reese 1954; Selby and Maitland 1967A, 1967B). This synergism is less when the two factors act sequentially, in either order.

It is not yet clear how the "C_1" enzyme functions. Siu (1954) jokingly called it a "hydrogen-bondase:" this describes, but does not explain the action. If this action is actually the separation of single macromolecules as originally proposed by Reese *et al.* (1950), this necessarily involves breaking hydrogen bonds and hydrating the hydroxyls, but the detailed mechanism is still only a matter of hypothesis. Reese (1956) has proposed that it splits a few bonds in the fiber of some different, but unspecified type, perhaps in the primary wall or the winding layer. He and Gilligan (1954) have also considered its possible relationship to the swelling factor of Marsh *et al.* (1953); swollen cotton fibers are hydrolyzed by "Cx" enzymes, but, as Reese says, we are still not sure how the C_1 enzyme functions to produce swelling if it does. Finally it may be sorbed very strongly, breaking up the original particle into smaller ones. Halliwell (1965, 1966) observed the breakdown of fibers into smaller and smaller fibers. King (1966) observed a fragmentation of microcrystalline cellulose into progressively smaller units, but considers the action with this substrate to be different. Since sorption and fragmentation involve formation and breaking of hydrogen bonds, Siu's "hydrogen-bondase" still comes to mind. However, in this case this first step in the enzymolysis of native cotton may not be enzymatic at all; Liu and King (1967) have proposed that it is effected by a protein which is not an enzyme.

Once the crystal structure is destroyed, either by the "C_1" system or by some nonenzymic treatment, the "Cx" system commences to function. There seem to be several "Cx" components in most cellulase systems, but whether they are actually produced by the organism or whether they are subsequent modification of a single enzyme so produced is still an unsettled question. Most of these "Cx" cellulases appear to attack the cellulose chain at random (Gilligan and Reese 1954; Nisizawa *et al.* 1963; Enger and Sleeper 1965), although a few instances have been cited indicating end-attack (Storvick and King 1960; King 1963). Besides the "C₁" and "Cx" enzymes, there is a cellobiase which converts the lower oligosaccharides to glucose (Reese and Levinson 1952).

SUMMARY

What I have said can be summed up in very few words; perhaps I should have said it in fewer words to start with. Cellulose is a β-1,4-glucan of considerable size; the macromolecules are held together with hydrogen bonds to form a highly ordered fibrillar structure. It is this molecular arrangement which makes it relatively inert to the digestive process and which causes it to play an almost exclusively structural role in foods, as in the rest of nature.

DISCUSSION

T. J. Schoch.—The question comes up as to the mechanism of synthesis of such materials as cellulose and granular starch where you have a solid surface. The problem resolves to this: in each case you are putting glucose units on to a polymer which itself consists of glucose units. It is an additive process: one after the other these glucose units get tied in to the solid surface. Now, how does this happen? Do you form a starchy material in the substrate surrounding the surface of the solid or cellulose phase and then deposit it onto the crystalline lattice? Or do you add successive glucose units to the solid surface itself? Now this has considerable connotations, for example as to whether the chain can fold. If you are adding a preformed polymer which may not be a complete cellulose molecule or a complete linear starch molecule from a starch solution surrounding that solid surface, then it may fold in the process of going in. If you are adding single glucose units at a time to a solid surface, then it is difficult to see how you can get any folding because the rest of the surface would be growing up right along with the single chain that you are synthesizing. Has anything been done on this? Has there been any thinking as to how that glucose unit gets into the cellulose fibril?

K. Ward.—There has been a good deal of thinking about it, but no very detailed hypothesis has come out. Colvin in his studies of bacterial cellulose has given some pretty interesting evidence that the microfibril grows as a unit, not as a single molecule but as a unit of approximately

30-50 angstroms across. In this case it is pretty hard to see how there can be a chain folding, on the other hand, Manley's picture of a possible spiral arrangement of a ribbon of cellulose formed by folding would permit growth, essentially of the microfibril, because the glucose is being added at the end of the ribbon, but first at one point of the periphery of the tip and then another. Manley describes a helix like an unwound soda straw in which the chain fold is produced by a single enzyme. I am inclined to follow Colvin's work.

J. L. Hickson.—Dr. Schoch, you have made an assumption that if molecules lie down on a surface that is more or less perfect that they will lie down in a perfect position. This does not bear out in fact; in such things as sucrose crystallization where there is no chance of polymer formation and yet the creation of imperfections in the inclusion of syrup is a very well demonstrated phenomenon. I don't think that you have to assume that, if molecules are laying down one at a time, they march up like Prussian soldiers.

S. M. Cantor.—I saw some figures recently on the prospective growth of disposable materials in the U.S. Market. The term disposables includes not just packaging but also clothes, and hospital linens. This projection is striking. So now, what attention is being paid to the problem of disposal utilization, perhaps moving from the structural to the storage type structure for potential feed use?

K. Ward.—Attention to utilization of disposables of this sort is recent and not as extensive as our efforts to get rid of processing wastes. Stream pollution is essentially a disposal problem and the attention paid to this is pretty large. This is more a matter of waste disposal and not necessarily of utilization of the waste. If we can get rid of it we will be satisfied. If we can use it as a result, so much the better.

E. S. Gordon.—Your comment concerning the use of cellulose by other organisms fascinates me. I understand that the termite doesn't have the capability of digesting cellulose, but there is a protozoon living in its intestinal tract that does have cellulase. This organism has a symbiotic relationship with the termite. Therefore it derives energy from the digestion of cellulose. Is this true? Do you have any comments about that?

K. Ward.—This is true. I recently saw a paper by Friedemann and co-workers in the Journal of Nutrition in which he showed that by putting cellulose regularly into the diet of a series of subjects (human), he increased the amount of cellulose that was digested. He was also increasing the population of cellulolytic bacteria that lived inside the human beings that were being studied. So apparently human beings too could "learn" to digest cellulose if it were absolutely necessary. This same article pointed out that cellulose is considered inert in the digestive process, but if you look over the literature, there is a large number of articles which point out that from 15 to 30, and in one case 80% of the crude fiber is actually digested. Now crude fiber is not cellulose entirely, but nevertheless he pointed out that digestibility is not zero.

H. W. Schultz.—I recall some studies using gamma radiation for cellulose degradation for use by ruminants. Would you care to comment?

K. Ward.—Radiation introduces a large number of carbonyl groups into the cellulose chain. These are points of action biologically as well as chemically. One might expect this to be more easily digested. There is probably a difference in types of carbonyl in this type of irradiation, but I really do not know enough to comment at length.

J. L. Hickson.—We do know a great deal about the nutrition of monogastric animals such as swine or poultry. It has been demonstrated that use of a purified, completely digestible diet not including cellulose results in feed inefficiency. Apparently some fiber is necessary for utilization of the feed product; the digestive tract seems to desire this.

BIBLIOGRAPHY

ADAMS, G. A., and BISHOP, C. T. 1955. Polysaccharides associated with alpha-cellulose. Tappi 38, 672–676.

BATTISTA, O. A. 1950. Hydrolysis and crystallization of cellulose. Ind. Eng. Chem. 42, 502–507.

BATTISTA, O. A., and SMITH, P. A. 1962. Microcrystalline cellulose. Ind. Eng. Chem. 54, No. 9, 20–29.

COWLING, E. B. 1963. Structural features of cellulose that influence its susceptibility to enzymatic hydrolysis. In Advances in Enzymic Hydrolysis of Cellulose and Related Materials, E. T. Reese (Editor). Macmillan Co., New York.

DICKEY, E. E., and WOLFROM, M. L. 1949. A polymer-homologous series of sugar acetates from the acetolysis of cellulose. J. Am. Chem. Soc. 71, 825–828.

DOLMETSCH, H., and DOLMETSCH, H. 1962. Evidence for a chainfolding in cellulose molecules. Kolloid-Z. 185, 106–119. (German).

ELLEFSEN, Ø., KRINGSTAD, K., and TØNNESON, B. A. 1964. Structure of cellulose as judged by x-ray methods. Norsk Skogind. 18, 419–429.

ENGER, M. D., and SLEEPER, B. P. 1965. Multiple cellulase system from Streptomyces antibioticus. J. Bacteriol. 89, 23–27.

FREUDENBERG, K., and BLOMQVIST, G. 1935. The hydrolysis of cellulose and its oligosaccharides. Ber. deut. chem. Ges. 68, 2070–2082. (German).

FREY-WYSSLING, A. 1959. The vegetable cell structure. Springer-Verlag, Berlin. (German).

GASCOIGNE, J. A., and GASCOIGNE, M. M. 1960. Biological Degradation of Cellulose. Butterworth, London.

GILLIGAN, W., and REESE, E. T. 1954. Evidence for multiple components in microbial cellulases. Can. J. Microbiol. 1, 90–107.

GORING, D. A. I., and TIMELL, T. E. 1962. Molecular weight of native celluloses. Tappi 45, 454–460.

HALLIWELL, G. 1965. Hydrolysis of fibrous cotton and reprecipitated cellulose by cellulolytic enzymes from soil micro-organisms. Biochem. J. 95, 270–281.

HALLIWELL, G. 1966. Solubilization of native and derived forms of cellulose by cell-free microbial enzymes. Biochem. J. 100, 315–320.

HAWORTH, W. N., LONG, C. W., and PLANT, J. H. G. 1927. The constitution of the disaccharides. XVI. Cellobiose. J. Chem. Soc., 2809–2814.

KING, K. W. 1963. Endwise degradation of cellulose. In Advances in Enzymic Hydrolysis of Cellulose and Related Materials, E. T. Reese (Editor). Macmillan Co., New York.

KING, K. W. 1966. Enzymic degradation of crystalline hydrocellulose. Biochem. Biophys. Res. Commun. 24, 295–298.

LIANG, C. Y., and MARCHESSAULT, R. H. 1959. Infrared spectra of crystalline polysaccharides. I. Hydrogen bonds in native celluloses. J. Polymer Sci. 37, 385–395.

LIU, T. H., and KING, K. W. 1967. Fragmentation during enzymic degradation of cellulose. Arch. Biochem. Biophys. 120, 462–464.

MANDELS, M., and REESE, E. T. 1964. Fungal cellulases and the microbial decomposition of cellulosic fabric. Develop. Ind. Microbiol. 5, 5–20.

MANLEY, R. ST. JOHN. 1964. Fine structure of native cellulose microfibrils. Nature 204, 1155–1157.

MARK, H. 1940. Intermicellar hole and tube system in fiber structure. J. Phys. Chem. 44, 764–788.

MARSH, P. B., BOLLENBACHER, K., BUTLER, M. L., and GUTHRIE, L. R. 1953. "S factor," a microbial enzyme which increases the swelling of cotton in alkali. Textile Res. J. 23, 878–888.

MARX-FIGINI, M., and PENZEL, E. 1965. Absolute molecular weight and molecular weight distribution in natural wood cellulose. Makromol. Chem. 87, 307–315. (German)

MARX-FIGINI, M., and SCHULZ, G. V. 1963. New research on mass and mass distribution of β glucosidic chains in natural cellulose. Makromol. Chem. 62, 49–65. (German)

MATSUZAKI, K., WARD, K., JR., and MURRAY, M. 1959. Mannose in cellulose and pulps. Tappi 42, No. 6, 474–476; No. 9, 128A–129A.

MEYER, K. H., and MISCH, L. 1937. Positions of the atoms in the new spatial model of cellulose. Helv. Chim. Acta 20, 232–244. (French)

MEYER, K. H., and VAN DER WYK, A. J. A. 1941. About the fine structure of the cellulose molecule. Z Elektrochem. 47, 353–360. (German)

MOREHEAD, F. F. 1950. Ultrasonic disintegration of cellulose fibers before and after acid hydrolysis. Textile Res. J. 20, 549–553.

NISIZAWA, K., HASHIMOTO, Y., and SHIBATA, Y. 1963. Specificities of some cellulases of the "random" type. In Advances in Enzymatic Hydrolysis of Cellulose and Related Materials, E. T. Reese (Editor). Macmillan Co., New York.

PAL, P. N., and GHOSH, B. L. 1965. Isolation, purification, and properties of cellulases from Aspergillus terreus and Penicillium variabile. Can. J. Biochem. 43, 81–90.

PURVES, C. B. 1954. Chemical nature of cellulose and its derivatives. A. Historical survey. B. Chain structure. In Cellulose and Cellulose

Derivatives, E. Ott, H. M. Spurlin, and M. W. Grafflin (Editors). Interscience Publishers, New York.

RÅNBY, B. G., and RIBI, E. 1950. On the fine structure of cellulose. Experientia. 6, 12–14.

RAPSON, W. H., and MORBEY, G. K. 1959. Highly purified cellulose from wood. Tappi 42, No. 2, 125–130.

REESE, E. T. 1956. Enzymatic Hydrolysis of Cellulose. Appl. Microbiol. 4, 39–45.

REESE, E. T. (Editor). 1963. Advances in Enzymic Hydrolysis of Cellulose and Related Materials. Macmillan Co., New York.

REESE, E. T., and GILLIGAN, W. 1954. Swelling factor in cellulose hydrolysis. Textile Res. J. 24, 663–669.

REESE, E. T., and LEVINSON, H. S. 1952. Comparative study of the breakdown of cellulose by microorganisms. Physiol. Plantarum 5, 345–366.

REESE, E. T., SEGAL, L., and TRIPP, V. W. 1957. Effect of cellulase on the degree of polymerization of cellulose and hydrocellulose. Textile Res. J. 27, 626–632.

REESE, E. T., SIU, R. G. H., and LEVINSON, H. S. 1950. Biological degradation of soluble cellulose derivatives and its relationship to the mechanism of cellulose hydrolysis. J. Bacteriol. 59, 485–497.

SELBY, K., and MAITLAND, C. C. 1967A. Components of Trichoderma viride cellulase. Arch. Biochem. Biophys. 118, 254–257.

SELBY, K., and MAITLAND, C. C. 1967B. Cellulase of Trichoderma viride. Biochem. J. 104, 716–724.

SELBY, K., MAITLAND, C. C., and THOMPSON, K. V. A. 1963. Degradation of cotton cellulose by the extracellular cellulase of Myrothecium verrucaria. Biochem. J. 88, 288–296.

SIU, R. G. H. 1954. Microbial degradation. In Cellulose and Cellulose Derivatives, E. Ott, H. M. Spurlin, and M. W. Grafflin (Editors). Interscience Publishers, New York.

SOOKNE, A. M., and HARRIS, M. 1954. End groups. In Cellulose and Cellulose Derivatives, E. Ott, H. M. Spurlin, and M. W. Grafflin (Editors). Interscience Publishers, New York.

STORVICK, W. O., and KING, K. W. 1960. Complexity and mode of action of the cellulase system of Cellvibrio gilvus. J. Biol. Chem. 235, 303–307.

TØNNESEN, B. A., and ELLEFSEN, Ø. 1960. Chain folding—a possibility to be considered in connection with the cellulose molecule? Norsk Skogind. 14, 266–269.

TRAUBERMAN, L. 1961. Crystalline cellulose: versatile new food ingredient. Food Eng. 33, No. 8, 44–47.

TUSING, T. W., PAYNTER, O. E., and BATTISTA, O. A. 1964. Birefringence of plant fibrous cellulose and microcrystalline cellulose in human stools freezer-stored immediately after evacuation. J. Agr. Food Chem. 12, 284–287.

WALSETH, C. S. 1952. Influence of the fine structure of cellulose on the action of cellulases. Tappi 35, 233–238.

WARD, K. JR., and MORAK, A. J. 1964. Reactions of cellulose. B.1 Hydrolysis. In Chemical Reactions of Polymers, E. M. Fettes (Editor). Interscience Publishers, New York.

WARD, K. JR., and MURRAY, M. L.   1959.   Alkaline extraction of spruce
  pulps. Tappi *42*, 17–20.
WARWICKER, J. O., JEFFRIES, R., COLBRAN, R. L., and ROBINSON, R. N.   1966.
  Review of Literature on the Effect of Caustic Soda and other Swelling
  Agents on the Fine Structure of Cotton.   Cotton Silk and Manmade Fibres
  Res. Assoc., Manchester.
WHITAKER, D. R.   1963.   Criteria for characterizing cellulases.   *In* Advances
  in Enzymic Hydrolysis of Cellulose and Related Materials, E. T. Reese
  (Editor).   Macmillan Co., New York.
WHITAKER, D. R., HANSON, K. R., and DATTA, P. K.   1963.   Improved pro-
  cedures for preparation and characterization of Myrothecium cellulase.
  Can. J. Biochem. Physiol. *41*, 667–670, 671–676, 697–705.
WISE, L. E.   1958.   What is wood cellulose—a semantic dilemma.   Tappi *41*,
  No. 9, 14A–22A.
WOLFROM, M. L., DACONS, J. C., and FIELDS, D. L.   1956.   The cellodextrins:
  preparation and properties.   Tappi *39*, 803–806.
ZECHMEISTER, L., and TOTH, G.   1931.   On the hydrolysis of cellulose and
  its products of hydrolysis.   Ber. Deut. Chem. Ges. *64*, 854–870.   (German)

CHAPTER 5

Roy L. Whistler | Pectins and Gums

INTRODUCTION

This title might be listed also as "Industrial Gums of Value in the Food Industry." It will be the intent of the review to present a brief summary of the industrial gums now used or of immediate potential usefulness to the food industry. Although the review will not include cellulose or starch and their derivatives, it must be emphasized that these polysaccharides and especially their improved derivatives are a major and continually growing segment of industrial gums devoted to food applications.

Naturally occurring food gums are polysaccharide hydrocolloids derived from land plants, seaweeds or microorganisms. Those from land plants are obtained as exudates or are extracted from plant tissues especially fruits or seeds. Some can be obtained from seed endosperm by simple dry milling.

Limitations on the length of this review preclude more than a brief discussion of a few of the gums used extensively, or of potential use, in the food industry.

PLANT EXTRACTIVE GUMS

These are gums which may be obtained by extraction of plant tissue. The oldest and still most important one in the food industry is pectin. Hemicelluloses, extractable by alkaline solution from the cell walls of all land plants, have a potential use as low cost gums of the arabic type.

Pectin

Pectin use continues to advance even at its present (1968) price of $2.20 per pound. Nearly 10 million pounds are produced annually in the United States with a slightly lesser production in Europe. Pectins are found universally in land plants where they occur principally in the intercellular, or middle lamella region and to a much smaller extent in the primary cell wall. Due to this intercellular occurrence, pectins greatly influence the texture of plants and particularly the texture of fruits and vegetables. A few examples of the percentages present in various plants are shown in Table 4.

Principal industrial sources in the United States are apple pomace and citrus peel with increasing amounts being produced from the latter. In Sweden and Russia pectin is produced from sugar beet pulp and in Germany, Bulgaria and Romania from sunflower heads.

Reviews on the chemistry (Whistler and Smart 1953; Worth 1967) and industry (Bender 1959) of pectin have appeared.

TABLE 4

PECTIN CONTENT OF SEVERAL PLANT TISSUES

Tissue	% Pectin
Potato	2.5
Tomato	3
Apple	5–7
Apple pomace	15–20
Carrot	10
Sunflower heads	25
Sugar beet pulp	15–20
Citrus albedo	30–35

Pectin has long been considered a triad (see for example Whistler and Smart 1953) of a $1 \rightarrow 4$-α-D-galacturonan in the partial methyl ester form, a branched L-arabinan and a $1 \rightarrow 4$-β-D-galactan. However, much work has been done and is still being done to refine knowledge of the fine structures of these glycans and to establish the manner of their interrelation.

TABLE 5

ACETYL CONTENT OF SEVERAL PECTINS

Pectin	% Acetyl
Cherry	0.18
Raspberry	0.25
Citrus	0.24
Strawberry	1.43
Sugar beet	2.50

Most pectins contain 9–12% ester methoxyl but range down to 0.2% for strawberry pectin. Acetyl groups are present in some pectins (McComb and McCready 1957) as illustrated in Table 5. Specific optical rotations to the sodium D-line are usually + 230 to + 250° but may be as low as + 216°. The highest specific rotation appears associated with the highest D-galacturonic acid content.

Since isolated pectins are a triad of at least three glycans, it was long understood that hydrolysis would yield D-galacturonic acid, D-galactose and L-arabinose. With the advent of chromatography, hydrolyzates were found to contain L-rhamnose (Aspinall and Canas-Rodriguez 1958; Aspinall and Fanshawe 1961; Carrao 1954; Coleman *et al.* 1955; Barrett and Northcote 1965), often D-xylose (Aspinall and Canas-Rodriguez 1958; Carrao 1954; Coleman *et al.* 1955; McCready and Gee 1960; Barrett and Northcote 1965), and sometimes L-fucose (Aspinall and Fanshawe 1961; Barrett and Northcote 1965), D-glucose (Aspinall and Canas-Rodriguez 1958), 2-*O*-methyl-L-fucose and 2-*O*-methyl-D-xylose (Aspinall and Canas-Rodriguez 1958; Aspinall and Fanshawe 1961; Barrett and North-cote 1965). It is likely that pectins from all sources contain D-galacturonic acid, D-galactose, L-arabinose, and L-rhamnose. Barrett and Northcote (1965) report that apple pectinic acid contains 87% D-galacturonic acid, 9.3% L-arabinose, 1.4% D-galactose, 1.2% L-rhamnose, 0.9% D-xylose, and traces of fucose, 2-*O*-methylfucose and 2-*O*-methylxylose.

Separation of the members of the pectin triad has not been easy, and some work has not been reproducible. In fact, only in recent years have reproducible homoglycans been obtained. Galacturonans, hydrolyzing only to D-galacturonic acid, have been obtained. One is obtained from the pectin removed from the bark of amabiles fir (Bhattacharjee and Timell 1965). For separation the pectin is passed down a cation exchange column in the acid form and the eluent ultracentrifuged. This results in precipitation of a pure galacturonan, while the supernatant contains D-galacturonic acid and neutral sugars. Structural analysis shows the polymer as a linear arrangement of $1 \rightarrow 4$-α-D-galacturonic acid units (Fig. 36). Another pure galacturonan is obtained from sunflower heads (Bishop 1955; Zitko and Bishop 1965, 1966) by fractionation of the deesterified pectin from 1% aqueous solution through addition of 2M aqueous sodium acetate. Methylation analysis and periodate oxidation of the carboxyl-reduced pectic acid gives data expected for a linear $1 \rightarrow 4$ linked α-D-galacturonic acid polymer. A second galacturonan separated by sodium acetate fractionation appears to contain covalently bonded neutral sugars.

Pectins and pectic materials have been separated on diethylaminoethylcellulose columns. In one separation absorbed pectic material was eluted from the column with a phosphate buffer at pH 6.1 to produce an arabinan and later by elution of the column with caustic solution an acidic polysaccharide was produced (Neukom *et al.* 1960). However, the pectic material had the unusually high arabinose content of 50%. Only 14% of the pectic material was eluted with buffer with the remainder appearing with the acidic polymer, which itself was a mixture of at least 3 substances. Sugar beet arabinan comes through the column with the pectic acid, which indicates that in this pectic material the arabinan is covalently bound to the galacturonan. Arabinan, galactan, and galacturonan have been said to be separated on a similar exchange column by using dilute acetate buffer at pH 5 (Hatanaka and Ozawa 1966).

Pectic material from white mustard, on extraction with 75% ethanol, yields an arabinan which is separable on a diethylaminoethylcellulose column. Better purification is obtained by precipitation with cetyltri-

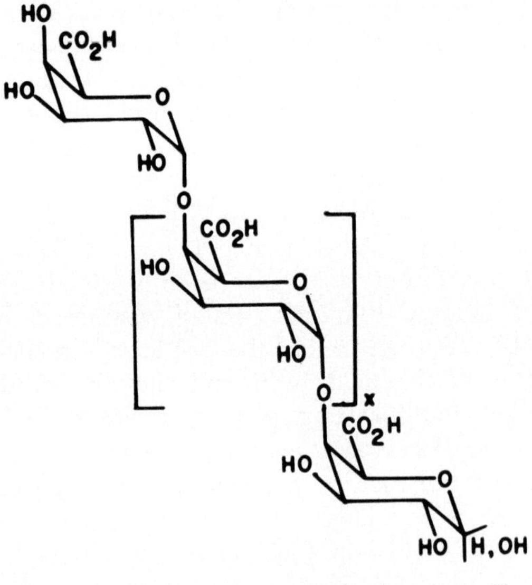

FIG. 36. GALACTURONAN FROM AMABILIS FIR

methylammonium hydroxide in the presence of sodium hydroxide (Hirst et al. 1965). This arabinan is similar to others (Hirst and Jones 1946; Hough and Powell 1960) except that it may contain more $1 \rightarrow 2$ links. The structures of none of the pectin associated arabinans can be written with precision at this time, but a number of structures are consistent with methylation analyses and with some fragmentation analyses. A likely representation (Gould et al. 1965) for white mustard arabinan is shown in Fig. 37.

While in a few instances part of the galacturonan has been obtained pure, the remainder seems to carry attached neutral sugars or even sugar polymers. It is difficult to deduce the exact structure but one suggestion (Gould et al. 1965) is that the linear D-galacturonic ester chain of white mustard pectin may have D-xylopyranosyl units attached by $1 \to 3$ linkages and L-arabinose or arabinan molecules attached by similar $1 \to 3$ linkages (see Fig. 38).

The suggestion that L-rhamnose is linked to the galacturonan chain agrees with the opinion of most recent investigators. Thus, 3-*O*-methyl-L-rhamnose (Aspinall and Fanshawe 1961; Bhattacharjee and Timell 1965), 3,4-di-*O*-methyl-L-rhamnose (Aspinall and Fanshawe 1961), and

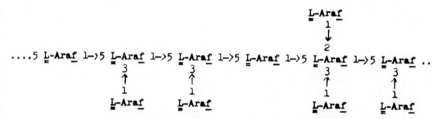

FIG. 37. A POSSIBLE STRUCTURE OF ARABINAN

2,3,4-tri-*O*-methyl-L-rhamnose (Aspinall and Canas-Rodriguez 1958; Aspinall and Fanshawe 1961) have been isolated from the methylated products of pectic substances but not from the separated arabinans or galactans. Consequently, L-rhamnose is assumed to occur in the galacturonan chain as nonreducing end units or as single units in side chains.

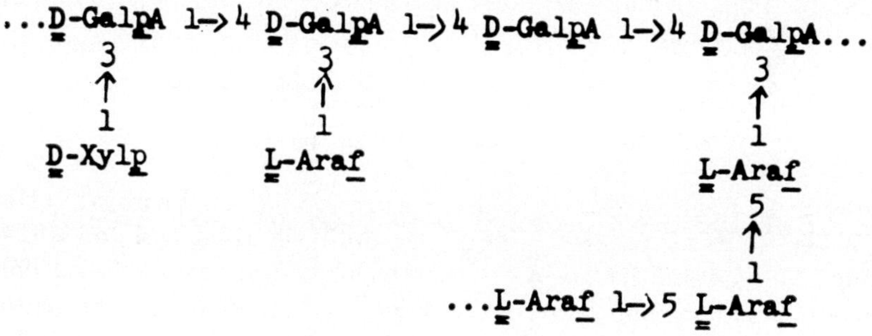

...D-GalpA $1 \to 4$ D-GalpA $1 \to 4$ D-GalpA $1 \to 4$ D-GalpA...

(structure diagram with branches: D-Xylp 1→3, L-Araf 1→3, and L-Araf 1→3 with L-Araf 5→1 ...L-Araf $1 \to 5$ L-Araf)

FIG. 38. SUGGESTED STRUCTURE OF WHITE MUSTARD POLYSACCHARIDE

Hydrolysis of a variety of pectins with acids and enzymes and acidic hydrolysis of the carboxyl reduced pectins has provided a polymer homologous series of oligosaccharides consisting of D-galacturonic acid or derived D-galactose units where linkages, determined by classical chemistry, are $1 \to 4$-α-D (for a partial list of oligosaccharides, see Aspinall and Fanshawe 1961; Bhattacharjee and Timell 1965; Bourne *et al.*

1967; Jermyn and Tomkins 1950; Jones and Reid 1954; Zitko *et al.* 1965).

The galactan component of pectic substances on investigation by methylation, periodate oxidation and other means is shown as a polymer of $1 \rightarrow 4$-β-D-galactopyranose units (Hirst and Jones 1946; Hough and Powell 1960). See Fig. 40. Evidence is obtained (Hough and Powell

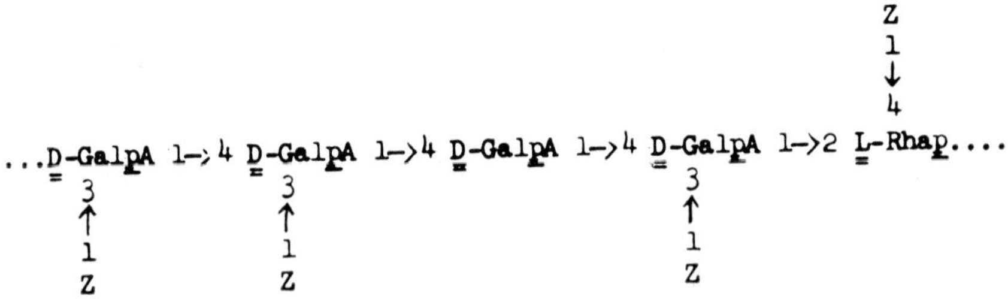

```
                                                                Z
                                                                1
                                                                ↓
                                                                4
...D-GalpA 1-›4 D-GalpA 1-›4 D-GalpA 1-›4 D-GalpA 1-›2 L-Rhap....
        3              3              3
        ↑              ↑              ↑
        1              1              1
        Z              Z              Z
```

FIG. 39. SUGGESTED STRUCTURE OF THE GALACTURONAN CHAIN WITH NEUTRAL
SUGARS ATTACHED

Z is a D-xylopyranosyl unit or an arabinan molecular unit.

1960) that part of the galactan in sugar beet pectin is branched, with branching points arising from units linked through positions 1, 3, and 6.

As indicated above, there is strong chemical evidence that in some pectins neutral sugars are linked to the galacturonan chain (see for example Aspinall and Canas-Rodriguez 1958; Apsinall and Fanshawe

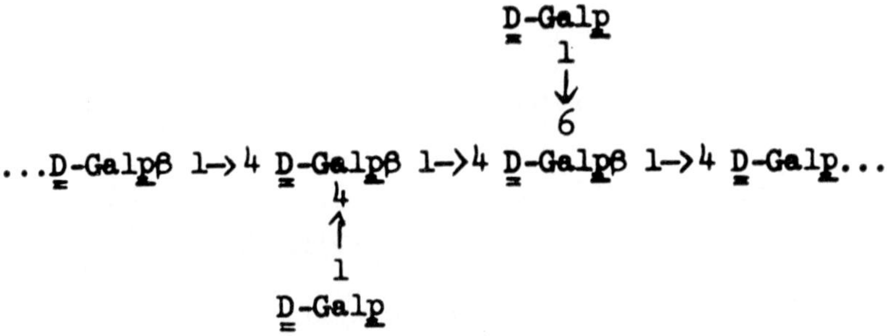

```
                          D-Galp
                            1
                            ↓
                            6
...D-Galpβ 1-›4 D-Galpβ 1-›4 D-Galpβ 1-›4 D-Galp...
                    4
                    ↑
                    1
                  D-Galp
```

FIG. 40. THE GALACTAN COMPONENT OF PECTIC SUBSTANCES

1961; McCready and Gee 1960; Bhattacharjee and Timell 1965; Zitko *et al.* 1965; Timell and Mian 1961). Neither the nature of these inter-component linkages nor the extent of such linkages is definitely known. Pectins undergo degradation so easily that doubt is cast on the signifi-cance that may be attached to the finding of separated pectin components. Certainly, pectins extracted with alkali or strong acids, or pectins subjected to other than neutral hydrogen ion concentrations are partially degraded. Even under neutral conditions, it is claimed that chain breakage can occur (Albersheim 1959; Albersheim *et al.* 1960). It is well-known that pectins are rapidly degraded by alkali even at room temperature (Vollmert 1950; Neukom and Deuel 1958). Degradation occurs by a β-elimination mechanism (Whistler and BeMiller 1958) as shown in Fig. 41.

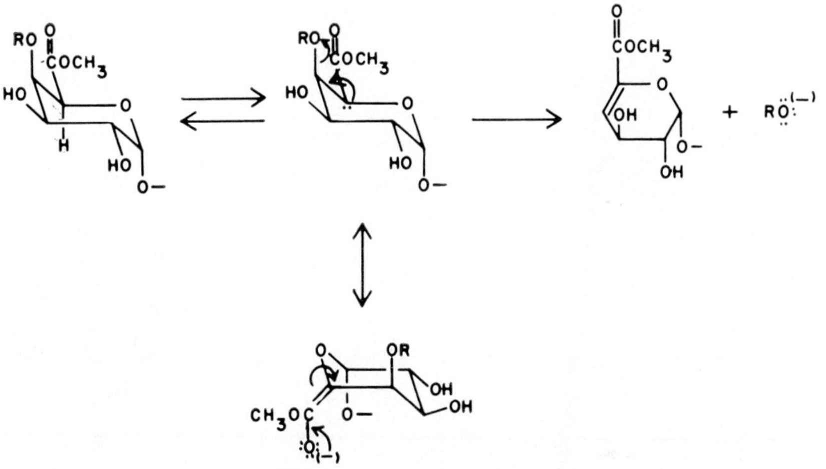

FIG 41. β-ELIMINATION MECHANISM IN ALKALINE DEGRADATION OF PECTIN

Mesomeric shifts incident to anion elimination at position C-4 proceed most easily with uronic carboxyl groups in the ester form. This is evident in the rapid depolymerization of pectins with a high methyl ester content. In these pectins, viscosity decreases with amazing rapidity under the influence of alkaline catalysis. Carboxylic acid groups do not readily undergo the required electronic shifts and consequently chain cleavage next to a free carboxyl group occurs much more slowly. Saponification of esters also proceeds under alkaline catalysis, but this reaction does not lead to significant loss in viscosity.

Uronic acid polymers such as the galacturonan of pectic substances are more resistant to acid hydrolysis than polysaccharides of neutral sugars. Feather and Harris (1965) have investigated the acid catalyzed hydrolysis of many glycopyranosides and suggest that steric hindrance to rotation about positions C-2, C-3, C-4, and C-5 may be a major rate determining factor.

Pectins are important industrial hydrocolloids. They have a high viscosity in water even when present at low concentrations and they have the remarkable property of producing strong gels either alone or in the presence of other components. These rheological properties result from the linear structure of the galacturonan which gives the molecules a large radius of gyration causing the molecules to sweep through large volumes of space with resultant interference with each other producing high viscosities even when concentrations are low. This is generally true for linear molecules as compared to branched molecules of equal molecular weight as illustrated by Fig. 42.

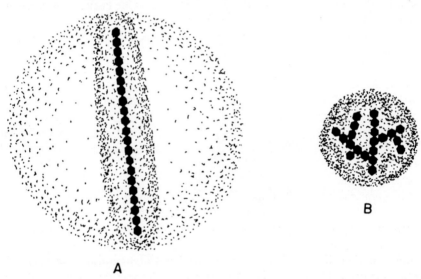

A

B

FIG. 42. A—A LINEAR MOLECULE WITH A LARGE RADIUS OF GYRATION; B—A BRANCHED MOLECULE WITH A SMALL RADIUS OF GYRATION

Stability of the solutions results from the molecules structural and charge effects. Protruding methyl ester groups, as well as neutral sugars or neutral polysaccharide molecules attached to the galacturonan chains, prevent the polysaccharide chains from extensively associating, but still allow optimum association creating an interlocked reticulum which is space filling and capable of holding immense volumes of water and low molecular weight solute molecules. Intermolecular bridging can take place through salt bridges produced between free carboxyl groups and polyvalent ions such as calcium. Thus, pectins can be gelled or gels can be firmed by polyvalent cations, among which calcium is most commonly selected.

Hemicelluloses

Hemicelluloses are potentially important industrial gums. It is likely that the future will see their broad application in the food industry. Hemicelluloses constitute about ¼ of perennial plants and about ⅓ of annual plants. The name hemicellulose was proposed by Schulze in 1891 to designate those polysaccharides extractable from plants by alkaline solution. The name seemed appropriate since these polysaccharides were found in close association with cellulose in the cell wall and were thought to be intermediates in cellulose biosynthesis. It is now known that hemicelluloses are not precursors of cellulose and have no part in cellulose biosynthesis, but rather represent a distinctly separate group of polysaccharides. These polysaccharides are independently produced in plants as structural components of the plant cell wall and make up a portion of the intercellular material called middle lamella. Most workers limit the term hemicellulose to designate cell wall polysaccharides of land plants, except cellulose and pectin, and classify hemicelluloses according to the type of sugars present. Thus, xylan is a polymer of xylose units, mannan of mannose units, and galactan of galactose units. Most hemicelluloses, however, are not homoglycans but are heteroglycans containing 2 to 4 and rarely 5 or 6 different types of sugar units.

A wide variety of hemicelluloses have been investigated and their properties evaluated. Many of these would be excellent industrial polysaccharides if they could be obtained at low cost. Annual plants offer an excellent industrial source of hemicelluloses. One large potential source is corn fiber produced by the Corn Wet Milling Industry. Corn fiber results from corn hull, a by-product produced in the wet milling of corn for starch, protein, and oil. The pericarp, or corn hull, which gives rise to most of the corn fiber, comprises 5 to 6% of the whole kernel dry substance. It is estimated that about 420 million pounds of corn fiber are produced each year which could yield up to 100 million pounds of hemicellulose.

Corn hulls, or industrial corn fiber, can be extracted with alkaline solution to remove hemicelluloses. The most economical alkaline solution for extraction is calcium hydroxide solution. The gum obtained is completely water soluble and is molecularly homogeneous. It is an acidic arabinoglucuronoxylan. Viscosity and other rheological properties of its solutions are similar in many respect to those of gum arabic.

Another hemicellulose now being produced in commercial quantities is an arabinogalactan. This water soluble highly branched polysaccharide is present in the wood of conifers and in the sap of at least one angiosperm, the maple tree. While most gymnosperm woods contain only small portions of this polysaccharide, the wood in the genus *Larix* may contain up to 25%. Larch wood is the source of the commercial arabinogalactan produced by the St. Regis Paper Company which markets it as the commercial gum, Stractan. The polysaccharide is isolated in good yield by hot water extraction of larch wood chips. It is a mixture of 2 structurally similar components possessing molecular weights of 100,000 and 160,000. Under mild hydrolysis, the L-arabinose units are removed.

SEAWEED GUMS

Gum extraction from seaweeds, originally practiced in oriental countries, has become an important source of hydrocolloids of great value to the food industry. Seaweed gums produced in largest volume are algin, carrageenan, and agar. Algin, which is sodium alginate, is extracted mainly from the brown seaweed *Macrocystis pyrifera*, the giant kelp growing in shallow water from Point Conception California down into southern Mexican waters. Carrageenan is obtained from the red seaweed, particularly *Chondrus crispus* growing in many shallow waters, but especially abundant along the northeast coast of the United States and Nova Scotia area. Agar use has not been expanding as extensively as those of the other seaweed gums. It is obtained from various species of *Gelidium* and is collected mainly from the Japanese coastal area. Other seaweed gums are harvested but do not yet find wide usage in foods, although Danish agar and laminaran may extend their application in the food area.

Agar

Agar and agaroides are obtained from various genera and species of red-purple seaweeds widely distributed throughout temperate zones. Agar is one of the oldest known seaweed gums. The principal plant source is *Gelidium*, although considerable amounts are made from *Gracilaria* and other species. Sea plants are collected by divers who must often work under the formidable hazards of jagged topography, strong currents, low illumination, and cold water. An expert diving team can produce between 40 and 320 kg of dried weed per day.

The soluble ⅔ portion of agar, agaran (agarose), isolated from *Gelidium amansii* is shown as an alternating copolymer of 3,6-anhydro-α-L-galactopyranosyl and β-D-galactopyranosyl units (Araki and Hirase 1960) joined by $1 \rightarrow 3$ and $1 \rightarrow 4$ linkages as shown in Fig. 43.

Structure of the remaining ⅓ is not clear. It is a branched, partially sulfated molecule.

Agar is a potent gel-forming agent, capable of gel formation at concentrations as low as 0.04%. Its aqueous gels have the unusual property of possessing a melting temperature, which varies with the concentration and quality of the agar used. Depending upon source, agar gels of 1.5% concentration may melt from 30° to 97°C.

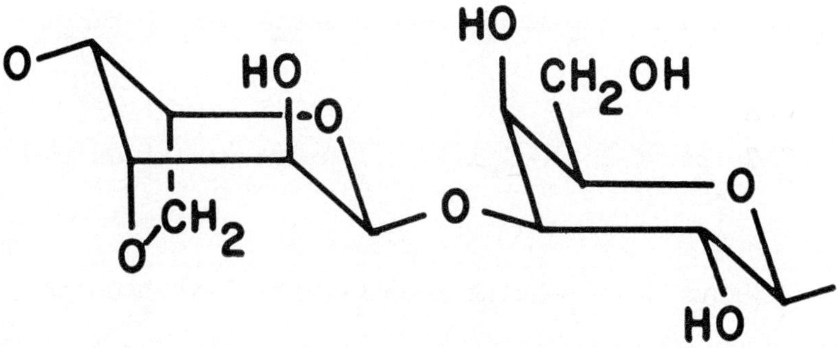

FIG. 43. AGARAN (AGAROSE) FROM *Gelidium Amansii*

Algin

A number of years after Nelson and Cretcher (1929) and others initiated the first fundamental work on the constitution of alginic acid which indicated that the polysaccharide was a linear polymer of β-D-mannuronic acid units, Fisher and Dörfel (1955) and Whistler and Kirby (1959) showed that L-guluronic acid was also present. D-mannuronic to L-guluronic acid ratios vary with seaweed source. Manganous salts specifically can be used to separate alginates from different brown seaweeds into fractions differing in sugar monomer ratio (McDowell 1958; Haug 1959). Separation probably results from the different cation exchange properties of the polymers for manganous II, cation.

Solubility of alginates in acidic solutions is shown to depend also on the uronic acid composition (Haug and Larsen 1963), but marked differences are observed between samples of nearly identical uronic acid composition. Recent work (Haug *et al.* 1966, 1967; Haug *et al.* 1967) shows that alginates from *Laminaria digitata* and possibly from other sources, contain blocks of D-mannuronic acid units, blocks of L-guluronic acid units, and regions with predominantly alternating uronic acid types. Further work is needed to completely clarify the algin structures. Presently, the linear molecule may be represented by Fig. 44.

Algins, like pectins, are gelled by polyvalent cations such as calcium. Significantly algins will selectively remove calcium, barium, and other similar cations from a solution where their ions are present with monovalent cations. Gelation of alginates by calcium is the basis of important food applications. If the alginic acid is transformed to the hydroxypropyl ester through reaction with propylene oxide, the product is no

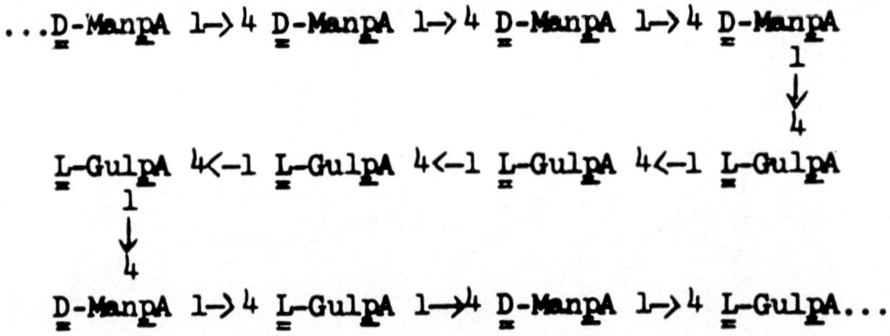

FIG. 44. A POSSIBLE STRUCTURE FOR ALGINATE FROM *Laminaria Digitata*

longer precipitated or gelled by polyvalent cations and is an effective ice cream stabilizer at concentrations from 0.1 to 0.5%.

Under acidic conditions, where the ionization of the carboxyl groups is repressed, alginic acid precipitates as a gel.

Carrageenan

A mixture of two sulfated polysaccharides occurs in the red seaweeds *Chondrus crispus* and *Gigartina stellata*, commonly known as Irish moss. These sea plants are found along the North Atlantic shores from Rhode Island northward to Newfoundland and from Norway south to the coast of North Africa. Large stands are reported near the Chilean coast and in the South Pacific. Commercial harvesting along the shores of Massachusetts, Maine, Nova Scotia, and Prince Edward Island provides the major supply for the American industry. Most of the plants are collected by raking or as beachweed left on the shore by tidal action.

A major difference among polysaccharides from red seaweeds is in their sulfate ester content. Agaran (agarose), the more soluble component of agar, contains no, or only a few sulfate groups, whereas the insoluble agaropectin contains 5 to 10% ester sulfate. Extractives as hypnean and furcellaran, from *Hyphnea* and *Furcellaria* respectively, form another group with from 12 to 16% ester sulfate. Carrageenans range in sulfate ester content from 20 to 36% depending on source, available nutrients, water conditions, and other environmental factors. These comparisons are shown in Table 6. While carrageenan is isolated from algal species within the order of Gigartinales, polysaccharides closely related to carrageenan have been extracted from species outside this order, according to enzyme classifications proposed by Yaphe (1959).

Early in the characterization of carrageenan, Smith *et al.* (1953, 1954), found that it could be fractionated into 2 portions by precipi-

TABLE 6

SULFATE ESTER CONTENT OF SELECTED SEAWEED POLYSACCHARIDES

Polysaccharide	Sulfate Ester %
Agar, Agaran	(little or none)
Agaropectin	5–10
Hypnean	12–16
Furcellaran	12–18
Carrageenan	20–36

tation with potassium chloride in the concentration range of 0.125–0.25M. The soluble fraction is termed lambda and the insoluble fraction kappa. The kappa fraction had a high content of 3,6-anhydro-D-galactose (Smith *et al.* 1955), while the soluble lambda portion from some samples of *Chondrus crispus* contained only traces of the anhydro sugar (Smith *et al.* 1955; Rees 1963; Dolan and Rees 1965). The biological precursor of the anhydro sugar may be D-galactose 6-sulfate, as L-galactose 6-sulfate was shown to be the precursor for 3,6-anhydro-L-galactose in porphyran, the galactan sulfate of the red algae, *Porphyra umbilicalis* (Rees 1961). Kappa carrageenan is very susceptible to hydrolysis due to its extremely acid labile 3,6-anhydro-D-galactopyranosyl linkages (O'Neill 1955A,B).

Using potassium chloride fractionation kappa and lambda carrageenans have been separated into fractions whose possible structures are represented in Fig. 45 (Anderson *et al.* 1965).

Black *et al.* (1965) show that caragalenans from many algal species do not give pure lambda fractions on potassium chloride fractionation. More recent potassium chloride fractionations have led to the conclusion that the two-fraction concept must be abandoned and that carrageenans be assumed to consist of a series of molecules of different chemical composition and solubility (Pernas *et al.* 1967).

One of the most interesting properties of kappa carrageenan, furcellaran and related polysaccharides is their capacity to form an insoluble gel in the presence of potassium ions. Bayley (1955) suggested that this could be due to the formation of a highly ordered ionic lattice into which only hydrated cations with a certain critical diameter could fit. However, Bayley assumed that the sulfate groups were regularly spaced

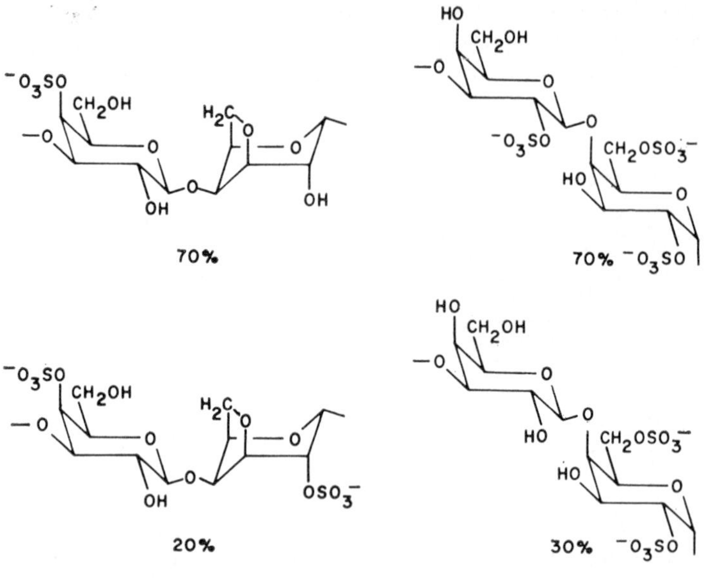

FIG. 45.   KAPPA (left) AND LAMBDA (right) CARRAGEENAN

along the polysaccharide chains. Painter (1965) believes that kappa carrageenan consists of α-D-(1 → 3) linked units of carrabiose (4-*O*-β-D-galactopyranosyl-3,6-anhydro-D-galactose) with sulfuric acid half-ester groups, more or less randomly distributed over all available hydroxyl groups, although they may be located mainly at positions C-4 as previously assumed. Consequently, the gelation of kappa carrageenan solutions on the addition of potassium ions must be a consequence of other factors. Since potassium ions are not highly hydrated their presence on the sulfate units could lower the solubility of the salt. The high solubility of the kappa fraction is not in any event due to the presence of 3,6-anhydro-D-galactose units which are less hydrophilic than D-galactose units. Whistler and Hirase (1961) show that introduction of 3,6-anhydro rings into D-glucopyranosyl units of amylose greatly decreases solubility, and leads to water insoluble products when 3,6-anhydro rings are present in more than half of the sugar units. Whistler *et al.* (1968) show that 3,6-anhydro rings can be introduced into starch sulfates by treatment with alkali (see Fig. 46). On resulfation the sulfated polysaccharide now containing numerous 3,6-anhydro rings does not gel when potassium salts are added to its aqueous solutions. Thus, carrageenan's characteristic of gelation on addition of potassium ions to its solutions while due, in large part, to a delicate solubility balance resulting from both the presence of less hydrophilic 3,6-anhydro rings and poorly hydrated potassium ion, must also involve other details of fine structure. It is important to note that cellulose,

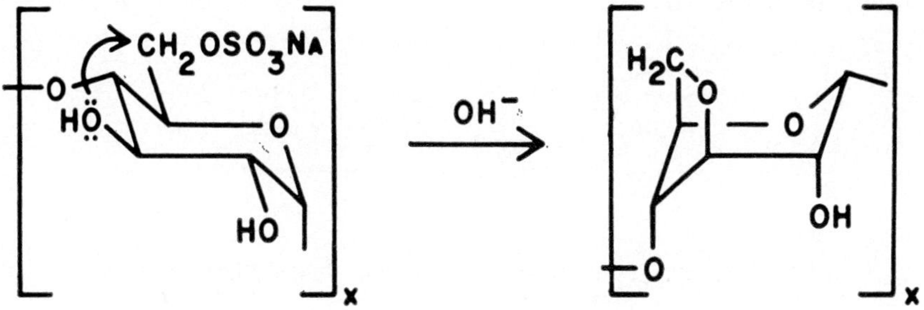

FIG. 46.   INTRODUCTION OF 3,6-ANHYDRO RINGS INTO STARCH SULFATES

sulfated to the high D.S. of about 2.5, will gel in solutions to which potassium ions are added. In this instance the Bayley philosophy of gelation may be entirely applicable.

Danish Agar (Furcellaran)

An interesting red seaweed occurring primarily in the waters near Denmark and Norway is *Furcellaria fastigiata*. The seaweed grows at depths of 20 to 30 ft and is gathered in trawl nets. Furcellaran is extracted by hot water and often bleached with hypochlorite. Aqueous solutions gel at about 40°C. Gelling properties are midway between those of agar and carrageenan. Somewhat like carrageenan dispersions, furcellaran solutions develop improved gel strength on addition of potassium ions. Addition of galactomannans also increases gel strength. The normally sharp fracturing furcellaran gel is made more elastic by incorporation of galactomannans. Because of these properties, jams, jellies, and marmalades are easily made with this excellent gelling agent. Boiling is unnecessary and high flavor retention results. The gum makes an excellent quick-set glaze and serves as a stabilizer in unboiled icing bases. Other bakery uses include pie fillings and special jellies.

When potassium chloride is added to an aqueous extract of *Furcellaria fastigiata* a rather homogeneous polysaccharide precipitates which is composed of D-galactose and 3,6-anhydro-D-galactose in the ratio of 1.3 : 1 and contains sulfate ester groups at a degree of substitution of about 0.4 (Kylin 1943; Clancey *et al.* 1960; Painter, 1960). Mercaptolysis produces a derivative of 4-O-β-D-galactopyranosyl-3,6-anhydro-D-galactose. The polysaccharide may consist, for the most part, of a chain in which this disaccharide unit repeats in some regular or nearly regular order, as in Fig. 47.

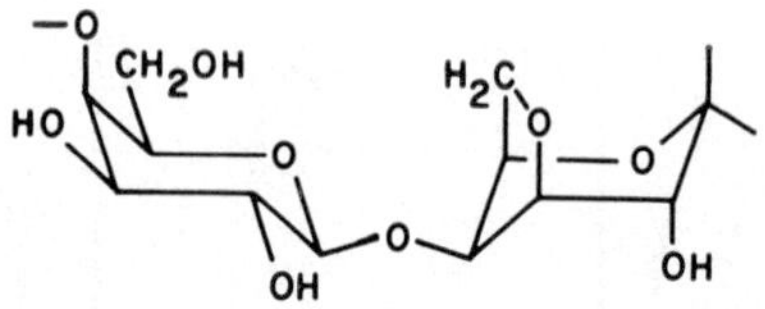

FIG. 47. FURCELLARAN FROM *Furcellaria Fastigiata*

A polysaccharide with similar properties has been isolated from *Hypnea specifera* by Clingman and Nunn (1959).

EXUDATE GUMS

Gums of the ancient world were largely plant exudates. Most plant families include species which exude gums to greater or lesser degrees. Plants which produce commercial gums are usually shrubs or low-growing trees from which gums exude, vermiform or tear-shaped, and may build up in thick layers. Harvesting is by hand picking, usually by native workers in countries where labor costs are low. Because labor costs are rising and because of contamination problems incident to exported adhesive exudates, industrial use and especially food use, while increasing, is not increasing commensurate with hydrocolloid demand. Perhaps partly because of these reasons, little scientific work has been done recently to elucidate further the structure and properties of exudate gums.

Gum Arabic

Gum arabic is one of the oldest known commercial gums having come down from antiquity. It is dried exudate from trees belonging to the genus *Acacia,* subfamily Mimosoideae, and family Leguminosae of which about 500 species exist. The trees usually grow in arid or semiarid regions distributed over tropical and subtropical areas of Africa, India, Australia, Central America, and southwest North America. The most important gum yielding area is the Republic of the Sudan, followed by West Africa. Large quantities are received from Nigeria, Tanzania, Morocco, and India.

Gum arabic is a slightly acidic, highly branched polysaccharide, producing aqueous solutions of comparatively low viscosity. The backbone

```
...D-Galpβ 1-)4 D-Galpβ 1-)4 D-Galpβ 1->6 D-Galpβ 1->4 D-Galp...
                   3                          3
                   ↑                          ↑
                   1β                         1β
   D-GlupAβ1->6 D-Galp        D-GlupAβ1->6 D-Galp
        4                          4
        ↑                          ↑
        1                          1α
   L-Araf                     L-Rhap
```

FIG. 48. SUGGESTED STRUCTURE OF GUM ARABIC

chain of the polysaccharide is composed of D-galactopyranose units joined by β-D-$(1 \to 4)$ and β-D-$(1 \to 6)$ linkages. Side chains composed of D-galactopyranose units are attached usually by β-D-$(1 \to 3)$ linkages. To these side chains L-rhamnopyranose or L-arabinofuranose residues are attached as end units. D-Glucuronic acid units are frequently attached by β-D-$(1 \to 6)$ linkages to D-galactose units and often L-arabinofuranose units are attached to the D-glucuronic acid units by $(1 \to 4)$ bonds. Most of the L-rhamnose is attached to D-glucuronopyranosyl units as 4-O-α-L-rhamnopyranosyl nonreducing terminal units (Aspinall *et al.* 1963; Aspinall and Young 1965). (See Fig. 48)

The molecular weight of gum arabic is in the range of 200,000 to 270,000. Gum arabic is extremely soluble in water, but because of its low molecular weight and branch structure must be employed in higher concentrations than most other gums to effect significant viscosity values.

The gum reacts with cationic polymers such as gelatin to form coacervates which have been used for microincapsulation.

About 55% of the total gum arabic imported into the United States is used in the food industry. It imparts desirable qualities through its influence over viscosity, body, and texture of foods. It is nontoxic, odorless, colorless, tasteless, and does not affect the flavor, odor or color of other food ingredients.

In confectionery, it is used extensively to prevent crystallization of sugar and because of its thickening power it is used as a glaze in candy products and is a component of chewing gums, cough drops, and candy lozenges. Originally, almost all gum drops were produced from gum arabic, but they are now made largely with thin boiling starches or pectins. Gum arabic is used as a stabilizer in dairy products, particularly ice creams, ices, and sherbets. It is widely used as a flavor fixative, particularly with spray-dried flavors. It is also employed as an emulsifier although it is not as effective as carrageenan.

Gum Ghatti

Gum ghatti (Indian Gum) is a water-soluble exudate from the tree *Anogeissus latifolia* found abundantly in the dry deciduous forests of India and to a lesser degree in Ceylon. It is harvested in large amounts and some 400 tons per year are imported into the United States.

Ghatti is a highly branched chain whose backbone is composed of D-galactopyranose units joined principally by $(1 \rightarrow 6)$ linkages. It contains L-arabinose, D-galactose, D-mannose, D-glucuronic acid and D-xylose units in the ratio $10 : 6 : 2 : 2 : 1$. The majority of the L-arabinose units is present in the furanose form and is present as nonreducing end units (Aspinall *et al.* 1965). Some of the L-arabinose units in both furanose and pyranose forms contain 3-O-substituents. D-Mannose residues are 2-O-substituted and the majority also contains 3-O- and 6-O-substituents.

Gum Karaya

Karaya gum is the commercially important exudate gum of *Sterculia urens*, a large deciduous tree growing in the dry, elevated areas of North and Central India. *Sterculia setigera* of tropical West Africa and *Sterculia caudata* of Northern Queensland, Australia produce similar gums although these are not collected for industrial use.

Gum from *S. urens* swells but does not dissolve in water. It consists of a partially acetylated polysaccharide with about 8% acetyl groups and 37% uronic acid residues. On deacetylation with sodium hydroxide solution the polysaccharide becomes water soluble. It has rather uniform molecular weight and is precipitable as its copper or cetyltrimethylammonium complex. It contains D-galactose, D-galacturonic acid, D-glucuronic acid, and L-rhamnose. Partial acid hydrolysis gives 2-O-(α-D-galactopyranosyluronic acid)-L-rhamnose, 4-O-(D-galactopyranosyluronic acid)-D-galactose, and a trisaccharide corresponding to O-(β-D-galactopyranosyluronic acid)-1 → 3-O-(α-D-galactopyranosyluronic acid)-1 → 2-L-rhamnose, Fig. 49 (Aspinall and Nasir-Ud-Din 1965). It is concluded that the majority of the D-galacturonic acid units and some of the L-rhamnose units represent branching points in the polysaccharide structure whereas D-glucuronic acid, or its 4-O-methyl derivative, are nonreducing end groups.

$$\text{D-Gal}p1 \rightarrow 2\alpha\text{-L-Rha}$$

$$\text{D-Gal}p\text{Al} \rightarrow 4\alpha\text{-D-Gal}$$

$$\text{D-Gal}p1 \rightarrow 3\beta\text{-D-Gal}p\text{Al} \rightarrow 2\alpha\text{-L-Rha}$$

FIG. 49. HYDROLYSIS PRODUCTS OF KARAYA GUM FROM *Sterculia urens*.

Tragacanth

Tragacanth gum is the exudate of several varieties of small shrub-like plants of the *Astragalus* species. These plants are common in sections of Asia Minor and in the semidesert mountainous regions of Iran, Syria, and Turkey. The gum is one of the oldest known emulsifiers. It exudes spontaneously from the shrubs and hardens into ribbons 2–4 in. in length and in flakes or brittle pieces 0.5–2 in. in diameter. The gum is collected by hand picking. In water the gum is extensively hydrated but does not completely dissolve. It is fairly resistant to hydrolysis and is consequently a frequent additive to salad dressings and emulsions where breakdown by acid would be detrimental to other less resistant additives. It has been used widely in hair lotions and hand creams and as an emulsifier for insecticides. Because it costs $1–$4 per pound and because its collection depends upon hand picking by natives, its usage has been decreasing as other lower cost natural and synthetic gums become available.

Tragacanth can be separated into two fractions: one soluble and one only swellable in water. Both fractions are complex. Their highly branched structure is supported by methylation work (James and Smith 1945A, B). Tragacanthic acid, the main fraction, is composed of 43% D-galacturonic acid, 40% D-xylose, 10% L-fucose, and 4% D-galactose, and the second fraction is composed of 75% L-arabinose, 12% D-galactose, 3% D-galacturonic acid, and a small amount of L-rhamnose. While no proper structures can be depicted at this time it can be said that tragacanthic acid is a main chain of $1 \rightarrow 4$-α-D-galacturonopyranosyl units with side chains attached at carbons C-3, perhaps as shown in Fig. 50.

The more neutral polymer must have a core of D-galactopyranose units joined by $1 \rightarrow 6$ linkages and a smaller proportion of $1 \rightarrow 3$ linkages. Attached to this central core are highly branched L-arabinofuranose units joined by $1 \rightarrow 2$, $1 \rightarrow 3$, and $1 \rightarrow 5$ linkages (Aspinall and Baillie 1963A, B).

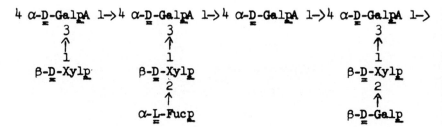

FIG. 50. SUGGESTED STRUCTURE FOR TRAGACANTHIC ACID

SEED GUMS

Endosperm gums of seeds are continually increasing in industrial importance.

Guar and Locust Bean Gums

Two important industrial gums are derived from the endosperm of seeds from leguminous plants. Locust bean gum is derived from the seeds of the tree *Ceratonia siliqua*. Locust beans have been used by man since his early beginnings. Culturing of locust bean trees was known before the Christian Era and in the first century A.D., Dioscorides referred to the curative properties of locust bean fruit. In ancient times the Arabs used the seed as a counterweight for gold and diamonds. The carob fruit is known all around the Mediterranean shore. Its goodness and richness is immortalized in the Bible as the food of John the Baptist.

Guar gum is derived from the seed of the guar plant, *Cyamopsis tetragonolobus*. This leguminous plant has been grown for centuries in India and Pakistan where it is one of the principal crops used both as food for humans and animals. It has either an erect or branched stem on which pods grow, containing the guar seeds which are approximately ⅛ in. in diameter. The seeds consist of 14–17% hull, 43–47% germ, and 35–42% endosperm, from which guar gum is produced by grinding.

Guar gum became commercially available in 1953 and has grown rapidly in the subsequent years until, presently, some 40 million pounds are consumed annually in the United States with usage continuing to increase. The principal source of guar seed is still India and Pakistan, although small quantities are being grown in the United States and in

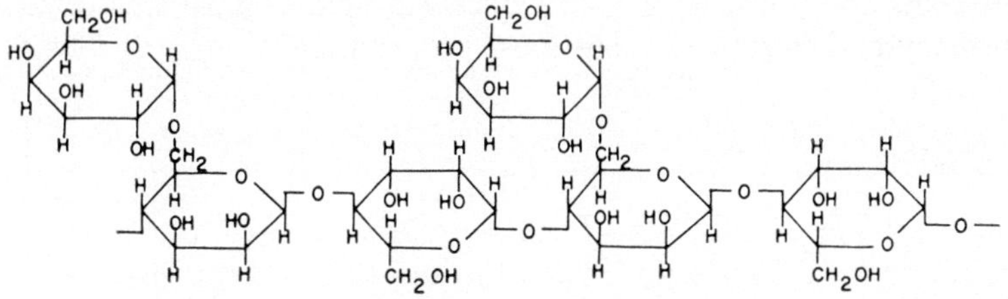

FIG. 51. GUARAN

some other parts of the world. The guar plant can be grown commercially in the same manner as other annual agricultural crops. It can be planted, cultivated and harvested by standard agricultural machinery. Since very little genetic work has been done to improve disease resistance and seed yield, it can be anticipated that significant improvements in productivity can be made by plant breeding. Because of its ease of cultivation by ordinary agricultural means, guar gum is replacing locust bean gum in the commercial markets of the world.

Both locust bean gum and guar gum are galactomannans. Both consist of a straight chain of D-mannose units joined by β-D($1 \rightarrow 4$) linkages, having α-D-galactopyranose units attached to this linear chain by $1 \rightarrow 6$ linkages. In the pure polysaccharide guaran, from guar, the ratio of D-galactose to D-mannose units is $1:2$; whereas, in locust bean gum the ratio varies from $1:3$ to $1:6$. The molecular weight of guaran is 220,000. Work by Whistler *et al.* (1948, 1950, 1951, 1952) has established the structure of guaran, Fig. 51.

Both gums form viscous, colloidal dispersions when hydrated in cold water. Because of the linear nature of the backbone chain, the gums form highly viscous solutions even at low concentrations. Guar gum is somewhat more viscous than locust bean gum but both have higher viscosities than most other natural gums. Since the polysaccharides are neutral, their viscosities are little affected by changes in pH. Furthermore, since the polysaccharides contain numerous α-D-galactopyranose unit side chains, the dissolved polysaccharides cannot fit together smoothly to produce aggregation and hence solutions of these gums are remarkably stable. The solutions are easily gelled by the addition of borate which acts as a cross-linking agent in developing a cohesive gel structure. Solutions of the gums display very little adhesive quality.

Guar gum is often used in foods as a thickener and as a binder. It is widely used as a stabilizer in ice cream to prevent graininess by controlling ice crystal growth. It acts in a similar way as a stabilizer in ice sherbets and pops. It is used extensively with other gums to provide consistency and improved rheological properties.

Tamarind Gum

Tamarind gum is obtained from the endosperm of the seeds of the tamarind tree, *Tamarindus indica,* a beautiful evergreen tree of the Middle East, particularly abundant in India. A good bearing tree may yield 500 lb of fruit in large flat pods 4–6 in. long. The seeds can be dry milled to produce an endosperm powder of high thickening power. The endosperm polysaccharide is composed of D-galactose, D-xylose and D-glucose in a $1:2:3$ molar ratio. It is hydrolized by commercial takadiastase (Khan and Mukherjee 1959) which with methylation data suggests a main chain of $(1 \to 4)\beta$-D-glucopyranosyl units to which are joined by $1 \to 6$ linkages D-galactopyranosyl and branches of one or more D-xylopyranosyl units, Fig. 52.

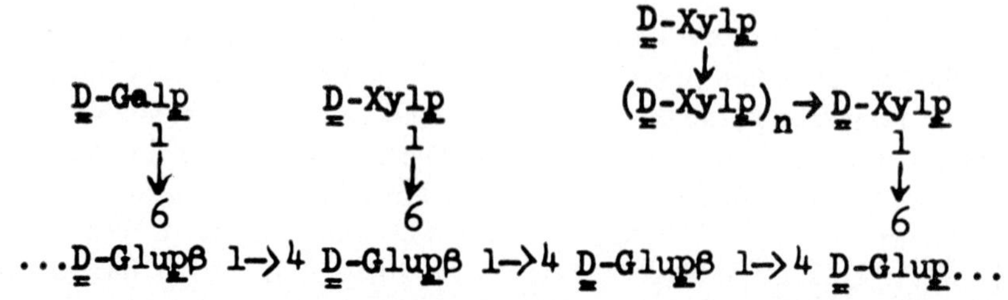

FIG. 52. STRUCTURE OF THE POLYSACCHARIDE FROM THE SEEDS OF *Tamarindus Indica*

Polysaccharide Derivatives

As stated at the beginning of this article, many polysaccharide derivatives are, and will continue to be produced for a variety of uses in the food industry. The potential for chemical modification of the properties of polysaccharides is almost unlimited. One has only to examine the multitude of modified starches and of water soluble cellulose derivatives to appreciate the magnitude of these modified polymers in the industrial field. As chemical skills become continually more sophisticated, it will be possible to further custom-modify polysaccharide structure and willfully introduce nuances of physical behavior that give polymers the proper shade of usefulness and adaptability which will bring them into new and more extensive application. The modification of the water soluble polysaccharide gums has a brilliant and growingly useful future in the food area.

BIBLIOGRAPHY

ALBERSHEIM, P. 1959. Instability of pectin in neutral solutions. Biochem. Biophys. Res. Cummun. *1*, 253–256.

ALBERSHEIM, P., NEUKOM, H., and DEUEL, H. 1960. Splitting of pectin chain molecules in neutral solutions. Arch. Biochem. Biophys. *90*, 46–51.

ANDERSON, N. S., DOLAN, T. C. S., and REES, D. A. 1965. Evidence for a common structural pattern in the polysaccharide sulfates of the *Rhodophyceae*. Nature *205*, 1060–1062.

ARAKI, C., and HIRASE, S. 1960. Chemical constitution of agar-agar. XXI. Reinvestigation of methylated agarose of *Gelidium amansii*. Bull. Chem. Soc. Japan *33*, 291–295.

ARAKI, C., and HIRASE, S. 1960. Studies of the chemical constitution of agar-agar. XXII. Partial methanolysis of methylated agarose of *Gelidium amansii*. Bull. Chem. Soc. Japan *33*, 597–600.

ASPINALL, G. O., and BAILLIE, J. 1963A. Gum tragacanth. Part I. Fractionation of the gum and the structure of tragacanthic acid. J. Chem. Soc., 1702–1714.

ASPINALL, G. O., and BAILLIE, J. 1963B. Gum tragacanth. Part II. The arabinogalactan. J. Chem. Soc., 1714–1721.

ASPINALL, G. O., and CANAS-RODRIGUEZ, A. 1958. Sisal pectic acid. J. Chem. Soc., 4020–4027.

ASPINALL, G. O., and FANSHAWE, R. S. 1961. Pectic substances from Lucerne (*Medicago sativia*). Part I. Pectic acid. J. Chem. Soc., 4215–4225.

ASPINALL, G. O., and NASIR-UD-DIN. 1965. Plant gums of the genus *Sterculia*. J. Chem. Soc., 2710–2720.

ASPINALL, G. O., and YOUNG, R. 1965. Further oligosaccharides from carboxyl-reduced gum arabic. J. Chem. Soc., 3003–3004.

ASPINALL, G. O., BHARANANDAN, V. P., and CHRISTENSEN, T. B. 1965. Gum ghatti (Indian gum). Part V. Degradation of the periodate-oxidized gum. J. Chem. Soc., 2677–2684.

ASPINALL, G. O., CHARLSON, A. J., HIRST, E. L., and YOUNG, R. 1963. Location of L-rhamnose residues in gum arabic. J. Chem. Soc., 1696–1702.

BARRETT, A. J., and NORTHCOTE, D. H. 1965. Apple fruit pectic substances. Biochem. J. 94, 617–627.

BAYLEY, S. T. 1955. X-ray and infrared studies on carrageenin. Biochim. Biophys. Acta. 17, 194–205.

BENDER, W. A. 1959. In Industrial Gums, R. L. Whistler (Editor). Academic Press, New York.

BHATTACHARJEE, S. S., and TIMELL, T. E. 1965. A study of the pectin present in the bark of amabilis fir (Abies amabilis). Can J. Chem. 43, 758–765.

BISHOP, C. T. 1955. Carbohydrates of sunflower heads. Can. J. Chem. 33, 1521–1529.

BLACK, W. A. P., BLAKEMORE, W. R., COLQUHOUN, J. A., and DEWAR, E. T. 1965. Evaluation of some red marine algae as a source of carrageenan and its K- and λ-components. J. Sci. Food Agr. 16, 573–585.

BOURNE, E. J., PRIDHAM, J. B., and WORTH, H. G. J. 1967. Pectic substances in cured and uncured tobacco. Phytochem. 6, 423–431.

CARRAO, A. 1954. Free pectin in persimmon fruit. Ann. Sper. Agrar. 8, 1675–1683. (Rome).

CLANCY, M. J., WALSH, K., DILLON, T., and O'COLLA, P. S. 1960. The gelatinous polysaccharide of Furcellaria fastigiata. Sci. Proc. Roy. Dublin Soc. Ser. A.1, 197–204.

CLINGMAN, A. L., and NUNN, J. R. 1959. Red seaweed polysaccharides. Part III. Polysaccharide from Hypnea specifera. J. Chem. Soc., 493–498.

COLEMAN, R. J., LENNY, J. F., COSCIA, A. T., and DI CARLO, F. J. 1955. Pectic acid from the mucilage of coffee cherries. Arch. Biochem. Biophys. 59, 157–164.

DOLAN, T. C. S., and REES, D. A. 1965. The carrageenans. Part II. The positions of the glycosidic linkages and sulfate esters in λ-carrageenan. J. Chem. Soc., 3534–3539.

FEATHER, M. S., and HARRIS, J. F. 1965. The acid catalyzed hydrolysis of glycopyranosides. J. Org. Chem. 30, 153–157.

FISCHER, F. G., and DÖRFEL, H. Z. 1955. The polyuronic acids of brown algae. Z. Physiol. Chem. 302, 186–203.

GOULD, S. E. B., REES, D. A., RICHARDSON, N. G., and STEELE, I. W. 1965. Pectic polysaccharides in the growth of plant cells: Molecular structural factors and their role in the germination of white mustard. Nature 208, 876–878.

HATANAKA, C., and OZAWA, J. 1966. Enzymic degradation of pectic acid. II. Chromatography of pectic substances on DEAE-cellulose. Nippon Nogeikagaku Kaishi. 40, 98–105.

HAUG, A. 1959. Fractionation of alginic acid. Acta Chem. Scand. 13, 601–603.

HAUG, A., and LARSEN, B. 1963. The solubility of alginate at low pH. Acta Chem. Scand. 17, 1653–1662.

HAUG, A., LARSEN, B., and SMIDSRØD, O. 1966. A study of the constitution of alginic acid by partial acid hydrolysis. Acta Chem. Scand. 20, 183–190.

HAUG, A., LARSEN, B., and SMIDSRØD, O. 1967. Studies on the sequence of uronic acid residues in alginic acid. Acta Chem. Scand. 21, 691–704.

HAUG, A., MYKLESTAD, S., LARSEN, B., and SMIDSRØD, O. 1967. Correlation between chemical structure and physical properties of alginates. Acta Chem. Scand. 21, 768–778.

HIRST, E. L., and JONES, J. K. N. 1946. The chemistry of pectic materials. Advan. Carbohydrate Chem. 2, 235–251.

HIRST, E. L., REES, D. A., and RICHARDSON, N. G. 1965. Seed polysaccharides and their role in germination. Biochem. J. 95, 453–458.

HOUGH, L., and POWELL, D. B. 1960. Methylation and periodate oxidation studies of the alkali-stable polysaccharide of sugar-beet pectin. J. Chem. Soc., 16–22.

JAMES, S. P., and SMITH, F. 1945A. Chemistry of gum tragacanth. I. Tragacanthic acid. J. Chem. Soc., 739–746.

JAMES, S. P., and SMITH, F. 1945B. Chemistry of gum tragacanth. III. J. Chem. Soc., 749–751.

JERMYN, M. A., and TOMKINS, R. G. 1950. The chromatographic examination of the products of the action of pectinase on pectin. Biochem. J. 47, 437–442.

JONES, J. K. N., and REID, W. W. 1954. The structure of the oligosaccharides produced by the enzymic breakdown of pectic acid. Part I. J. Chem. Soc., 1361–1365.

KHAN, N. A., and MUKHERJEE, B. D. 1959. The polysaccharide in tamarind seed kernel. Chem. & Ind. 1413–1414.

KYLIN, H. 1943. Biochemistry of Rhodophyceae. Kgl. Fysiograf. Sällskap. Lund, Förh. 13, 51–63.

McCOMB, E. A., and McCREADY, R. M. 1957. Determination of acetyl in pectin in acetylated carbohydrate polymers. Hydroxamic acid reaction. Anal. Chem. 29, 819–821.

McCREADY, R. M., and GEE, M. 1960. Determination of pectic substances by paper chromatography. J. Agr. Food Chem. 8, 510–513.

McDOWELL, R. H. 1958. A method for the fractionation of alginates. Chem. & Ind., 1401–1402.

NELSON, W. L., and CRETCHER, L. H. 1929. Alginic acid from Macrocystis pyrifera. J. Am. Chem. Soc. 51, 1914–1922.

NELSON, W. L., and CRETCHER, L. H. 1930. Isolation and identification of D-mannuronic acid lactone from the Marcocystis pyrifera. J. Am. Chem. Soc. 52, 2130–2132.

NELSON, W. L., and CRETCHER, L. H. 1932. Properties of D-mannuronic acid lactone. J. Am. Chem. Soc. 54, 3409–3412.

NEUKOM, H., and DEUEL, H. 1958. Alkaline degradation of pectin. Chem. & Ind., 683.

NEUKOM, H., DEUEL, H., HERI, W. J., and KÜNDIG, W. 1960. Chromatographic fractionation of polysaccharides on cellulose-anion exchangers. Helv. Chim. Acta 43, 64–71.

O'NEILL, A. N. 1955A. 3,6-Anhydro-D-galactose as a constitutent of K-carrageenin. J. Am. Chem. Soc. 77, 2837–2839.

O'Neill, A. N. 1955B. Derivatives of 4-O-β-D-galactopyranosyl-3,6-anhydro-D-galactose from K-carrageenin. J. Am. Chem. Soc. 77, 6324–6326.

Painter, T. J. 1960. The polysaccharides of *Furcellaria fastigiata* and partial mercaptolysis of a gel-fraction. Can. J. Chem. 38, 112–118.

Painter, T. J. 1965. Fifth International Seaweed Symposium. Pergammon Press. Halifax, Nova Scotia.

Pernas, A. T., Smidsrød, O., Larsen, B., and Haug, A. 1967. Chemical heterogeneity of carrageenans as shown by fractional precipitation with potassium chloride. Acta. Chem. Scand. 21, 98–110.

Rees, D. A. 1961. Enzymic synthesis of 3,6-anhydro-L-galactose within porphyran from L-galactose 6-sulfate units. Biochem. J. 81, 347–352.

Rees, D. A. 1963. The carrageenan system of polysaccharides. Part I. The relation between the K- and λ-components. J. Chem. Soc., 1821–1832.

Smith, D. B., and Cook, W. H. 1953. Fractionation of carrageenin. Arch. Biochem. Biophys. 45, 232–233.

Smith, D. B., Cook, W. H., and Neal, J. L. 1954. Physical studies on carrageenin and carrageenin fractions. Arch. Biochem. Biophys. 53, 192–204.

Smith, D. B., O'Neill, A. N., and Perlin, A. S. 1955. Studies of the heterogeneity of carrageenin. Can. J. Chem. 33, 1352–1360.

Timell, T. E., and Mian, A. J. 1961. A study of pectin present in the inner bark of white birch (*Betula papyrifera* Marsh). Tappi. 44, 788–793.

Vollmert, B. 1950. Alkaline degradation of pectin. Makromol. Chem. 5, 110–127.

Whistler, R. L., and Ahmed, Z. F. 1950. The structure of guaran. J. Am. Chem. Soc. 72, 2524.

Whistler, R. L., and BeMiller, J. N. 1958. Alkaline degradation of polysaccharides. Advan. Carbohydrate Chem. 13, 289–329.

Whistler, R. L., and Durso, D. F. 1951. The isolation and characterization of two crystalline disaccharides from partial acid hydrolysis of guaran. J. Am. Chem. Soc. 73, 4189–4190.

Whistler, R. L., and Durso, D. F. 1952. A new crystalline trisaccharide from partial acid hydrolysis of guaran and the structure of guaran. J. Am. Chem. Soc. 74, 5140–5141.

Whistler, R. L., and Heyne, E. 1948. Chemical compositions and properties of guar polysaccharide. J. Am. Chem. Soc. 70, 2249–2252.

Whistler, R. L., and Hirase, S. 1961. Introduction of 3,6-anhydro rings into amylose and characterization of the products. J. Org. Chem. 26, 4600–4605.

Whistler, R. L., and Kirby, K. W. 1959. The composition of alginic acid in *Macrocystis pyrifera*. Z. Physiol. Chem. 314, 46–48.

Whistler, R. L., and Smart, C. L. 1953. Polysaccharide Chemistry. Academic Press, New York.

Whistler, R. L., and Smith, C. G. 1952. A crystalline mannotriose from the enzymatic hydrolysis of guaran. J. Am. Chem. Soc. 74, 3795–3796.

Whistler, R. L., and Zimmerman Stein, J. 1951. A crystalline mannobiose from the enzymatic hydrolysis of guaran. J. Am. Chem. Soc. 73, 4187–4188.

WHISTLER, R. L., LI, T. K., and DVONCH, W. 1948. Branched structure of guaran. J. Am. Chem. Soc. 70, 3144.

WHISTLER, R. L., UNRAU, D. G., and RUFFINI, G. 1968. Preparation and properties of a new series of starch sulfates. Arch. Biochem. Biophys. 126, 647–652.

WORTH, H. G. J. 1967. The chemistry and biochemistry of pectic substances. Chem. Rev. 67, 465–473.

YAPHE, W. 1959. The determination of K-carrageenin as a factor in the classification of the Rhodophycae. Can. J. Botany. 37, 751–757.

ZITKO, V., and BISHOP, C. T. 1965. Fractionation of pectins from sunflowers, sugar beets, apples and citrus fruits. Can. J. Chem. 43, 3206–3214.

ZITKO, V., and BISHOP, C. T. 1966. Structure of the galactomannan from sunflower pectic acid. Can. J. Chem. 44, 1275–1282.

ZITKO, V., ROSIK, J., and KUBALA, J. 1965. Pectic acid from wild apples (Malus sylvestris). Collection Czech. Chem. Commun. 30, 3902–3908.

CHAPTER 6

Ronald A. Pieringer | # Glycolipids[1,2]

INTRODUCTION

In an attempt to keep within the boundaries of allotted time and space, the subject of glycolipids will be approached by considering just two classes of glycolipids: the sphingoglycolipids and the glyceride glycolipids. The sphingoglycolipids have been the subject of a number of recent excellent reviews (Carter *et al.* 1965; Ledeen 1966; Stoffyn 1966; Wiegandt 1968). However, the glyceride glycolipids of animal origin have not been previously reviewed. For these reasons the glyceride glycolipids will be discussed in greater depth than the sphingoglycolipids.

GLYCERIDE GLYCOLIPIDS

The discussion of glyceride glycolipids will include those glycolipids in which a carbohydrate is in direct glycosidic linkage with the glycerol moiety of a diglyceride (or the closely related monoalkyl ether derivative). Glycolipids of this type have been isolated from forms of life ranging from bacteria and plants to nerve tissues of animals.

Glyceride Glycolipids of Animals

The glyceride glycolipid of animal origin, first reported by Steim and Benson (1963), Norton and Brotz (1963), and Rouser *et al.* (1963) was shown to be a constituent of brain and to have the structure of a monogalactosyl diglyceride (Fig. 53). These initial findings have been confirmed by data in so many other publications (Rumsby and Gray 1965; Pelick *et al.* 1965; Wells and Dittmer 1966; Rumsby 1967; Steim 1967) from a number of laboratories that there can be little doubt as to the existence of monogalactosyl diglyceride in brain and some other nerve tissues. The detailed structure of the galactolipid has been shown conclusively by Steim (1967), who used the techniques of cochromatography, infrared spectroscopy, nuclear magnetic resonance, and the

[1] Supported by grants from The United Cerebral Palsy Research and Educational Foundation and The National Institute of Allergy and Infectious Diseases, United States Public Health Service.
[2] The helpful criticisms and suggestions made by Dr. Leonard N. Norcia after reading this manuscript are greatly appreciated.

specific β-galactosidase, to be 1,2-diacyl-3-(β-D-galactopyranosyl)-*sn*-glycerol.[3] Rumsby (1967) has also contributed to the detailed characterization of the lipid. Work from our laboratory has demonstrated that the monogalactosyl diglyceride of brain is biosynthesized from UDP-galactose and 1,2-diglyceride in the presence of a particulate enzyme from rat brain (Wenger *et al.* 1967; Subba Rao *et al.* 1968). The biosynthesized monogalactosyl diglyceride has been shown to have the same structure as the monogalactosyl diglyceride isolated from brain.

Unfortunately the presence of a monoalkyl-monoacyl glyceryl galactoside (Norton and Brotz 1963; Rumsby 1967) in addition to the diacyl glyceryl galactoside (monogalactosyl diglyceride) in brain has not been

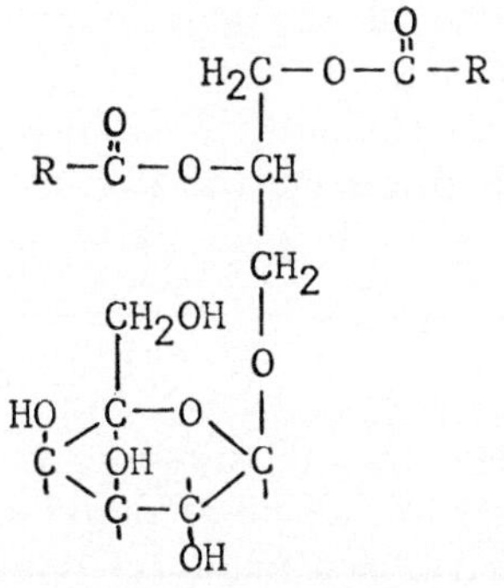

FIG. 53. 1,2-DIACYL-3-(β-D-galactopyranosyl) - *SN* - GLYCEROL (MONOGALACTO-SYL DIGLYCERIDE) OF BRAIN

settled to the same degree of certainty that the existence of monogalactosyl diglyceride has been established. Norton and Brotz (1963) and Rumsby (1967) present convincing results which indicate that galactose of a glyceride galactolipid fraction from sheep or bovine brain remains soluble in organic solvents after saponification. This fact together with fatty acid ester to galactose ratios and other data led Norton and Brotz (1963) and Rumsby (1967) to conclude that there does exist a monoalkyl ether derivative of monogalactosyl diglyceride in brain. However, Wells and Dittmer (1966), who developed an excellent microanalytical technique for the determination of brain lipids, could not detect a monoalkyl ether derivative of galactosyl diglyceride

[3] The stereospecific number nomenclature system (IUPAC-IUB Commission on Nomenclature. 1967. The nomenclature of lipids. Biochem. *6*, 3287–3292.) is employed in this paper for the glyceride glycolipids.

at a concentration equivalent to 0.05 μmole per gram wet weight of rat brain. On this basis they ruled out the presence of significant quantities of monoalkyl ether analog. The impasse between the two viewpoints seems difficult to resolve. However, 1 obvious difference between the 2 findings is that Wells and Dittmer used rat brain while Norton and Brotz, and Rumsby used bovine and sheep brain as the source of the lipid.

Galactosyl diglyceride of animal origin has been found to be restricted (with possibly one exception) to nerve tissues. Steim and Benson (1963) could not detect this lipid in mammary gland, intestine, spleen, liver, and kidney. Steim (1967) in a later paper confirmed these results except that trace amounts may have been detected in kidney. The identification of the galactolipid in kidney was quite tentative and may be subject to change.

The major fatty acid constituents of the monogalactosyl diglyceride of sheep brain were palmitic acid (56.42%), stearic acid (17.96%), and myristic acid (10.26%) (Rumsby 1967), and of beef spinal cord monogalactosyl diglyceride were palmitic acid (53.2%), and oleic acid (22.3%) (Steim 1967). No hydroxy fatty acids were detected as a component of galactosyl glyceride even though hydroxy fatty acids are normally found in cerebrosides and sulfatides isolated from brain.

Studies on the isolation of galactosyl diglyceride have been limited to the tissue level, and have been found to comprise about 0.4 to 0.6% of total lipids (Steim 1967; Rumsby 1967) of nerve tissue. No work as yet has been done on the exact cellular location of galactosyl diglyceride. Steim (1967) and Norton and Brotz (1963) found the lipid in white matter, but not in gray matter. Since myelin is a major component of the former portion of the brain, these results suggest that the galactolipid may be an integral part of myelin. Evidence supporting this view is that the concentration of monogalactosyl diglyceride in brain of rats (Wells and Dittmer 1967) and the activity of the enzyme responsible for its synthesis (Wenger et al. 1968) increases dramatically during the period of most active myelination (in rat about 10 to 20 days old) (Fig. 54).

A glycolipid having the chromatographic properties of a digalactosyl diglyceride (Fig. 55) has been found in extracts of human brain (Rouser et al. 1967). The detailed structure of this isolated galactolipid was not reported. However, in vitro studies on the biosynthesis of glyceride glycolipids have revealed that brain tissues have the enzymatic capability to synthesize digalactosyl diglyceride (Wenger et al. 1967; Subba Rao et al. 1968). The galactolipid is formed from UDP-galactose and

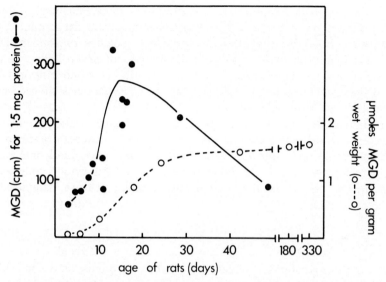

FIG. 54. EFFECT OF AGE OF RAT ON CONCENTRATION OF GALACTOSYL DIGLYCERIDE AND ON ACTIVITY OF ENZYME RESPONSIBLE FOR THE SYNTHESIS OF MONOGALACTOSYL DIGLYCERIDE IN ISOLATED BRAIN OF RATS

Concentration of MGD from Wells and Dittmer. Enzyme activity from Wenger, Petitpas, and Pieringer.

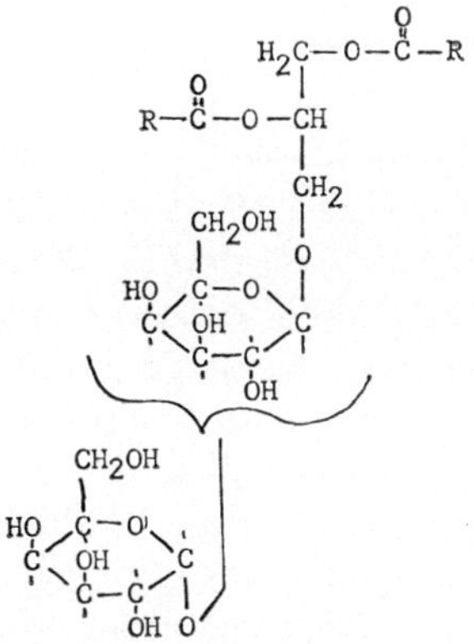

FIG. 55. 1,2-DIACYL-3-(α-D - GALACTOPYRANOSYL - β - D - GALACTOPYRANOSYL) - SN-GLYCEROL (DIGALACTOSYL DIGLYCERIDE) OF BRAIN

1,2-diacyl-3-(β-D-galactopyranosyl)-sn-glycerol (monogalactosyl diglyceride) in the presence of a particulate fraction from rat brain. The galactosidic bond between the two galactose moieties of the galactolipid has been shown by specific galactosidase action to have an α-stereochemical configuration. While the point of attachment of the acetal linkage between the two galactose moieties has not as yet been elucidated, the remainder of the structure of the biosynthesized galactosyl diglyceride has been characterized as 1,2-diacyl-3-(α-D-galactopyranosyl-β-D-galactopyranosyl)-sn-glycerol. It is interesting to note that the digalactosyl diglyceride may be the only glycolipid of animal origin to have an aldehydic bond with an α-configuration. An α-ketosidic linkage occurs in the attachment of sialic acid to gangliosides (Wiegandt 1968).

Glyceride Glycolipids of Plants

Of all the glycolipids the interest of the nutritionist and food biochemist may be greatest in the glyceride glycolipids of plants. The reason for this supposition is that the glyceride glycolipids are a major lipid component of plants which in turn can serve as a major source of dietary glycolipid for animals. Glyceride glycolipids, for example, comprise 23, 20, and 13% of the total lipids of red clover (Weenirk 1961), runner-bean leaves (Sastry and Kates 1964), and alfalfa (O'Brien and Benson 1964), respectively. As in brain the glyceride glycolipid of plants has thus far been found to contain only galactose as the carbohydrate component. Monogalactosyl and digalactosyl diglyceride were first discovered in wheat flour by Carter et al. (1956) and have been completely characterized as 1,2-diacyl-3-(β-D-galactopyranosyl)-sn-glycerol and 1,2-diacyl-3-(6-α-D-galactopyranosyl-β-D-galactopyranosyl)-sn-glycerol, respectively (Carter et al. 1961 A, B). Similar structures have been found for the galactosyl diglycerides of spinach (Zill and Harmon 1962) and runner-bean leaves (Sastry and Kates 1964).

Light energy appears to have a profound effect on the concentration of galactosyl diglycerides in many plant tissues. Cabbage was found to have higher concentrations of galactolipids in those tissues exposed to the highest amount of light (Nichols 1963). Wintermans (1960) found more galactolipids in the green leaves of *Sambus nigra* and *Phaseolus vulgaris* than in the yellow leaves. Some of the more striking results on the effect of light on the concentration of galactosyl diglycerides have been observed in more elementary cells such as those of the photosynthetic microorganism, *Euglena gracilis*. On exposure of dark-grown cells of *Euglena gracilis* to constant light for periods up to 160 hr, the cells accumulated galactosyl diglycerides and chlorophyll simultaneously at almost linear rates (Rosenberg and Gouaux 1967). The concentration of monogalactosyl diglyceride rose from 2 μmoles per 100 mg of the dark-grown cells to 27 μmoles in the fully green cells; the digalactosyl diglyceride concentration increased from 1 to 11 μmoles. Galactosyl diglycerides with different fatty acid constituents accumulated under these conditions to varying degrees. Predominant at the end of the greening process were the digalactosyl diglyceride with 16 : 2 fatty acids and the monogalactosyl diglycerides with 16 : 2, 16 : 3, 16 : 4, 18 : 2, and 18 : 4 fatty acids. This study and other studies which have demonstrated that chloroplasts of photosynthesizing cells contain large quantities of galactosyl diglycerides (Allen and Good 1965; Benson 1964) suggest that galactosyl diglycerides may have a function in the assembly of chloroplasts. The relatively high content of linolenic acid in the galactolipids of a variety of chloroplasts of higher plants has led to the proposal that the galactolipids of the chloroplasts function in the electron transport of photosynthesis (Bloch *et al.* 1967).

The biosynthesis of mono- and digalactosyl diglycerides (and possibly higher homologs) has been investigated by Neufeld and Hall (1964), and Ongun and Mudd (1968). Chloroplasts of spinach catalyzed the formation of the two galactolipids from UDP-galactose and diglyceride; monogalactosyl diglyceride functioned as an intermediary metabolite in the reaction. The general structures of the biosynthesized galactolipids were determined by cochromatography, but the detailed structures have not as yet been determined. Although etiolated pea leaves were able to carry out the synthesis, light increased the activity of the biosynthetic enzymes within 11 hr by a factor of two- to threefold (Ongun and Mudd 1968).

A galactolipid composed of glycerol, galactose, and fatty acid in the molar ratio of 1 : 1 : 3 has been isolated from homogenates of spinach leaves (Heinz 1965, 1967A, B). The galactolipid is formed enzymatically at an optimum pH of 4.6 by the transfer of a fatty acyl group from either galactosyl diglyceride or a phospholipid to monogalactosyl diglyceride. The 3rd fatty acid is esterified mainly to the 6-hydroxyl position of the galactose moiety. However, the 3 or 4 positions also have been found in ester linkage with the fatty acid.

The unique sulfolipid of plants has been characterized as 1,2-diacyl-3-(α-D-6-sulfo-6-deoxyglucopyranosyl)-*sn*-glycerol (Benson 1963). The sulfur containing constituent is a sulfonic acid (not sulfuric acid) group. The biosynthesis of this glycolipid with its intriguing carbon to sulfur bond has not as yet been elucidated. The sulfolipid, similar to the galactosyl diglycerides of plants, is concentrated in the chloroplast fraction, usually contains a relatively large amount of linoleic acid (except in alfalfa) (O'Brien and Benson 1964), and is found in higher concentrations in light-grown cells of *Euglena gracilis* than in dark-grown cells (Rosenberg 1963).

Glyceride Glycolipids of Bacteria

One of the areas of most active research in the field of glycolipids in recent years has been that area involving the determination of the structure and metabolism of the glyceride glycolipids of bacteria. The numerous studies undertaken have established a trend which indicates that the glyceride glycolipids are the principal glycolipid of all Gram-positive bacteria, but are absent from Gram-negative bacteria (Shaw and Baddiley 1968). The segregation of these glycolipids between two morphologically different types of bacteria; the location of the glycolipid in the membranes of Gram-positive bacteria (Vorbeck and Marinetti 1965; Cohen and Panos 1966; Bishop *et al.* 1967); the high probability of the location of the enzymes responsible for the biosynthesis of the lipids (except dimannosyl diglyceride of *M. lysodeikticus*) in membrane-containing fractions of the cells (Kaufman *et al.* 1965; Lennarz and Talamo 1966; Pieringer 1968); and the molecular conformation that the diglycosyl diglycerides can assume (Brundish *et al.* 1967) have led to the speculation that the glycosyl diglycerides might have some role in the passage of molecules into and out of the cell, or in the biosynthesis of nonlipid constituents.

Depending on the bacterial source the percent of glyceride glycolipid of the total lipid may be small (a few percent) as in lactobacilli and staphylococci or it may be quite significant (greater than 40%) in *Microbacterium lacticum* and *Mycoplasma laidlawii B* (Shaw and Baddiley 1968). The structures of the glyceride glycolipids of bacterial origin are more variable than those of animal and plant origin. The carbohydrate moiety of the monoglycosyl diglycerides may be glucose, galactose, or mannose. The combinations of these carbohydrates in the diglycosyl diglycerides thus far characterized are α-D-*mannosyl* (1 → 3)-α-D-mannosyl- found in *M. lysodeikticus* (Lennarz and Talamo 1966), α-D-galactosyl-(1 → 2)-α-D-glucosyl- found in *Pneumococcus* types I and XIV (Brundish *et al.* 1965A; and Kaufman *et al.* 1965), β-D-glucosyl-(1 → 6)-β-D-glucosyl- found in *Staphylococcus lactis* (Brundish *et al.* 1967), α-D-glucosyl-(1 → 2)-α-D-glucosyl- found in *S. faecalis* (Pieringer

1968; Shaw and Baddiley 1968), and β-D-galactosyl-$(1 \rightarrow 6)$-β-D-galactosyl- found in *Arthrobacter globiformis* (Walker and Bastl, cited by Shaw and Baddiley 1968). (The review by Shaw and Baddiley (1968) is recommended for a more comprehensive listing of the above disaccharide combinations found in other bacteria.) Biosynthetic studies, which demonstrated a specific requirement for 1,2-diglycerides as substrates (Lennarz and Talamo 1966; Pieringer 1968), and chemical analysis (Brundish *et al.* 1965B) have established that the carbohydrate is linked to the 3-position and the fatty acids to the 1- and 2- positions of the glycerol moiety. It has been observed that in all naturally occurring phospholipids the nonlipid group is linked to an alcohol function at the 3-position of the phospholipid (Hanahan 1960). Apparently this generalization can also be applied to the structure of the glycosyl diglycerides of animal, plant, and bacterial origin. Biosynthetic studies have been employed to significant advantage in the identification of the position of attachment of the outer hexose of the diglycosyl diglycerides. In the case of the diglucosyl diglyceride of S. *faecalis* it was possible to prepare this glucolipid with *only* the *internal* glucose labeled uniformly with ^{14}C (Pieringer 1968). Because the 2 glucose units could be differentiated from one another, the demonstration of the linkage as $1 \rightarrow 2$ through the use of periodate oxidation, sodium borohydride reduction, acid hydrolysis and paper chromatography was relatively facile.

SPHINGOGLYCOLIPIDS

Cerebrosides

At the focal point of the structure of the glycosphingosides is, of course, the long chain base sphingosine (2-amino-*trans*-4-octadecene-1,3-diol) or some closely related derivative (e.g., dihydrosphingosine, C_{20}-dihydrosphingosine, or phytosphingosine). Attachment of the base to a long chain fatty acid through an amide linkage produces a compound referred to as ceramide. Ceramides in glycosidic linkage at the primary alcohol function of the sphingosine moiety with mono- or oligosaccharides comprise that family of compounds that share the common name of glycosyl ceramides. The glycosyl ceramides have been isolated from animal and plant sources and contain from 1 to 4 monosaccharide units (Fig. 56). Ceramide-monohexosides (cerebrosides) of animal origin contain either glucose or galactose, and mostly C_{18}-sphingosine and some C_{18}-dihydrosphingosine (Carter *et al.* 1965). The galactose containing cerebrosides have been isolated from human kidney, brain, and spinal cord, while the glucose-containing cerebrosides have been found in human serum, spleen, and liver. In the genetically linked Gaucher's disease extremely high concentrations of glucocerebroside accumulate in the spleen (Philippart and Menkes 1967). The genetic defect appears to be the lack of a β-glucosidase (Brady et al. 1965; Patrick 1965).

The ceramide dihexosides have either 2 galactose units or 1 each of glucose and galactose. The ratio glucose to galactose in ceramide trihexosides is 1 to 2. A trihexoside having the structure galactosyl-$(1 \rightarrow 4)$-galactosyl-$(1 \rightarrow 4)$-glucosyl-ceramide has been isolated from the kidney of a patient with Fabry's disease (Sweeley and Klionsky, cited in Carter et al. 1965).

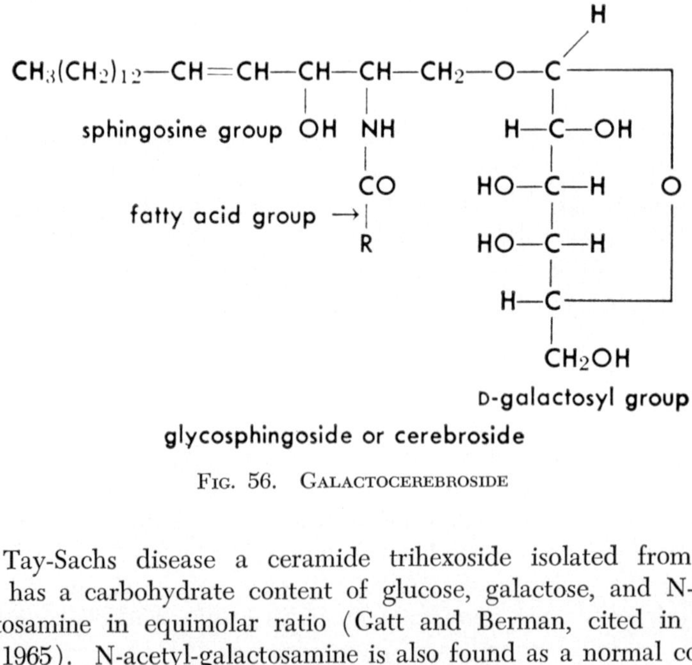

FIG. 56. GALACTOCEREBROSIDE

In Tay-Sachs disease a ceramide trihexoside isolated from brain tissue has a carbohydrate content of glucose, galactose, and N-acetyl-galactosamine in equimolar ratio (Gatt and Berman, cited in Carter et al. 1965). N-acetyl-galactosamine is also found as a normal constituent in combination with galactose and glucose in a ratio of $1 : 2 : 1$ respectively in ceramide tetrahexosides (globosides) (Martensson; Rapport et al., cited in Carter et al. 1965). The structure of the fatty acid moieties of ceramide glycosides varies depending to some degree upon the organ from which the compound is isolated. For example, kidney cerebrosides have predominantly hydroxy fatty acids with 22 and 24 carbon-chain lengths; however, the fatty acids of spleen cerebrosides are mostly nonhydroxylated. In brain the fatty acid composition of cerebrosides ranges in chain length from C_{16} to C_{26} in both hydroxylated and nonhydroxylated types of fatty acid.

Sulfatides

Sulfatides are usually thought of as sulfated (sulfuric acid group) derivatives of galactosyl ceramide (Fig. 57). The sulfate moiety has been shown to exist in the 3-position of the galactosyl moiety (Stoffyn 1966). Stoffyn (1966) has conclusively shown that the galactose is in

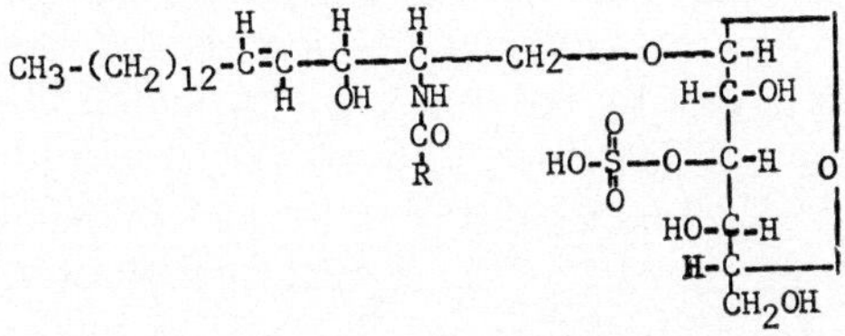

FIG. 57. SULFATIDE

β-glycosidic linkage to the ceramide portion of the molecule. In addition to the ceramide monohexoside sulfate, a ceramide dihexoside sulfate containing glucose and galactose has also been isolated (Stoffyn 1968). Sulfatides appear in higher than normal concentrations in the white matter of brain in the demyelinating disease, metachromatic leukodystrophy (Stoffyn 1966). The deficiency of a cerebroside sulfatase may be the cause of the accumulation of the sulfatides in this disease.

Gangliosides

Glycosyl ceramides containing sialic acids (acylated neuraminic acids) (Fig. 58) are commonly designated gangliosides. Of the dozen or so gangliosides discussed by experts in recent reviews (Carter *et al.* 1965;

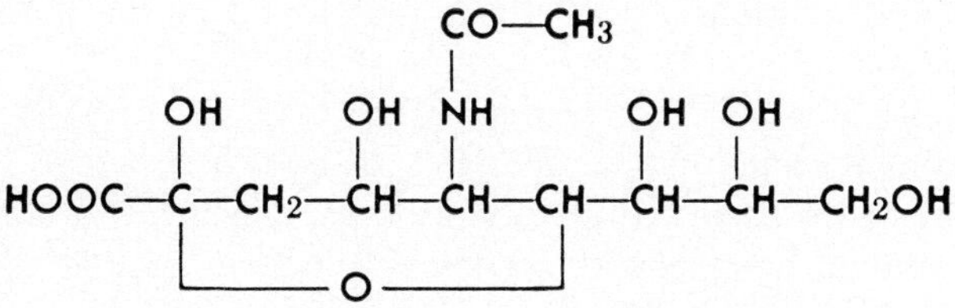

β-D (—)-N-acetylneuraminic acid, sialic acid

FIG. 58. N-ACETYLNEURAMINIC ACID

Ledeen 1966; Wiegandt 1968), the major structural difference of these gangliosides occurs in the carbohydrate moiety. It is this structural variation that forms the basis of the physicochemical separation, the nomenclature, and the immunochemical properties of gangliosides. The carbohydrate moiety linked to the ceramide portion may contain up to four monosaccharide units. The hexose units most often found are glucose, galactose, and N-acetylgalactosamine (in nerve tissues) or N-acetylglucosamine (in other tissues) (Ledeen 1966). The tetrasaccharide-ganglioside appears to be the most prevalent ganglioside. However, mono-, di-, and trisaccharide-containing gangliosides are also found. The four predominant gangliosides isolated from brain contain a common tetrasaccharide structure, galactosyl-(β,1 → 3)-N-acetylgalactosaminyl-(β,1 → 4)-galactosyl-(β,1 → 4)-glucosyl-(1 → 1)-ceramide to which is bound 1 to 3 molecules of sialic acid (Wiegandt 1968). The sialic acid, which in the case of these brain gangliosides is an N-acetyl neuraminic acid, may be linked to the internal galactose in a 2 → 3 ketosidic bond to form a monosialoganglioside. The disialogangliosides are of two varieties. In one the second sialic acid is attached in a 2 → 8 ketosidic linkage to the sialic acid of monosialoganglioside. In the other the second sialic acid is attached to the terminal galactose in a 2 → 3 linkage. The 2 types of linkage of the sialic acid of the 2 disialogangliosides are combined in the trisialoganglioside shown below:

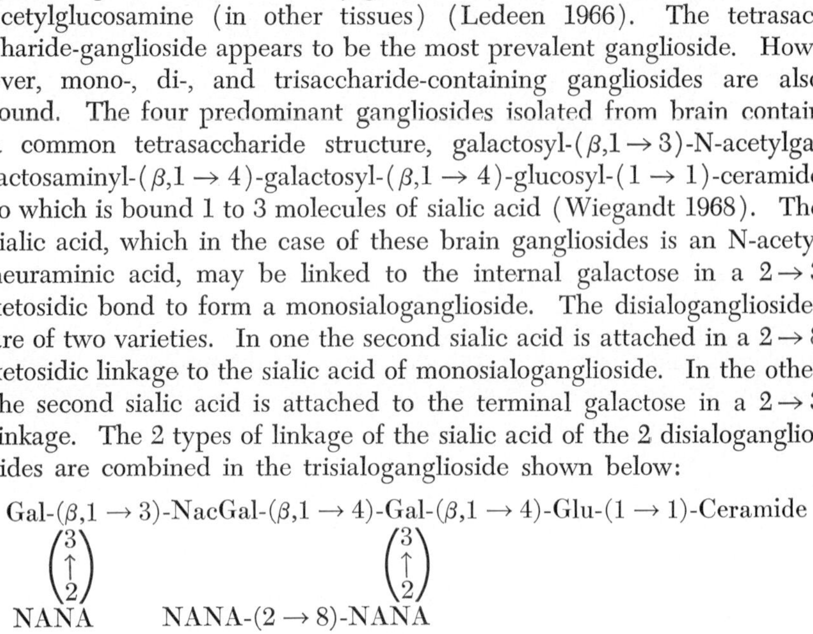

Gal-(β,1 → 3)-NacGal-(β,1 → 4)-Gal-(β,1 → 4)-Glu-(1 → 1)-Ceramide

$$\left(\begin{array}{c}3\\ \uparrow \\ 2\end{array}\right) \qquad\qquad \left(\begin{array}{c}3\\ \uparrow \\ 2\end{array}\right)$$

NANA   NANA-(2 → 8)-NANA

trisialoganglioside

The most striking changes in ganglioside concentration from the normal appears to occur in Tay-Sachs disease. A monosialoganglioside, in which the sialic acid is attached to the galactose unit of N-acetylgalactosaminyl-galactosyl-glucosyl-ceramide, accumulates in white and gray matter in concentrations 2 to 3 times that found in the normal (Wiegandt 1968).

Gangliosides isolated from sources other than nerve tissues have not as yet been as thoroughly characterized as those from brain. However, the structure of the main ganglioside of visceral system contains a single sialic acid unit in linkage with galactosyl-glucosyl-ceramide (Wiegandt 1968). In horse erythrocytes the sialic acid is an N-glycosylneuraminic acid (Klenk and Padberg; and Yamakawa and Suzuki, cited in Ledeen 1966) rather than the N-acetyl derivative.

The sphingosine component of gangliosides of the brain changes in structure with age. The gangliosides of fetal brain contain mostly C_{18}-sphingosine; however, in adult brain C_{20}-sphingosine greatly predominates (Svennerholm cited in Wiegandt 1968). Small amounts of C_{18}- and C_{20}-dihydrosphingosines have also been detected (Sambasivarao and McCluer 1964).

The brain gangliosides all have stearic acid as the major fatty acid constituent. Analysis of brain gangliosides by Sambasivarao and Mc-Cluer (1964) have shown the fatty acid compositions to be 86–95% stearic acid, 1–2.5% palmitic acid, and 2–12% arachidonic acid. It is interesting to note that the gangliosides of spleen have lignoceric acid as the major fatty acid (Wiegandt 1968).

DISCUSSION

D. F. Farkas.—Have any feeding studies been carried out in glyceride glycolipids? Do you have any data on their caloric density, or the level you can feed them in a diet as a source of calories and fat, particularly unsaturated fatty acids?

R. A. Pieringer.—No, there have been no studies done on that level. One could calculate the caloric content from the lipids that have been isolated from plants but there have been no published data.

BIBLIOGRAPHY

ALLEN, C. F., and GOOD, P. 1965. Plant lipids. J. Am. Oil Chemists' Soc. 42, 610–614.

BENSON, A. A. 1963. The plant sulfolipid. Advan. Lipid Res. 1, 387–394.

BENSON, A. A. 1964. Plant membrane lipids. Ann. Rev. Plant Physiol. 15, 1–16.

BISHOP, D. G., RUTBERG, L., and SAMUELSSON, B. 1967. The chemical composition of the cytoplasmic membrane of Bacillus subtilis. Europ. J. Biochem. 2, 448–453.

BLOCH, K., CONSTANTOPOULOS, G., KENYON, C., and NAJAI, J. 1967. Lipid metabolism of algae in the light and in the dark. In Biochemistry of Chloroplasts. Vol. II, T. W. Goodwin (Editor). Academic Press, New York.

BRADY, R. O., KANFER, J. N., SHAPIRO, D. 1965. Metabolism of glucocerebosides. II. Evidence of an enzymatic deficiency in Gaucher's disease. Biochem. Biophys. Res. Commun. 18, 221–225.

BRUNDISH, D. E., SHAW, N., and BADDILEY, J. 1965A. The glycolipids from a rough strain of Pneumococcus type I. Biochem. Biophys. Res. Commun. 18, 308–311.

BRUNDISH, D. E., SHAW, N., and BADDILEY, J. 1965B. The glycolipids from the non-capsulated strain of Pneumococcus I-192R, A.T.C.C.12213. Biochem. J. 97, 158–165.

112   CARBOHYDRATES AND THEIR ROLES

BRUNDISH, D. E., SHAW, N., and BADDILEY, J. 1967. The structure and possible function of the glycolipid from Staphylococcus lactis I 3. Biochem. J. 105, 885–889.

CARTER, H. E., HENDRY, R. A., and STANACEV, N. Z. 1961A. Wheat flour lipids: III. Structure of the mono- and digalactosyl glycerol lipids. J. Lipid Res. 2, 223–227.

CARTER, H. E., JOHNSON, P., and WEBER, E. J. 1965. Glycolipids. Ann. Rev. Biochem. 34, 109–142.

CARTER, H. E., McCLUER, R. H., and SLIFER, E. D. 1956. Lipids of wheat flour: I. Characterization of galactosylglycerol components. J. Am. Chem. Soc. 78, 3735–3738.

CARTER, H. E. et al. 1961B. Wheat flour lipids: II. Isolation and characterization of glycolipids of wheat flour and other plant sources. J. Lipid Res. 2, 215–222.

COHEN, M., and PANOS, C. 1966. Membrane lipid composition of Streptococcus pyogenes and derived L form. Biochemistry 5, 2385–2392.

HANAHAN, D. J. 1960. Lipide Chemistry, John Wiley and Sons, New York.

HEINZ, E. 1965. Enzymatic acylation of monogalactosyl diglyceride by isolated chloroplasts. Z. Naturforsch. 20b, 83. (German)

HEINZ, E. 1967A. Acylgalactosyl diglyceride from leaf homogenates. Biochim. Biophys. Acta. 144, 321–332. (German)

HEINZ, E. 1967B. On the enzymatic formation of acylgalactosyl diglyceride. Biochim. Biophys. Acta. 144, 333–343. (German)

KAUFMAN, B., KUNDIG, F. D., DISTLER, J., and ROSEMAN, S. 1965. Enzymatic synthesis and structure of two glycolipids from type XIV Pneumococcus. Biochem. Biophys. Res. Commun. 18, 312–318.

LEDEEN, R. 1966. The chemistry of gangliosides: A review. J. Am. Oil Chemists' Soc. 43, 57–66.

LENNARZ, W. J., and TALAMO, B. 1966. The chemical characterization and enzymatic synthesis of mannolipids in Micrococcus lysodeikticus. J. Biol. Chem. 241, 2707–2719.

NEUFELD, E. F., and HALL, C. W. 1964. Formation of galactolipids by chloroplasts. Biochem. Biophys. Res. Commun. 14, 503–508.

NICHOLS, B. W. 1963. Separation of the lipids of photosynthetic tissues; improvements in analysis by thin-layer chromatography. Biochim. Biophys. Acta. 70, 417–422.

NORTON, W. T., and BROTZ, M. 1963. New galactolipids of brain: A mono-alkyl-monoacyl-glyceryl galactoside and cerebroside fatty acid esters. Biochem. Biophys. Res. Commun. 12, 198–203.

O'BRIEN, J. S., and BENSON, A. A. 1964. Isolation and fatty acid composition of the plant sulfolipid and galactolipids. J. Lipid Res. 5, 432–436.

ONGUN, A., and MUDD, J. B. 1968. Biosynthesis of galactolipids in plants. J. Biol. Chem. 243, 1558–1566.

PATRICK, A. D. 1965. A deficiency of glucocerebrosidase in Gaucher's disease. Biochem. J. 97, 17c–18c.

PELICK, N. et al. 1965. Some practical aspects of thin-layer chromatography of lipids. J. Am. Oil Chemists' Soc. 42, 393–399.

PHILIPPART, M., and MENKES, J. H. 1967. Isolation and characterization of the principal cerebral glycolipids in the infantile and adult forms of

Gaucher's disease. *In* Inborn Disorders of Sphingolipid Metabolism, S. M. Aronson, and B. W. Volk (Editors). Pergamon Press, New York.

PIERINGER, R. A.   1968.   The metabolism of glyceride glycolipids: I. Biosynthesis of monoglucosyl diglyceride and diglucosyl diglyceride by glucosyltransferase pathways in *Streptococcus faecalis*. J. Biol. Chem. *243*, 4894–4903.

ROSENBERG, A.   1963.   A comparison of lipid patterns in photosynthesizing and nonphotosynthesizing cells of *Euglena gracilis*. Biochemistry *2*, 1148–1154.

ROSENBERG, A., and GOUAUX, J.   1967.   Quantitative and compositional changes in monogalactosyl and digalactosyl diglycerides during light-induced formation of chloroplasts in *Euglena gracilis*. J. Lipid Res. *8*, 80–83.

ROUSER, G., KRITCHEVSKY, G., HELLER, D., and LIEBER, E.   1963.   Lipid composition of beef brain, beef liver, and sea anemone: two approaches to quantitative fractionation of complex lipid mixtures. J. Am. Oil Chemists' Soc. *42*, 393–399.

ROUSER, G., KRITCHEVSKY, G., SIMON, G., and NELSON, G. J.   1967.   Quantitative analysis of brain and spinach leaf lipids employing silicic acid column chromatography and acetone for elution of glycolipids. Lipids *2*, 37–40.

RUMSBY, M. G.   1967.   Preparation and characterization of a glycerogalactolipid fraction from sheep brain. J. Neurochem. *14*, 733–741.

RUMSBY, M. G., and GRAY, I. K.   1965.   A monogalactolipid component in extracts of sheep brain. J. Neurochem. *12*, 1005–1006.

SAMBASIVARAO, K., and MCCLUER, R. H.   1964.   Lipid components of gangliosides. J. Lipid Res. *5*, 103–108.

SASTRY, P. S., and KATES, M.   1964.   Lipid components of leaves. V. Galactolipids, cerebrosides, and lecithin of runner-bean leaves. Biochemistry *3*, 1271–1280.

SHAW, N., and BADDILEY, J.   1968.   Structure and distribution of glycosyl diglycerides in bacteria. Nature *217*, 142–144.

STEIM, J. M.   1967.   Monogalactosyl diglyceride: a new neurolipid. Biochim. Biophys. Acta. *144*, 118–126.

STEIM, J. M., and BENSON, A. A.   1963.   Galactosyl diglyceride in brain. Federation Proc. *22*, 299.

STOFFYN, P. J.   1966.   The structure and chemistry of sulfatides. J. Am. Oil Chemists' Soc. *43*, 69–74.

STOFFYN, A., STOFFYN, P., and MARTENSSON, E.   1968.   Structure of kidney ceramide dihexoside sulfate. Biochim. Biophys. Acta. *152*, 353–357.

SUBBA RAO, K., WENGER, D. A., and PIERINGER, R. A.   1968.   Metabolism of digalactosyl diglyceride in brain. Federation Proc. *27*, 346.

VORBECK, M., and MARINETTI, G. V.   1965.   Intracellular distribution and characterization of the lipids of *Streptococcus faecalis* (ATCC 9790). Biochemistry *4*, 296–305.

WEENIRK, R. O.   1961.   Acetone-soluble lipids of grasses and other forage plants. I. Galactolipides of red clover (*Trifolium pratense*) leaves. J. Sci. Food Agr. *12*, 34–38.

WELLS, M. A., and DITTMER, J. C. 1966. A microanalytical technique for the quantitative determination of twenty-four classes of brain lipids. Biochemistry 5, 3405–3418.

WELLS, M. A., and DITTMER, J. C. 1967. A comprehensive study of the postnatal changes in the concentration of the lipids of developing rat brain. Biochemistry 6, 3169–3175.

WENGER, D. A., PETITPAS, J. W., and PIERINGER, R. A. 1967. Biosynthesis of galactosyl diglyceride in brain. Federation Proc. 26, 675.

WENGER, D. A., PETITPAS, J. W., and PIERINGER, R. A. 1968. The metabolism of glyceride glycolipids: II. biosynthesis of monogalactosyl diglyceride from UDP-galactose and diglyceride in brain. Biochemistry 7, 3700–3707.

WIEGANDT, H. 1968. The structure and the function of gangliosides. Angewandte Chem. 7, 87–96.

WINTERMANS, J. F. G. M. 1960. Concentrations of phosphatides and glycolipides in leaves and chloroplasts. Biochim. Biophys. Acta. 44, 49–54.

ZILL, L. P., and HARMON, E. A. 1962. Lipids of photosynthetic tissue. I. Silicic acid chromatography of the lipids from whole leaves and chloroplasts. Biochim. Biophys. Acta. 57, 573–583.

CHAPTER 7

A. Neuberger and
R. D. Marshall

Aspects of the Structure of Glycoproteins

INTRODUCTION

Many proteins contain only the "standard" amino acids, i.e., the set of 20 amino acids for which codons exist in the genome. According to present belief, which is well founded, the nature and sequence of such amino acid residues is directly and in general unambiguously determined by the base sequence in the nucleic acid controlling the synthesis of that particular protein. However, there are amino acids found in many proteins for which no codons exist and which probably arise from modifications of "standard" amino acids. These modifications are most likely produced after the formation of the peptide chains is completed. Examples of such situations are the occurrence of hydroxyproline (Stetten 1948) and of hydroxylysine (Sinex and Van Slyke 1955) in collagen, the presence of the desmosines in elastin or the occurrence of ϵ-N-methyl-lysine and ϵ-N-dimethyl-lysine in histones and in flagellar proteins of *Salmonella*. These changes clearly arise from the action of specific enzymes modifying certain amino acid residues probably in a completed peptide chain. Similar "postsynthetic" changes are the acetylation of terminal amino groups of some proteins or the phosphorylation of certain β-hydroxy-α-amino acid residues observed in casein and other phospho-proteins.

Proteins containing relatively large groups of nonamino acid type have generally been called conjugated proteins. Thus lipoproteins may consist of up to 80 or 90% of their mass of cholesterol, triglycerides, and phospholipids. Nuclear proteins isolated from cell nuclei contain large amounts of deoxyribonucleic acid. But with these types of compounds the association between the protein and nonpeptide moiety is not effected through covalent bonds, but the binding occurs through Coulomb forces or hydrophobic interactions. The position is different with the glycoproteins.

We shall call glycoproteins substances in which sugar residues are linked to a peptide chain by clearly defined covalent bonds. Three types of such covalent bonding are known and these we shall discuss in some detail. However, we might first briefly mention that there is as yet no universal agreement about the exact definition of glycoproteins. Gottschalk (1966) has fully discussed the history of this argument on nomenclature of glycoproteins and his own definition excludes such substances as chondroitin 4-sulphate, chondroitin 6-sulphate, and dermatan sulphate. On the other hand he includes soluble blood group substances under the heading of glycoprotein. All these types of substances contain a relatively large number of sugar residues and on a weight basis the carbohydrate fraction is much larger than the peptide moiety. A good case can certainly be made for such a division as Gottschalk has proposed, but to us it seems somewhat artificial. A rational classification may have to wait until our knowledge is more complete than it is at present, but in the meantime we may call glycoproteins substances which contain peptide chains made by the ordinary processes of protein synthesis in the cell and to which carbohydrate moieties have been added by the formation of covalent bonds. This definition takes no account of the quantitative proportions of amino acids on the one hand and sugar moieties on the other. We thus include in our definition all mucopolysaccharides which contain covalently bound peptide.

A SUBDIVISION OF GLYCOPROTEINS

The list of glycoproteins described in the literature is at present increasing at a fairly high rate. They vary greatly in molecular weight; thus they include substances such as avidin (mol wt $17 \cdot 10^3$; Melamed and Green 1963), hen's egg albumin (mol wt $45 \cdot 10^3$), γ-G-globulin (mol wt $150 \cdot 10^3$), Le^a-substance (mol wt $1000 \cdot 10^3$, Pusztai and Morgan 1963), and cervical mucin (mol wt $4000 \cdot 10^3$; Gibbons 1959). They also vary greatly in the type of sugar they contain; thus the soluble collagen has only the two sugars, glucose and galactose; egg albumin contains only mannose and N-acetyl glucosamine. On the other hand γ-G-globulin contains mannose, galactose, fucose, N-acetyl glucosamine, and sialic acid. Other glycoproteins contain both glucosamine and galactosamine or only galactosamine. Some glycoproteins contain their sugar residues in the form of one oligosaccharide attached to a specific amino acid residue in the protein backbone. Examples are hen's egg albumin and ox ribonuclease B (Plummer and Hirs 1963, 1964). In other cases, such as hen's ovomucoid (see Melamed 1966) and fetuin (Spiro and Spiro 1962), there are three carbohydrate prosthetic groups. In fibrinogen, caeruloplasmin and thyroglobulin the number of prosthetic groups is between 7 and 23 (see Table 7). In the blood group substances the number of heterosaccharide groups per protein molecule is about 350 and the submaxillary gland glycoprotein of ox has more than 3,000 prosthetic groups per protein unit of molecular weight of $4 \cdot 10^6$

(Marshall and Neuberger 1968). Any classification of glycoproteins will have certain shortcomings, but we have recently proposed a division of glycoproteins based on the type of carbohydrate-peptide bond present and this division seemed to have certain satisfactory features (Marshall and Neuberger 1968). There are three groups according to this criterion, at least on the basis of our present knowledge. The first group (A) comprises the glycoproteins with N-glycosidic linkages involving the amide group of asparagine residues. The second group (B) consists of glycoproteins in which the amino-acid backbone is linked to the heterosaccharide prosthetic group by an O-glycosidic linkage involving serine and/or threonine. The third group (C) comprises a small number of glycoproteins where the linkage is through an alkali-stable O-

TABLE 7

CARBOHYDRATE CONTENT OF SOME ANIMAL GLYCOPROTEINS[1]

Protein	Mol. Wt.	No. of Carbohydrate Moieties per Mole	Types of Sugars[2]
Egg albumin	45×10^3	1	M, Gm.
Ribonuclease B	14.7×10^3	1	M, Gm.
Thyroglobulin	660×10^3	23	M, Ga, F, Gm, S
A-Substance	1000×10^3	350	Ga, F, Gm, Gam
OSM	1000×10^3	800	Gam, S
Tropocollagen	320×10^3	∽3	G, Ga

[1] See Marshall and Neuberger (1968) for original references.
[2] F = fucose, G = glucose, Ga = galactose, Gm = glucosamine, Gam = galactosamine, M = Mannose, S = Sialic acid.

glycosidic bond to the hydroxyl group of δ-hydroxylysine. This classification is not completely satisfactory, as a myeloma protein (Dawson and Clamp 1968) and rabbit γ-G-globulin (Smyth and Utsumi 1967) have recently been shown to contain linkages of both types A and B.

GROUP A GLYCOPROTEINS
(ASPARAGINE-N-ACETYL GLUCOSAMINE LINKAGE)

This group comprises a large number of more or less globular proteins such as egg albumin, ovomucoid, fetuin, caeruloplasmin, transferrin, ox ribonuclease B, and γ-G-globulin. The proteins mentioned so far have molecular weights below 200,000, but certain other proteins of higher molecular weight, such as fibrinogen and thyroglobulin also appear to belong to this group. It is likely that urinary glycoprotein (Tamm and Horsfall protein) and several plant glycoproteins are also of this type. The sugar involved in the N-glycosidic linkage is almost invariably N-acetyl glucosamine and the nitrogen atom is supplied by

the amide group of an asparagine residue. The heterosaccharide residues in this group contain mannose, often galactose, but never or rarely
glucose. In some of them fucose is present. Amongst amino sugars
glucosamine predominates, but galactosamine in addition to glucosamine is present in some of the proteins of this group; examples of this
are fetuin, thyrotropin, and other protein hormones produced by the
anterior pituitary gland. Sialic acid is present in a large proportion of
the glycoproteins of this group, but sometimes only in relatively small
amounts. There are usually few heterosaccharide residues per mole of
protein and only a small proportion of aspartic acid (asparagine)
residues are linked to sugar (Marshall and Neuberger 1968). But orosomucoid is an exception; in this protein 40% of the aspartic acid groups
are linked to heterosaccharides. The number of sugar residues per
heterosaccharide in general varies between about 7 and 25–30. Detailed
structures of the heterosaccharide residues have been determined only
in a very few cases, but with many other heterosaccharides at least
certain linkages between the sugar residues have been established.
The available evidence suggests that most of the heterosaccharides
have one or more branch points.

The N-glycosidic linkage found in this group of glycoproteins is
relatively stable to acid, heating at 100°C with 1 or 2N-mineral acid
for 1 or 2 hr being required for more than 90% of hydrolysis to occur
(Neuberger and Marshall 1966). In order to demonstrate this linkage in
a glycoprotein it is desirable to isolate, after appropriate enzymic hydrolysis, a glycopeptide which has at most only traces of amino acids
other than asparagine or aspartic acid. On further hydrolysis with acid,
ammonia should be liberated in equivalent amounts to that of aspartic
acid.

Specificity of Attachment of Carbohydrate Chain to the Peptide Moiety

It has already been mentioned that in most glycoproteins containing
N-glycosidic linkages only a few of the asparagine residues are joined
to carbohydrate chains. The sequence of amino acids in this region is
known for a number of glycoproteins (Table 8). There is no one
preferred sequence around the asparagine group involved in the linkage, but it is of interest that in a number of them a threonine or a
serine residue is in a position next but one on the C-terminal side to
this asparagine residue.

Certain titration studies may be relevant. In our earlier studies we

TABLE 8

AMINO ACID SEQUENCES AROUND THE ASPARAGINE-CARBOHYDRATE LINKING MOIETY

Protein	Amino Acid Sequence (1)
Egg albumin (2)	Glu. Glu. Lys. Tyr. ASN. Leu. Thr. Ser. Val. Leu
Ox γ-G-globulin (3) Lys. Pro. Arg.	Glu. Glu. Gln. Phe. ASN. (Ser. Thr.)
Ox ribonuclease B (4) Met. Met. Lys. Ser. Arg.	ASN. Leu. Thr. Lys.
Ovotransferrin ⎱ (5)	⎧Ile. His. ASN. Arg. Thr. Gly. Thr.
Serum transferrin ⎰	⎩ Ala. ASN. Leu. Thr. Gly.
Stem bromelain (6) Asn. Glu. Ser. Asn.	ASN. Glu. Ser. Ser.
Mouse myeloma L chain (7) Ala. Ser. Gln.	ASN. Ile. Ser. Asn. Asn. Leu.
Fibrinogen (8)	ASN. Lys. Thr. Ser.
	Val. Gly. Glu. ASN. Arg.
	Glu. ASN.
Human thyroglobulin (9)	Ala. Leu. Glu. ASN. Ala. Thr. Arg.

1. The asparagine residue which is linked to N-acetyl-D-glucosamine is indicated in capitals (ASN).
2. Cunningham et al. 1963; Neuberger and Marshall 1966.
3. Nolan and Smith 1962.
4. Plummer and Hirs 1964.
5. Williams 1968.
6. Takahashi et al. 1967.
7. Coleman et al. 1967; Melchers 1968.
8. Mester et al. 1967; Iwanaga et al. 1968.
9. Rawitch et al. 1968.

titrated a glycopeptide from egg albumin, the sequence of the amino acids being

carbohydrate
|
— Asn. Leu. Thr. Ser —

As well as the expected groups titrating as amino and carboxyl terminal moieties, one with a pK of about 9.5 was found. From later work it appeared that this group was the imido group constituting the linkage. When glycopeptide, containing aspartic acid as the only amino acid, or the synthetic asparagine-N-acetyl glucosamine model compound was titrated, the pK of this group was now about 2.5 units higher (Fletcher 1965; Montgomery et al. 1965).

It seemed reasonable to suggest (Marshall 1967) that with a glyco-peptide containing several amino acids the hydroxyl group of the threonine residue, which is next but one to the asparagine residue, facilitates the dissociation of the imido hydrogen atom through hydrogen bonding. It is indeed possible to make a model of the form suggested and in this the distance between the two relevant atoms (2.5–2.6 Å) is that which is generally accepted for a hydrogen bond (Fig. 59). The model shown has the nonpolar side chain of leucine directed away from the carbohydrate moiety.

This problem might be further discussed with respect to ox ribonuclease. This enzyme, of which the amino acid sequence is known, exists in 2 forms, 1 with and 1 without carbohydrate, ribonucleases B

and A respectively.   For the latter the three-dimensional structure has
also been determined.   Ribonuclease contains 10 asparagine and 7
glutamine residues, and only 1 of the former, number 34, is linked to a
carbohydrate group in ribonuclease B.   Inspection of the crystallographic structure indicates that the asparagine residues are sterically

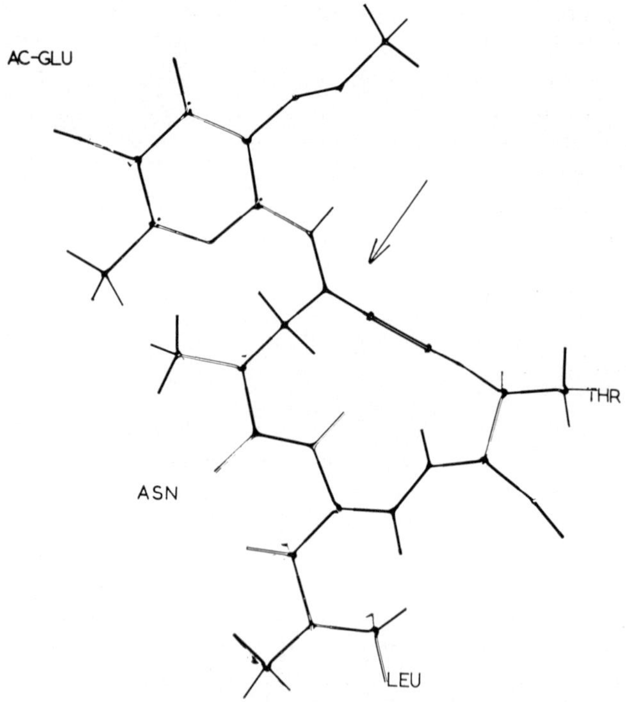

Fig. 59. The Sequence of Amino Acids (. . . . Asn. Leu. Thr.
. . . .) in the Region of the Carbohydrate Moeity,
Represented by N-Acetyl Glucosamine (AC-GLU),
of Hen's Egg Albumin

The carbonyl oxygen of the β-aspartamido group involved in the
carbohydrate-peptide bond is indicated by an arrow and is hydrogen bonded to the β-hydroxyl group of the threonine residue.

accessible to a varying degree and that the residue substituted is somewhat less accessible than are some other asparagine groups.   The
crystallographic model does not incorporate the hydrogen bonded structure postulated in Fig. 59, but the conformation of ribonuclease in
aqueous solution is unlikely to be fixed in the same way as in the

crystal. A protein in solution is likely to exist in several conformations, differing slightly in energy, and one or more of these may contain the structure postulated.

In ribonuclease A the asparagine residue number 34 is the only one which is followed in a position next but one by a β-hydroxy-α-amino acid residue. Experimental facts indicate that this structure endows the β-amide group of asparagine with increased acidity. It is therefore proposed that it is this structural feature which is the basis for the specificity of the transferase which adds an N-acetyl-β-glucosaminyl residue to the asparagine group. It is also suggested that this may be a requisite feature for all glycoproteins of this group (Table 8). This type of sequence is relatively rare in proteins which do not contain carbohydrate, but it may be found in a few proteins such as trypsinogen (Mikeš et al. 1966) and elastase (Hartley et al. 1965).

GROUP B.
GLYCOSIDIC LINKAGES INVOLVING THREONINE AND/OR SERINE

The carbohydrate-peptide linkage in glycoproteins of this type is frequently fairly labile under alkaline conditions, splitting occurring by means of a β-elimination reaction. However, if the β-hydroxy-α-amino acid which is linked to the sugar is itself N- or C- terminal, the β-elimination reaction does not readily occur. It seems likely also that, if a glutamic or aspartic acid residue were closely adjacent to a linkage of this type, the stability of the bond under alkaline conditions would be much greater.

The group comprises the soluble blood group substances, the submaxillary gland proteins as well as chondroitin-4-sulfate and certain other mucopolysaccharides. The latter have been called proteoglycans. In the first subgroup the glycosidic linkage is between an N-acetyl galactosamine and the hydroxyl group of serine or threonine, but with many of the proteoglycans the linkage involves a xylose residue (Rodén and Armand 1966; Lindahl and Rodén 1966). Perhaps such a rigid division will be found to be unsatisfactory, for a recent report (Greiling et al. 1968) describes the presence of the asparagine-N acetyl glucosamine linkage in keratan sulfate of the cornea.

Glycoproteins of this group appear to have a very large proportion of their serine or threonine residues glycosylated, so that it is not known whether the substitution of the hydroxyl group of the β-hydroxy acids occurs by a random process.

GROUP C. GLYCOSIDIC BONDS INVOLVING δ-HYDROXYLYSINE

It is now established that soluble collagen contains sugar and Butler and Cunningham (1966) have clearly established that the linkage is between the hydroxyl group of hydroxylysine and C_1 of a galactosyl group. More recently Cunningham and Ford (1968) have extended this work; they have shown that the sugar moiety is present either as monosaccharide, i.e., galactose only, or as a disaccharide containing both a glucosyl and a galactosyl residue. This type is present in both soluble and insoluble collagen and it is thus unlikely, but not impossible, to be involved in such cross linkages as those responsible for the conversion of the soluble collagen into the insoluble form.

A second type of glycopeptide of much higher molecular weight, contained no hydroxylysine, and was found only in insoluble collagen. This material was heterogenous and may have contained impurities arising from contaminating other glycoproteins. However, the larger part appeared to be associated with insoluble collagen and may play a part in the cross-linking or maturation of this protein.

A similar glycopeptide was isolated by Spiro (1967) from a highly purified preparation of basement membrane of ox kidney glomeruli. The biosynthesis of this type of linkage, i.e., one involving an O-glycosidic bond between galactose and hydroxylysine, has recently been investigated by Rosenbloom et al. (1968). Their results indicate that the synthesis consists of three distinct stages. The first step is the assembly of amino acid residues on the ribosomes. Hydroxylation of proline and lysine residues to give hydroxyproline and hydroxylysine appears to occur mainly or entirely after the peptide chains are released from the ribosomes. This is then followed by glycosylation of certain hydroxylysine residues and only then are the collagen molecules extruded from the cell.

In contrast to the O-glycosidic linkage to peptide bound serine and threonine, the glycosidic linkage to hydroxylysine is stable to alkali. In the latter case there are no carbonyl groups in suitable positions to make β-elimination possible. All the workers in this field have in fact observed that no hydroxylysine is liberated from the appropriate glycopeptide by heating with aqueous alkali. The linkage is also stable to acid. It has been known for some time that glucosaminides containing a free amino group are very stable to acid hydrolysis and this has been ascribed by Moggridge and Neuberger (1938) to the repulsion exerted by the positively charged ammonium group on an approaching proton. As protonation of the glycosidic oxygen is assumed to precede fission and as the distance between the 2-amino nitrogen in the hydroxylysine and the glycosidic oxygen is approximately the same as the corresponding distance in the glucosaminides, we should expect that hydrolysis by acid will also be impeded. However the experimental results obtained with the natural glycopeptide do not bear out this prediction. It seemed worthwhile to study this problem further.

Graham and Neuberger (1968) synthesized, as a model for the hydroxylysine glycoside, 2-aminoethyl-β-D-glucopyranoside. The synthesis involved condensation between 2(p-nitrobenzyloxy carbonylamino) ethanol and 2,3,4,6-tetra-O-acetyl-D-glucosyl bromide followed by removal of the protecting groups. The glucopyranoside was obtained as the crystalline hydrochloride and had the expected properties.

TABLE 9

RATES OF HYDROLYSIS OF SOME GLYCOPYRANOSIDES IN $2N$ HCL AT 80°C

Glycopyranoside	$[10^3 \times k \ (\text{min}^{-1})]$
Ethyl β-D-Glucopyranoside	19.2[1]
2-Aminoethyl β-D-glucopyranoside	4.3[2]
Methyl α-D-glucosaminide	0.09[3]

[1] Overend *et al.* 1962.
[2] Graham and Neuberger 1968.
[3] In 2.5N HCl; Moggridge and Neuberger 1938.

The results of kinetic experiments shown in Table 9 indicate that the protonated aminoethyl glycopyranoside is only slightly less easily hydrolyzed than is the β-ethyl glucopyranoside; the corresponding glucosaminide is about 60 times more stable. We have tried to rationalize these findings in the following manner. It is now generally assumed that hydrolysis of glycosides by acid is a unimolecular reaction involving preliminary protonation, most probably of the glycosidic oxygen. Fission then occurs between C_1 of the pyranoside and the glycosidic oxygen yielding in the first place a cyclic carbonium ion.

Protonation will be hindered by the existence of the positive charge both with the methyl glucosaminide and with the aminoethyl glycopyranoside. However the actual fission of the C-O bond will be greatly facilitated in the glycoside which contains the ammonium group in the aglycone, as cleavage will relieve the electrostatic repulsion between the two positive charges (Fig. 60A). With the glucosaminide however, fission will produce in the first place a dication in which the unfavorable Coulomb interaction is still retained (Fig 60B).

This interpretation would lead to the conclusion that a free amino group will hinder acid hydrolysis only if it is present in the "glycone," but not significantly if it is present in the aglycone; it might be used in structural studies on heterosaccharides. If in an N-acetylated hetero-saccharide the acetyl groups were removed, e.g., by hydrazinolysis, the resulting deacetylated product could then be subjected to hydrolysis by acid. The glycosidic linkages surviving the treatment would be only those in which the amino group is present in the "glycone."

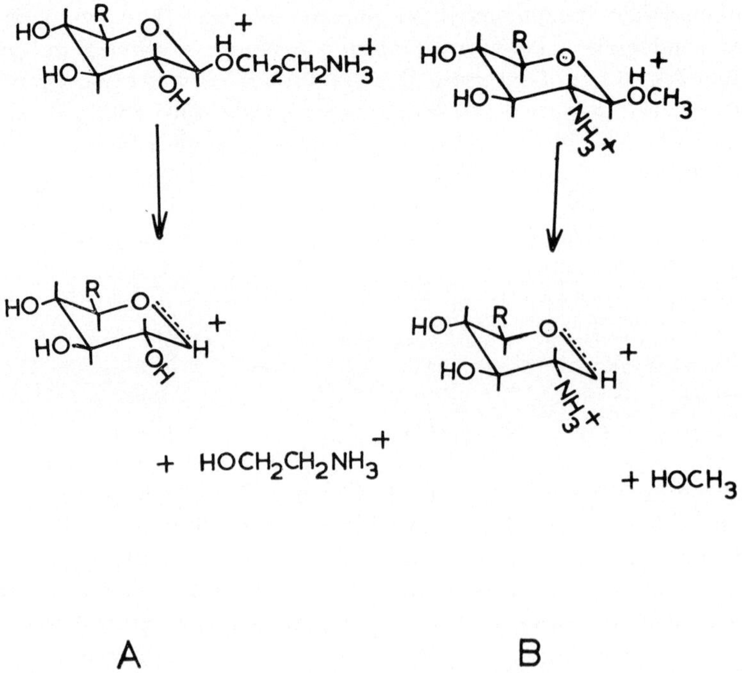

A   B

Fig. 60. A Representation of the Reaction Mechanism in the Hydrolysis of (A) Aminoethyl Glucoside and (B) Methyl Glucosaminide

$R = CH_2OH$.

An alternative explanation could be based on the assumption that the protonation which is preliminary for cleavage of the glycosidic bond occurs on the ring oxygen (Bunton *et al.* 1955). If in the 2-aminoethyl glucopyranoside a dication is cleaved which contains positive charges on the amino group and on the ring oxygen repulsion between the two charges is significantly less. On the other hand, with the alkyl gluco-saminides a shift of the proton from the glycosidic oxygen to the ring oxygen will not alter the distances between the two positive charges to the same extent. It should be noted that the two explanations are not mutually exclusive.

HETEROGENEITY OF CARBOHYDRATE PROSTHETIC GROUPS

Cunningham and his colleagues (Cunningham *et al.* 1965) were the first to separate different glycopeptides varying both in mannose and glucosamine content from enzymic hydrolysis products of egg albumin, and his findings have been confirmed (Bhoyroo and Marshall 1965; Levvy *et al.* 1966). In this case there is only one heterosaccharide prosthetic group for each protein molecule and thus these results suggest that the number of sugar residues in a given heterosaccharide may vary. With immunoglobulins and related myeloma proteins, Dawson and Clamp (1968; see also Clamp *et al.* 1968) have recently described a similar situation and they distinguish between central and peripheral heterogeneity. To quote from their paper, "in central heterogeneity the glycopeptides have totally different structures, whereas in peripheral heterogeneity the fundamental or core structure is the same, but there are small differences in the terminal residues." Similar findings have been reported for other proteins (e.g., orosomucoid, ovomucoid, and the blood group substances discussed later). These findings are not unexpected. The sequence of amino acid residues in a given protein is clearly defined by the base sequence of the DNA molecule concerned with the biosynthesis of that particular protein. Any change in amino acid sequence requires substitution of a base or a deletion resulting either in amino acid replacement in the protein or a shift in the reading frame. In other words the effects of a genetic change is direct and requires only the processes of transcription and translation. The position is different when we consider changes in the sequence of sugar residues in the heterosaccharide portion of glycoproteins. We may assume that this is determined directly by the action of certain glycosyl transferases which transfer a particular sugar to a certain position of an "unfinished" glycoprotein. These enzymes have themselves an amino acid sequence, which is determined by the appropriate gene, i.e., the base sequence of the DNA. However, the action of the transferase on its substrate will also depend on other factors, such as the position of amino acid residues near the active site, the nature of sugar residues already attached to the protein, the specificity of the transferase with respect to the sugar to be transferred, the ionic environment of the cell, and the presence of inhibitors or activators of the enzyme. There is therefore no logically compelling reason to assume that the exact structure of the heterosaccharide side chains of glycoproteins will show the same uniformity which we expect to find in the amino acid sequence of the protein, for the latter is the direct result of the DNA structure. We may thus expect that the action of the transferases may not be always quantitative and some of the prosthetic groups may be incomplete with regard to the sugar residues situated in the outer positions or periphery of the heterosaccharide moieties. This may explain the findings mentioned above with immunoglobulins and egg albumin. Another type of heterogeneity may result from the possibility that the transferases may not always be absolutely specific with regard to the sugar to be transferred.

We have thus far tacitly assumed that each sugar residue is transferred one by one to an unfinished glycoprotein and that the enzyme "recognizes" the substrate site by interacting in some way with properly placed amino acid residues and sugar residues already present in the unfinished glycoprotein. There is quite strong suggestive evidence that this concept is generally valid, but this may not necessarily apply in all cases. We have recently pointed out that the admittedly slight possibility that 2-acetamido-1-(β-L-aspartamido)-1,2-dideoxy-β-D-glucose might be incorporated through a specific tRNA and activating enzyme has not been excluded, neither has a mechanism in which the sugar is added at some stage between aminoacyl-tRNA and ribosome-bound polypeptide (Marshall and Neuberger 1968).

The close relationship between genetic characteristics and relevant chemical structure is beautifully illustrated by the work on blood group substances, which has recently been reviewed by Watkins (1966A, B). In the A,B,O and Lewis (Le) system at least the immunological specificity appears to be determined entirely by the terminal residues on the carbohydrate chains. All the evidence leads to a scheme which is shown in an abbreviated form in Table 10. A glycoprotein having prosthetic

TABLE 10

IMMUNOLOGICAL DETERMINANTS OF A,B,O BLOOD GROUP SUBSTANCES

Precursor Substance	β-D-Gal-(1 → 3)—β-D-N Ac Glu —
H-Substance	α-L-Fuc-(1 → 2)—β-D-Gal-(1 → 3)—β-D-N Ac Glu —
A-Substance	α-L-Fuc-(1 → 2)—β-D-Gal-(1 → 3)—β-D-N Ac Glu —
	α-D-N Ac Gal-(1 → 3)╱
B-Substance	α-L-Fuc-(1 → 2)—β-D-Gal-(1 → 3)—β-D-N Ac Glu —
	α-D-Gal-(1 → 3)╱
Le[a]-Substance	β-D-Gal-(1 → 3)—β-D-N Ac Glu —
	α-L-Fuc-(1 → 4)
Le[b]-Substance	α-L-Fuc-(1 → 2)—β-D-Gal-(1 → 3)—β-D-N Ac Glu —
	α-L-Fuc-(1 → 4)

groups identical with those characteristic of the precursor substance is assumed to be the substrate for the action of genetically distinct trans-ferases. A transferase, a product of the H gene, transfers an α-fucose to the 2-position of the galactose residue producing the H-substance. The latter is then the substrate for two other genetically controlled transferases, which transfer either an α-N-acetyl galactosamine residue or an α-galactosyl residue to the 3-position of the subterminal β-galactose residue of the H chain. These two transferases are under the control of the A and B genes respectively, and produce the corresponding A and B substances. In order to produce A B specificity, both types of residue have to be added. The experimental facts suggest that A B individuals have a glycoprotein which has both A and B structures on the same molecule, but on different carbohydrate chains. Both anti-A globulins and anti-B globulins can thus interact independently with the same glycoprotein, but the precursor substance can also be acted on by another type of transferase. Under the influence of the Le gene a glycosyltransferase catalyses the transfer of an α-fucose residue to the 4-position of the penultimate N-acetyl-β-D-glucosaminyl residue of the precursor substance producing Le^a activity. If both fucosyl transferases are present in the same individual, one acting on the galactosyl residue and one acting on the N-acetyl-glucosaminyl residue, the Le^b structure is produced, a structure in which the characteristics of the H gene and of the Le gene may be present in the same carbohydrate chain (Marr et al. 1967). Thus a new immunological determinant is created with a specificity of its own. The position is different from that of the A B substance where the immunological determinants are found in different carbohydrate chains. The antigenic activity of the Le^b type is of a new character in spite of the fact that structurally it arises by the addition of H and Le features. This is due to the fact that the antibody-combining site is larger than the structures under consideration.

The ability to secrete A and B substances is also under the influence of a pair of allelic genes Se and se, which are independent of the ABO genes. Most secretors of blood group A or B substances also show H activity, suggesting that not all carbohydrate chains are substituted in the 3-position of the preterminal β-galactose residue. As Le activity is independent of the ABO system, it seems likely that the same glyco-protein may have carbohydrate chains both with Le^a and Le^b structural determinants in addition to A or B determinants. If this idea is accepted it follows that in the blood group substances the action of the genetically controlled transferases is not necessarily quantitative, i.e., not all positions in a glycoprotein molecule which are potential acceptor sites for a sugar residue are in fact substituted. This situation is basically different from that found in the amino acid sequence of a protein where a deletion or substitution of a residue can occur only by a process of mutation.

The question can be raised as to whether the transferases for fucose, galactose, and N-acetyl galactosamine, which are important in defining blood group specificity, are also involved in the biosynthesis of carbohydrate chains of glycoproteins other than blood group substances. There is no obvious reason why this should not be the case. However, in the blood group substances, the immunologically important residues are α-linked, whilst in other glycoproteins so far examined, β-linkages seem to predominate. It is not impossible therefore that genetically defined immunological differences might occur which in some cases are parallel to blood group specificity. However, the relatively small number of carbohydrate chains in many glycoproteins might make it difficult to recognize by immunological means such differences if indeed they exist. The possibility might also be considered whether immunological tissue specificity, which is important in tissue transplantation, may in part be caused by the action of genetically distinct transferases producing a specificity possibly altogether separate from those observed with red cell systems.

SUMMARY

Glycoproteins may be defined as macromolecules which have a polypeptide backbone to which are attached one or more carbohydrate moieties. This class of biological macromolecules may be subdivided into groups based on the nature of the carbohydrate-peptide bond which predominates, but some overlapping in such a division occurs. One may also classify these substances according to whether they contain large carbohydrate moieties with a predominantly regular repeating pattern, such as occurs in the so-called proteoglycans, or smaller heterosaccharides of less regular structure, as in the serum proteins and the soluble blood group substances. It is probable that a high degree of heterogeneity of the carbohydrate units of glycoproteins is of common occurrence.

DISCUSSION

M. L. Wolfrom.—Comment on the "original" abbreviations for monosaccharides. F and S are used by chemists to indicate fluorine and sulfur.

A. Neuberger.—I'm no authority on abbreviations, nor would I defend them.

W. A. Gortner.—Peroxidases from animal sources do not appear to be glycoprotein, but all those from plant sources that have been looked at contain substantial amounts of carbohydrate materials. What is the characterization of the linkage or the nature of the carbohydrate from these plant peroxidases? Can you comment or speculate on their possible role in terms of activity or specificity of these important enzymes?

A. Neuberger.—First, I don't know very much about the detailed chemistry of the plant peroxidases. I know they contain a variety of sugars. In order to show the type of linkage they contain you have to have reasonable amounts of these glycoproteins. Preferably you have to isolate glycopeptides and I'm not aware that it has been done. How about the function of the sugars? We who work with these materials have been hoping for some time that a clearly defined function could be assigned to the sugar part of the glycoproteins. They might be required for transporting these proteins from inside the cell to the outside environment. Some evidence supports this, but is not necessarily the right explanation. There is also the possibility that in order for the protein to get away from the ribosome, there has to be a handle and the sugar part may provide the handle. Some of the proteins may lose the sugar residue. In other cases the sugar may be required for the protein to assume a particular conformation. In fibrinogen, recently, it has apparently been shown that certain individuals showed defects in their clotting process. All the various clotting factors seemed to be present in normal amounts, but fibrinogen, while having the correct amino acid content, was grossly deficient in carbohydrate. Apparently in the polymerization which occurs after the proteolytic activity of the first step in fibrin formation, the carbohydrate is required to get a really stable polymer.

BIBLIOGRAPHY

Bhoyroo, V. D., and Marshall, R. D. 1965. Is the heterosaccharide of egg albumin polymorphous? Biochem. J. 97, 18P–19P.

Bunton, C. A., Lewis, T. A., Llewellyn, D. R., and Vernon, C. A. 1955. The acid catalysed hydrolysis of α- and β-Methyl and α- and β-Phenyl D-Glucopyranosides. J. Chem. Soc., 4419–4423.

Butler, W. T., and Cunningham, L. W. 1966. Evidence for the linkage of a disaccharide to hydroxylysine in tropocollagen. J. Biol. Chem. 241, 3882–3888.

Clamp, J. R., Dawson, G., and Spragg, B. P. 1968. Central and peripheral heterogeneity of oligosaccharide units from glycoproteins. Biochem. J. 106, 16P.

COLEMAN, T. J., MARSHALL, R. D., and POTTER, M. 1967. Preparation of a glycopeptide from an immunoglobulin kappa polypeptide chain from the mouse. Biochim. Biophys. Acta. *147*, 396–398.

CUNNINGHAM, L. W., CLOUSE, R. W., and FORD, J. D. 1963. Heterogeneity of the carbohydrate moiety of crystalline ovalbumin. Biochim. Biophys. Acta. *78*, 379–381.

CUNNINGHAM, L. W., and FORD, J. D. 1968. A comparison of glycopeptides derived from soluble and insoluble collagens. J. Biol. Chem. *243*, 2390–2398.

CUNNINGHAM, L. W., FORD, J. D., and RAINEY, J. M. 1965. Heterogeneity of β-aspartyl-oligosaccharides from ovalbumin. Biochim. Biophys. Acta. *101*, 233–235.

DAWSON, G., and CLAMP, J. R. 1968. Investigation on the oligosaccharide units of an A myeloma globulin. Biochem. J. *107*, 341–352.

FLETCHER, A. P. 1965. Ph.D. Thesis, Structural studies on glycoproteins. University of London.

GIBBONS, R. A. 1959. Chemical Properties of two mucoids from bovine cervical mucine. Biochem. J. *73*, 207–217.

GOTTSCHALK, A. 1966. Definition of glycoproteins. *In* Glycoproteins, A. Gottschalk (Editor). Elsevier Publishing Co., New York.

GRAHAM, E. R. B., and NEUBERGER, A. 1968. The synthesis and some properties of 2-aminoethyl β-D-glucopyranoside. J. Chem. Soc. (C), 1638.

GREILING, H., STUHLSATZ, R., KISTERS, R., and PLAGEMANN, H. 1968. Abstr. 5th Meeting Fed. European Biochem. Soc., Prague, 155.

HARTLEY, B. S., BROWN, J. R., KAUFMAN, D. L., and SMILLIE, L. B. 1965. Evolutionary similarities between pancreatic proteolytic enzymes. Nature (London) *207*, 1157–1159.

IWANAGA, S. *et al.* 1968. Amino acid sequence of the N-terminal part of γ-chain in human fibrinogen. Biochim. Biophys. Acta. *160*, 280–283.

LEVVY, G. A., CONCHIE, J., and HAY, A. J. 1966. The heterogeneity of ovalbumin glycopeptide. Biochim. Biophys. Acta. *130*, 150–154.

LINDAHL, U., and RODÉN, L. 1966. The chondroitin 4-sulphate-protein linkage. J. Biol. Chem. *241*, 2113–2119.

MARR, A. M. S., DONALD, A. S. R., WATKINS, W. M., and MORGAN, W. T. J. 1967. Molecular and genetic aspects of human blood-group Le[b] specificity. Nature (London) *215*, 1345–1349.

MARSHALL, R. D. 1967. 2-Acetamido-1-(β-L-aspartamido)-1,2-dideoxy-β-D-glucose and glycoproteins. Intern. Congr. Biochemistry, 7th, Tokyo. Colloq. XIV, No. 4.

MARSHALL, R. D., and NEUBERGER, A. 1968. Metabolism of glycoproteins and blood group substances. *In* Carbohydrate Metabolism and Its Disorders, Vol. 1, F. Dickens, P. J. Randle, and W. J. Whelan (Editors). Academic Press, New York.

MELAMED, M. D. 1966. Ovomucoid. *In* Glycoproteins, A. Gottschalk (Editor). Elsevier Publishing Co., New York.

MELAMED, M. D., and GREEN, N. M. 1963. Avidin 2. Purification and composition. Biochem. J. *89*, 591–599.

MELCHERS, F. 1968. Max-Planck-Institut für Molekulare Genetik, 1 Berlin 33, W. Germany.

MESTER, L., MOCZÀR, E., and SZABADOS, L. 1967. Glucoside and amino acid sequences in glycopeptides isolated from fibrinogen. Compt. Rend. 265, 877–880. (French)

MIKEŠ, O., HOLEYŠOVSKÝ, V., TOMÁŠEK, V., and ŠORM, F. 1966. Covalent structure of bovine trypsinogen. The position of the remaining amides. Biochem. Biophys. Res. Commun. 24, 346–352.

MOGGRIDGE, R. C. G., and NEUBERGER, A. 1938. Methylglucosaminide: its structure and the kinetics of its hydrolysis by acids. J. Chem. Soc., 745.

MONTGOMERY, R., LEE, Y. C., and WU, Y-C. 1965. Glycopeptides from ovalbumin. Preparation, properties and partial hydrolysis of the asparaginyl carbohydrate. Biochemistry 4, 566–577.

NEUBERGER, A., and MARSHALL, R. D. 1966. The N-acylglycosylamine linkage involving the amide-N of asparagine. In Glycoproteins, A. Gottschalk (Editor). Elsevier Publishing Co., New York.

NOLAN, C., and SMITH, E. L. 1962. Glycopeptides. III. Isolation and properties of glycopeptides from a bovine globulin of colostrum and from fraction II.3 of human globulin. J. Biol. Chem. 237, 453–458.

OVEREND, W. G., REES, C. W., and SEQUEIRA, J. S. 1962. Reactions at position 1 of carbohydrates. Part III. The acid catalysed hydrolysis of glucosides. J. Chem. Soc., 3429–3440.

PLUMMER, T. H., and HIRS, C. H. W. 1963. The isolation of ribonuclease B, a glycoprotein, from bovine pancreatic juice. J. Biol. Chem. 238, 1396–1401.

PLUMMER, T. H., and HIRS, C. H. W. 1964. On the structure of bovine pancreatic ribonuclease B. Isolation of glycopeptides. J. Biol. Chem. 239, 2530–2538.

PUSZTAI, A., and MORGAN, W. T. J. 1963. Studies in immunochemistry 2. The amino acid composition of human blood group A, B, H and Le^a specific substances. Biochem. J. 88, 546–555.

RAWITCH, A. B., LIAO, T. -H., and PIERCE, J. G. 1968. The amino acid sequence of a tryptic glycopeptide from human thyroglobulin. Biochim. Biophys. Acta. 160, 360.

RODÉN, L., and ARMAND, G. 1966. Structure of the chondroitin 4-sulphate-protein linkage region. J. Biol. Chem. 241, 65.

ROSENBLOOM, J., BLUMENKRANTZ, N., and PROCKOP, D. J. 1968. Sequential hydroxylation of lysine and glycosylation of hydroxylysine during the biosynthesis of collagen in isolated cartilage. Biochem. Biophys. Res. Commun. 31, 792–797.

SINEX, F. M., and VAN SLYKE, D. D. 1955. The source and state of hydroxylysine during the biosynthesis of collagen. J. Biol. Chem. 216, 245–250.

SMYTH, D. S., and UTSUMI, S. 1967. Structure at the linkage region in rabbit immunoglobulin-G. Nature (London) 216, 332–335.

SPIRO, R. G. 1967. Studies on the renal glomerular basement membrane. J. Biol. Chem. 242, 1915–1922.

SPIRO, M. J., and SPIRO, R. G. 1962. Composition of the peptide portion of fetuin. J. Biol. Chem. 237, 1507–1510.

STETTEN, M. R.   1948.   Some aspects of the metabolism of hydroxyproline studied with the aid of isotopic nitrogen.   J. Biol. Chem. *181,* 31–37.

TAKAHASHI, Y. Y., KUZUYA, M., SUZUKI, A., and MURACHI, T.   1967.   Structure of the glycopeptide from Stem Bromelain.   Intern. Congr. Biochemistry, 7th, Tokyo. Abstr. *D23.*

WATKINS, W. M.   1966A.   Blood group specific substances.   *In* Glycoproteins, A. Gottschalk (Editor).   Elsevier Publishing Co., New York.

WATKINS, W. M.   1966B.   Blood group substances.   Science. *152,* 172–181.

WILLIAMS, J.   1968.   A comparison of glycopeptides from the ovotransferrin and serum transferrin of the hen.   Biochem. J. *108,* 57–67.

Advances in Analytical Methodology

CHAPTER 8

Rex Montgomery

The Chemical Analysis of Carbohydrates[1]

INTRODUCTION

The biochemist, with an extract of some tissue, the food chemist, with a sample of a competitor's product, and the research chemist, with a crude polysaccharide for structural elucidation, are all faced at an early stage of their investigations with the problem of deciding what is the chemical composition of their particular material or mixture. Perhaps the food chemist is in the most difficult position because he not only faces the cunning of nature in the components of the food product, but also the scheming modifications that his fellow man has played upon the natural compounds. In most cases the material under

TABLE 11

TYPICAL ANALYSIS OF COMMERCIAL CARBOHYDRATE M

Component	Liquid Form (%)	Dried Form (%)
Solids	65	97
Carbohydrates	55	84
Ash	6	4
Fiber	1	1
Fat	0.5	0.5
pH	5.5	3.7

study is a mixture that may give an analysis by the methods described in official handbooks, such as the Official Methods of Analysis of the Association of Official Agricultural Chemists, AOAC, something like the material M represented in Table 11. This information is of some value but tells little about the nature of the individual components, for which knowledge a fractionation procedure must usually be applied. The mixture may be fractionated on the basis of molecular size, electrical

[1] Supported in part by research grants HE06717 and GM14013 from the National Institutes of Health.

charge, solubility, complex formation, or some specific property, such
as the formation of an insoluble derivative (Whistler and Wolfrom
1964B). Concentrating at this point on the carbohydrate-containing
fractions from such a separation, a more detailed study would include
hydrolysis of each fraction to the component monosaccharides or their
derivatives, which are subsequently analyzed quantitatively for more
complete characterization of the original material.

The general procedures of carbohydrate analysis are exemplified by
the determination of the composition of a polysaccharide. Any particular problem may be easier than this, maybe simply one of analyzing
for sucrose, glucose, or cellulose fiber, or it may be more difficult than
usual, as in the case of determining the composition of a glycoprotein.
If it is decided eventually to complete the characterization and to
elucidate the structure of the carbohydrate-containing polymer, the
procedure, which is general for all biopolymers, involves analytical
methods at nearly every step that leads to the primary structure and
follows the outline shown in Fig. 61. An end-group analysis of the
molecule under study gives a measure of the average molecular size,
since there is only one reducing group, or potential reducing-end, in
each carbohydrate molecule. A measure of the degree of branching
is calculated from the analysis of the nonreducing terminal ends (Smith
and Montgomery 1959). The types of glycosidic linkages may be established by classical methylation or periodate oxidation studies, but the
true sequence of residues rests principally upon analyses of the oligosaccharides resulting from partial hydrolysis of the polysaccharide. At
the various steps in the investigation, both physical and chemical
methods of analysis may be involved. The physical methods analyze
the material under study, without chemical modification, by such
techniques as optical rotation, nmr, infrared and ultraviolet spectroscopy,
and molecular weight determinations. These are considered in greater
detail in Chap. 9. The chemical procedures may involve enzymic
catalysis (see Chap. 10) or other types of reactions. It is the latter
that are discussed at this time.

CHEMICAL ANALYSES

Many methods have been used for the chemical analysis of carbohydrates. Good summaries of these procedures are to be found in the
literature (Bates *et al.* 1942; Dische 1955; Paech and Tracy 1955; Smith
and Montgomery 1959; Stanek *et al.* 1963; Whistler and Wolfrom 1962).
It seems therefore more relevant at this time to discuss the general
principles, merits and precautions of the methods as applied to poly-

saccharides than to compile a detailed list of procedures. The general types of analyses are given in Table 12. They depend principally upon the reducing properties of the hemiacetal function, which may

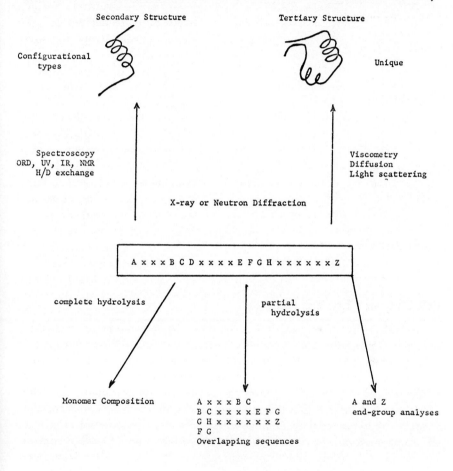

FIG. 61. GENERAL APPROACH TO THE ELUCIDATION OF THE STRUCTURE OF A BIOPOLYMER

react stoichiometrically but most commonly does not, and the complicated degradation of the carbohydrate in acid solution to furfural derivatives that can react with phenols or amines to give colored products. More specific reactions are based upon the products of periodate oxidation of the sugars or reactions of particular functional groups, such as the carboxyl group in the uronic acids.

TABLE 12

GENERAL TYPES OF ANALYTICAL PROCEDURES

Type of Sugar	Method	Examples of Method
Simple sugars		
	Stoichiometric reducing sugar reactions	Hypoiodite (Hirst *et al.* 1949)
	Nonstoichiometric reducing sugar reactions	Ferricyanide (Park & Johnson 1949) Dinitrosalicylate (Miller 1959)
	Colorimetric, by acid degradation	Phenol (Dubois *et al.* 1956) Anthrone (Dimler *et al.* 1952)
	Specific colorimetric reaction	Thiobarbituric acid for 2-deoxy sugars (Waravdekar & Saslaw 1959)
	Chromatographic	Gas-liquid chromatography of volatile derivative (Bishop 1964; Sweeley *et al.* 1963) Thin-layer chromatography (Mangold *et al.* 1964)
Polysaccharides		
	Colorimetric reactions	Phenol (Dubois *et al.* 1956)
Sugar alcohols		
	Oxidative	Periodate (Lambert & Neish 1950)
	Chromatographic	Gas-liquid chromatography of volatile derivative (Sweeley *et al.* 1963)
Uronic acids		
	Decarboxylation	CO_2 evolution (Barker *et al.* 1958)
	Colorimetric	Carbazole (Dische 1947).

Reliability of Chemical Analyses

The reliability of any analytical method depends upon its accuracy, precision, sensitivity, and specificity (Anastassiadis and Common 1968). The accuracy reflects the agreement between the experimental values and the true value, wheras the precision relates the experimental value to the average of an infinite number of the same determinations. The sensitivity tells how little of a material can be determined and also how small a difference can be demonstrated between two concentrations of the same material. For example, the determination of carbohydrates by the phenol-sulfuric acid procedure (Dubois *et al.* 1956) is precise to ± 2 to $\pm 3\%$, with a sensitivity of around 1 μg down to 5 to 10 μg of sugar. Since it is a nonstoichiometric colorimetric procedure, the accuracy of the method depends upon the construction of a standard curve and if the component being analyzed is available in pure form then the accuracy of the method is equal to the precision. As may be seen from Fig. 63, the color yield from different sugars is not the same.

It is not uncommon to see carbohydrate analyses reported as total hexose or in terms of some other sugar. This is a little more specific than the type of analysis given in Table 11 and is usually meant to refer to an equivalent amount of some selected sugar. The nonstoichiometric reactions involved in colorimetric analyses of sugars make it nearly impossible to determine the absolute amount of a heteroglycan or carbohydrate mixture without prior hydrolysis and determination of the ratio of monosaccharide in the sample. Even a homoglycan such as starch is best determined from a standard curve of that sample, which requires, of course, that the dry weight of the sample be known. In passing, it may be noted that the scientist with a biologically-active and heat-sensitive material finds it difficult to analyze for water without denaturation of the sample.

Analysis of Polysaccharides

The ideal analytical procedure for a polysaccharide would permit an accurate, precise, and specific microdetermination in the presence of other materials, which could include other carbohydrates. Such a specificity is rarely found in chemical methods, the closest approach being perhaps the determination of starch by its reaction with iodine, but even here we have variations due to molecular size (Szejtli *et al.* 1968) and source (Whistler and Wolfrom 1964A), as well as interference from a few other polysaccharides (Ehrenthal *et al.* 1954). There are a number of reasons for this lack of specificity. Polysaccharides are not unique chemical entities. Their biosynthesis does not involve a template as far as is known, with the result that a molecularly dispersed product is obtained in each case. Furthermore, the monosaccharide units are not so chemically different as to impart individual character to the family of molecules that each polysaccharide represents. However, the ratio of the monosaccharides and the linkages between them are characteristic of each polysaccharide, those from similar species often showing clear distinctions, such as the alginates from the *Laminariaciae*.

The differences in composition are determined by hydrolysis of the polysaccharide for analysis of the component building units. Unfortunately, hydrolysis does not yield the component sugars quantitatively because of the side reactions of acid degradation to produce furfurals and humins. The optimum yield of monosaccharides is obtained by different conditions for each kind of polysaccharide and it is wise to establish these conditions in each case if enough material is available. Customarily an unknown polysaccharide is hydrolyzed at 95°-100° with 1-2 N sulfuric acid for a length of time that varies from 1 hr and on. Insoluble materials may be dissolved initially in 72% sulfuric acid at 0°-5° for a period of time, followed by dilution to 1-2 N [H^+] and heating at 100°.

Glycosidic bonds are hydrolyzed with acid at rates that depend upon the sugar, the anomeric form, and the aglycone (BeMiller 1967). The relative rates of hydrolysis of methyl glycopyranosides are shown in Table 13. Since most polysaccharides have more than one type of glycosidic bond and many have more than one kind of sugar, the optimum conditions for hydrolysis of the polysaccharide must be a compromise between the relative rates of hydrolysis of the glycosidic bond-types and the rates at which the liberated sugars are destroyed. For example, the D-fructofuranosyl residue in the oligosaccharide stachyose is liberated more rapidly than the pyranose residues, but fructose also degrades at a more rapid rate than D-glucose and D-galactose so that the greatest yield of the mixture of sugars may be obtained by a compromise between the optimum condition for complete hydrolysis and the minimum amount of degradation. In such cases it is best to

TABLE 13
RELATIVE RATES OF HYDROLYSIS OF GLYCOSIDIC BONDS

Methyl Pyranoside of	Relative Rate[1]
α-D-Glucose	1.0
β-D-Glucose	1.9
α-D-Mannose	2.4
α-D-Galactose	5.2
β-D-Xylose	9.1
β-D-Glucuronic acid	0.62

Source: BeMiller, J. N. (1967).
[1] Glycoside hydrolyzed with 0.5 N H^+ at 75°.

analyze for the different types of sugars separately. For example, in gum arabic the L-arabinose and L-rhamnose residues are readily released leaving a more acid-resistant core of D-galactose and D-glucuronic acid residues. It would be difficult to find a single condition of hydrolysis that would give a mixture of these four sugars in the same proportion as they are present in the native gum.

The reactions of hydrolysis and acid degradation present problems in estimating the correction factors to be applied to the analytical results from the acid hydrolysates. Some attempts have been made to determine these correction factors by including an internal standard sugar in the hydrolysis of the polysaccharide, but it is recognized that such a free sugar degrades continuously, which is not necessarily true for the sugars in the polysaccharide until they are liberated. The sugar glycosides are better internal controls.

The problem of the hydrolysis of a carbohydrate polymer is a difficult one and its solution may be possible only by the use of suitable enzymes, particularly in the case of the polyuronides, which are difficult to hydrolyze and relatively easily decarboxylated by acids.

Analysis of Monosaccharides

Given a mixture of monosaccharides, such as that from the hydrolysate of a heteroglycan, analysis of the components is relatively simple and quantitative. Such analyses may proceed by paper chromatography, ion exchange chromatography, thin-layer chromatography, gas-liquid chromatography, or paper electrophoresis. Each sugar so separated by these micro methods is then usually determined by some chemical method. For example, the spots or bands from paper or thin layer chromatography, located by spraying a marker strip, are extracted and analyzed by some colorimetric method (Hay *et al.* 1962; Mangold *et al.* 1964). The fractions eluted from columns can be analyzed in a continuous manner and the resulting peak areas integrated for conversion to the weights of the corresponding sugars (see Fig. 62 and 63, Brummel *et al.* unpublished work). The elegant methods of gas-liquid chromatography separate the more volatile acetate or trimethylsilyl derivatives of the sugars, which are then detected by such methods as thermal conductivity, flame ionization, or other such physical methods (Bishop 1964; Sweeley *et al.* 1963).

Each of these procedures can give excellent quantitative separations and reproducible results if the usual care required for analytical work is taken. It is necessary to use reliable standard sugars for the preparation of standard curves. Fortunately, pure sugars are commercially available that meet the National Research Council specifications (1967). Preservatives, keto acids, aldehydes and certain metal ions may cause appreciable interference in the colorimetric methods. For example, some aldehydes, such as formaldehyde, and $Cu(II)$, $Fe(II)$, $Fe(III)$, or chloride ions interfere in the phenol sulfuric acid procedure (Montgomery 1961; Brummel *et al.* unpublished work). Some examples are given in Tables 14 and 15. In some cases the interference of the metal ion can be removed by including an excess of EDTA in the sugar solution before analysis or during the hydrolysis of the carbohydrate. Thus, the interference of $Fe(III)$ in the determination of sialic acid in ovomucoid is eliminated by adding a large excess of EDTA during the initial hydrolysis. EDTA has no similar effect on the determination of hexosamine or in the phenol-sulfuric acid reaction. Likewise, calcium ion interferes in the 3,5-dinitrosalicylic acid procedure for the determination of reducing sugars (Robyt and Whelan 1965). It has been noted that aqueous pyridine acetate buffers, when used for an ion exchange chromatography of carbohydrates, give a product that reacts like a sugar in the phenol-sulfuric acid test.

The resolving power of the gas-liquid chromatographic procedure is

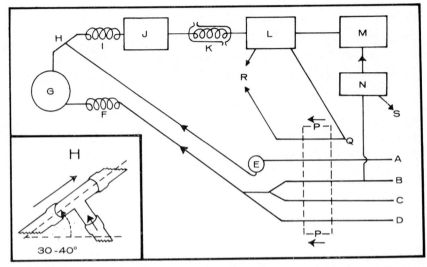

FIG. 62.   FLOW DIAGRAM OF AUTOANALYSIS OF CARBOHYDRATES USING THE PHENOL-
SULFURIC ACID REAGENTS

A. 0.056-in. Acidflex tubing for the concentrated sulfuric acid.
B. 0.020-in. or 0.0075-in. clear standard tubing for the column effluent.
C. 0.020-in. or 0.025-in. Solvaflex tubing for 2% (w/v) phenol and 0.2% (v/v)
   ARW7 or 1% (w/v) phenol and 0.1% ARW7 respectively.
D. 0.035-in. clear standard for air segmentation.
E. Pulse suppressor.
F. Glass mixing coil.
G. Time delay coil—about 100 ft of Intramedic PE 240 polyethylene tubing.
H. T-junction for mixing the concentrated sulfuric with the segmented stream.
I. Glass mixing coil.
J. Glass coil heating bath—96°.
K. Water jacketed cooling coil or two mixing coils in sequence.
L. 490 mμ colorimeter with a 15 mm tubular flow cell.
M. Three point recorder.
N. Standard peptide manifold and apparatus for ninhydrin color production on base
   hydrolyzed and unhydrolyzed column effluent.
P. One speed proportioning pump and roller assembly.
Q. 0.056-in. Acidflex pull-through.
R. Drain.
S. To fraction collector.

such that the different sugar anomers are often separated.   This must
be taken into account when the area under the curve of each peak is
integrated for the construction of a standard response curve.   The areas
for the anomeric forms must be added together.   The multiplicity of
sugar forms can be overcome by reducing the sugars first to their corresponding alditols, which are chromatographed as their acetate or
trimethylsilyl derivative as usual (Sawardeker *et al.* 1965).   Unfortu-

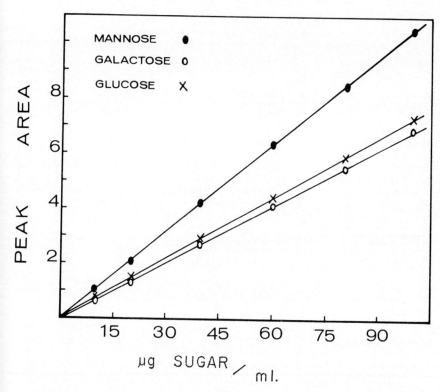

FIG. 63. STANDARD CURVES FOR D-MANNOSE, D-GLUCOSE AND D-GALAC-
TOSE BY THE PHENOL-SULFURIC ACID REACTION FROM THE AUTOANALYZER
(Fig. 62).

TABLE 14

CHROMOGENS FROM THE REACTION OF KETO ACIDS, ALDEHYDES AND KETONES
WITH THE PHENOL-SULFURIC ACID REAGENTS

Compound	Max[1] (nm)	$\log_{10} E_{1\,cm}^{mmole[2]}$	Glucose Equiv.[3] %
Glucose	489	3.24	100
2-Keto-D-gluconate	489	3.14	70
Pyruvate	477	2.83	49
2-Keto-glutarate	489	0.93	1
Acetaldehyde	477	1.41	
Acetone	351	1.66	
2-Butanone	342	0.96	

Source: Montgomery (1961).
[1] Only the maximum absorbance closest to 489 nm is reported.
[2] $\log_{10} E_{1\,cm}^{mmole}$ represents the absorbancy of 1 mmole of material dissolved in water (2 ml) and to which is added 80% aq phenol (0.10 ml) and conc. H_2SO_4 (5 ml).
[3] Glucose equivalent is the apparent glucose content, expressed as a percentage of the material when the absorbancy of the reaction is read at 489 nm.

TABLE 15

EFFECT OF METAL IONS ON SUGAR ANALYSES[1]

Analytical Procedure	Metal Ion				
	Cu(II)	Mn(II)	Fe(II)	Fe(III)	Co(II)
Hexose (Dubois 1956)	132[2](14.4[3]) 125 (0.9) 120 (0.1)	[4]	[4]	142(0.9) 125(0.1)	[4]
N-Acetyl-D-glucosamine (Morgan and Elson 1934)	84 (2.1) 85 (1.4) 83 (0.4)	83(2.1) 85(1.4) 94(0.4)		[4]	[4]
D-Glucosamine (Elson and Morgan 1933) (Winzler 1955)	57 (0.7) 68 (0.5) 89 (0.1)	72(0.7) 86(0.5) 96(0.1)		93(1.0) [4]	62(1.0) 87(0.7) 75(0.1)
Sialic acid[5] (Warren 1959)	[4]	[4]		69(2.9) 94(0.03)	[4]

[1] Brummel, Gerbeck and Montgomery, unpublished work.
[2] Percent of the value obtained in the absence of the metal ion.
[3] Mole ratio of metal ion to residue being analyzed.
[4] Analysis is unaffected at this level.
[5] Heated with 0.1 N H_2SO_4 for 1.5 hr at 80° with metal ion before analysis.

nately, the ketoses give two alditols, not necessarily in equimolar amounts.

Isotopic Dilution

Attempts to reduce the above mentioned problems have been made by applying the technique of isotopic dilution (Graham and Neuberger 1968). For the microanalysis of a biopolymer containing carbohydrate, known amounts of radioactively-labeled sugar glycosides are added to the material before hydrolysis. The hydrolysate is then fractionated by any of the chromatographic techinques listed above, and the separated sugars are determined by a microchemical method as well as being counted for their radioactivity. From the dilution of the radioactivity and the loss of total counts, one may calculate the amount of each "cold" sugar originally present in the biopolymer. Alternatively, the radioactive sugars, each of known specific activity, may be added to the hydrolysate which by separation and determination of the specific activity of each isolated sugar fraction can then give the required information of the composition. Table 16 shows the results obtained in the estimation of D-mannose in egg albumin (Graham and Neuberger 1968).

Carbohydrate Derivatives

For many years the constituents of foods have included derivatives of the natural polysaccharides. Such groups as methyl-, hydroxyethyl-,

TABLE 16

ESTIMATION OF MANNOSE IN EGG ALBUMIN BY RADIOISOTOPE DILUTION

D-Mannose μg	Methyl α-D-Mannopyranoside μg	Egg Albumin Mg	Estimated Unlabeled Mannose Content μg
408[1]	558	—	537 508
408[1]	—	22.85	458 458
403	556[1]	—	403
	556[1]	22.85	430 435

Source: Graham and Neuberger (1968).
[1] ^{14}C-labeled compound.

hydroxypropyl-, and carboxymethyl-, have been covalently linked through ether linkages to cellulose, starch, various natural gums, and mucilages giving products with unique properties. Analyses for these substituent groups may follow classical techniques as in the case of hydroxyalkyl-determination (Jullander and Lagerstrom 1963), but their actual location on the polysaccharide chain presents more of a problem because suitable reference compounds are frequently lacking. The methyl derivatives give little problem. The hydroxyalkyl compounds have been separated by gas-liquid chromatography (Lott and Brobst 1966) and it would seem that electrophoresis or ion exchange chromatography could separate the anionic polymer residues. However, their identification and estimation still await further basic synthetic work.

FUTURE POSSIBILITIES

As may have been gathered from the foregoing, it is easy to analyze chemically for complex carbohydrates with an accuracy of 80-90%. The last 5% of accuracy may not be obtainable unless new reactions are described that facilitate the hydrolysis of glycosidic bonds and reduce the concomitant degradation of the simple sugars so released. Even with the methods presently available it is possible to reduce the errors significantly by the approaches already discussed, but both time and more sophisticated instrumentation are required for these analytical procedures. As is evidenced by the fact that all the natural carbohydrates are biodegradable, the necessary enzymes must exist for cleaving the polysaccharides. It would seem that the ideal methods of analysis must await the necessary advances in enzymology.

DISCUSSION

R. A. Scanlan.—Is there any reason given for different rates of hydrolysis of glycosidic bonds? For example, galactoside hydrolysis is five times as great as glucoside.

R. Montgomery.—I will refer this question to Jim BeMiller, who has just completed an excellent review on the subject (BeMiller 1967).

J. N. BeMiller.—This is the subject of a two hour lecture. Basically it is a difference in the reaction mechanism. We are talking about enthalpy and entropy differences. As you go from the glycoside to a protonated conjugate acid form, heterolytic cleavage occurs to form the carbonium ion at C-1. During this reaction the conformation of the molecule has to change from a normal chair form to a half chair form in which the C-1 end of the molecule is planar. This required that all hydroxyl groups move with respect to one another and some of them become eclipsed. This accounts for the differences in rates.

P. Muneta.—Have you done anything on the possible interferences of the polyphenols with the reducing sugar reactions?

R. Montgomery.—In the sense that the polyphenols would produce a color under the acidic conditions of the colorimetric reaction this color would probably be different from that with say anthrone or phenol; its absorbance maximum may be different and there could be an interference.

P. Muneta.—What about the copper reagents, which depend on the reduction of Cu(II)? These phenols could act as reducing agents too.

R. Montgomery.—This is correct, but we have done nothing in this regard. I would expect interference.

A. Pittet.—On the separation of the monosaccharides on paper, what sort of recoveries do you get after elution?

R. Montgomery.—Within 95% recovery. We cut out the band or spot and soak this in water for 30 min.

A. S. Perlin.—Would you comment on the hydrolysis of glycoproteins? What about the recovery of sugars there? What losses do you expect in the presence of protein because the protein content will vary enormously through the range that Dr. Neuberger gave?

R. Montgomery.—The interference of the peptides and amino acids following the hydrolysis of the glycoprotein is quite extensive. One needs to consider carefully the optimum conditions of hydrolysis for each oligomer. The problems need individual consideration and re-evaluation by other methods, for example, isotopic dilution (Graham and Neuberger 1968). A whole chapter could be written on the determination of mannose in ovalbumin.

A. Neuberger.—It was the problem of estimation of sugars in glycoproteins that led to development of the isotope dilution technique, which still is not the complete answer. Even if you use methyl α-D-mannoside as an internal standard you must assume that this methyl glycoside is hydrolyzed at a rate which is approximately similar to the mannose residue in the protein. This is an approximation. We have now extended the method to amino sugars. Results are probably correct to within 5%.

BIBLIOGRAPHY

ANASTASSIADES, P. A., and COMMON, R. H. 1968. Some aspects of the reliability of chemical analysis. Anal. Biochem. 22, 409–423.

BARKER, S. A., FOSTER, A. B., SIDDIQUI, I. R., and STACEY, M. 1958. Uronic acid determination. Talanta 1, 216–218.

BATES, F. J., and ASSOCIATES. 1942. Polarimetry and saccharimetry of the sugars. Natl. Bur. Std., Circ. C440.

BEMILLER, J. N. 1967. Acid-catalyzed hydrolysis of glycosides. Advan. Carbohydrate Chem. 22, 25–108.

BISHOP, C. T. 1964. Gas-liquid chromatography of carbohydrate derivatives. Advan. Carbohydrate Chem. 19, 95–147.

BRUMMEL, M., GERBECK, C., and MONTGOMERY, R. Unpublished work.

BRUMMEL, M., MAYER, H., and MONTGOMERY, R. Unpublished work.

DIMLER, R. J., SCHAEFER, W. C., WISE, C. S., and RIST, C. E. 1952. Quantitative paper chromatography of D-glucose and its oligosaccharides. Anal. Chem. 24, 1411–1414.

DISCHE, Z. 1947. A new specific color reaction of hexuronic acids. J. Biol. Chem. 167, 189–198.

DISCHE, Z. 1955. New color reactions for determination of sugars in polysaccharides. Methods Biochem. Anal. 2, 313–358.

DUBOIS, M. et al. 1956. Colorimetric method for determination of sugars and related substances. Anal. Chem. 28, 350–356.

EHRENTHAL, I., MONTGOMERY, R., and SMITH, F. 1954. The carbohydrates of gramineae. II. The constitution of the hemicelluloses of wheat straw and corn cobs. J. Am. Chem. Soc. 76, 5509–5514.

ELSON, L. A., and MORGAN, W. T. J. 1933. A colorimetric method for the determination of glucosamine and chondrosamine. Biochem. J. 27, 1824–1828.

GRAHAM, E. R. B., and NEUBERGER, A. 1968. The estimation of galactose, mannose and fucose in glycoproteins by radioisotope dilution. Biochem. J. 106, 593–600.

HAY, G. W., LEWIS, B. A., and SMITH, F. 1962. Thin-film chromatography in the study of carbohydrates. J. Chromatog. 11, 479–486.

HIRST, E. L., HOUGH, L., and JONES, J. K. N. 1949. Quantitative analysis of mixtures of sugars by the method of partition chromatography. Part II. The separation and determination of methylated aldoses. J. Chem. Soc., 928–933.

JULLANDER, I., and LAGERSTROM, O. 1963. Ethyl ethers and their analytical determination. In Methods in Carbohydrate Chemistry. Vol. 3, Whistler, R. L., and Wolfrom, M. L. (Editors). Academic Press, New York.

LAMBERT, M., and NEISH, A. C. 1950. A rapid method for estimation of glycerol in fermentation solutions. Can. J. Res. B28, 83–89.

LOTT, C. E., and BROBST, K. M. 1966. Gas chromatographic investigation of hydroxylethyl amylose hydrolyzates. Anal. Chem. 38, 1767–1770.

MANGOLD, H. K., SCHMID, H. H. O., and STAHL, E. 1964. Thin-layer chroma tography (TLC). Methods Biochem. Anal. *12*, 393–451.

MILLER, G. L. 1959. Use of 3:5-dinitrosalicylic acid reagent for determination of reducing sugar. Anal. Chem *31*, 426–428.

MONTGOMERY, R. 1961. Further studies of the phenol-sulfuric acid reagent for carbohydrates. Biochem. Biophys. Acta. *48*, 591–593.

MORGAN, W. T. J., and ELSON, L. A. 1934. A colorimetric method for the determination of N-acetyl glucosamine and N-acetyl chondrosamine. Biochem. J. *28*, 988–995.

NATIONAL RESEARCH COUNCIL. 1967. Specifications and criteria for biochemical compounds, 2nd Edition, Publ. *1344*. Washington D.C.

PAECH, K., and TRACY, M. V. (Editors). 1955. *In* Modern Methods of Plant Analysis. Springer, Berlin.

PARK, J. T., and JOHNSON, M. J. 1949. A submicrodetermination of glucose. J. Biol. Chem. *181*, 149–151.

ROBYT, J. F., and WHELAN, W. J. 1965. Anomalous reduction of alkaline 3,5-dinitrosalicylate by oligosaccharides and its bearing on amylase studies. Biochem. J. *95*, 10P–11P.

SAWARDEKER, J. S., SLONEKER, J. H., and JEANES, A. 1965. Quantitative determination of monosaccharides as their alditol acetates by gas-liquid chromatography. Anal. Chem. 37, 1602–1604.

SMITH, F., and MONTGOMERY, R. 1959. Analytical procedures. *In* Chemistry of Plant Gums and Mucilages. Reinhold Publishing Co., New York.

STANEK, J., CERNEY, M., KOCOUREK, J., and PACAK, J. 1963. Sugar analysis. *In* The monosaccharides. Academic Press, New York.

SWEELEY, C. C., BENTLEY, R., MAKITA, M., and WELLS, W. W. 1963. Gas-liquid chromatography of trimethylsilyl derivatives of sugars and related substances. J. Am. Chem. Soc. 85, 2497–2507.

SZEJTLI, J., RICHTER, M., and AUGUSTAT, S. 1968. Molecular configuration of amylose and its complexes in aqueous solutions. Part IV. Determination of DP of amylose by measuring the concentration of free iodine in solution of amylose-iodine complex. Biopolymers *6*, 27–41.

WARAVDEKAR, V. S., and SASLAW, L. D. 1959. A sensitive colorimetric method for the estimation of 2-deoxy sugars with the use of malonaldehyde-thiobarbituric acid reaction. J. Biol. Chem. *234*, 1945–1950.

WARREN, L. 1959. The thiobarbituric acid assay of sialic acids. J. Biol. Chem. *234*, 1971–1975.

WHISTLER, R. L., and WOLFROM, M. L. (Editors). 1962. Methods in Carbohydrate Chemistry, Vol. 1, Analysis and Preparation of Sugars. Academic Press, New York.

WHISTLER, R. L., and WOLFROM, M. L. (Editors). 1964A. Methods in Carbohydrate Chemistry, Vol. IV, Starch. Academic Press, New York.

WHISTLER, R. L., and WOLFROM, M. L. (Editors). 1964B. Methods in Carbohydrate Chemistry, Vol. V, General Polysaccharides. Academic Press, New York.

WINZLER, R. J. 1955. Determination of serum glycoproteins. Methods Biochem. Anal. *2*, 279–311.

CHAPTER 9

David R. Lineback | **Physical Methods**[1]

INTRODUCTION

The analysis of carbohydrates implies not only the determination of the amount of carbohydrate present in a reaction mixture, a polysaccharide hydrolyzate, a food product or a biological sample but often the determination of the identity, configuration or conformation of the carbohydrate components. This information must commonly be obtained from very small samples. Sensitive analytical techniques are required to meet these needs. In the past decade, advances and improvements in the application of x-ray crystallography, optical rotatory dispersion, circular dichroism, nuclear magnetic resonance spectroscopy, mass spectrometry, infrared spectroscopy, gas-liquid chromatography and thin-layer chromatography to carbohydrate chemistry have made very sensitive analytical techniques available to research workers in this area. This presentation will be limited to a consideration of only three of these techniques: gas-liquid chromatography (glc), nuclear magnetic resonance (nmr) spectroscopy and mass spectrometry. Space does not permit a comprehensive treatment of even these three techniques but excellent reviews of the application of gas-liquid chromatography (Kircher 1962; Bishop 1964), mass spectrometry (Budzikiewicz *et al.* 1964; Hanessian 1964; Heyns *et al.* 1966; Kochetkov and Chizhov 1966; DeJongh in press; DeJongh *et al.* in press), and nmr spectroscopy (Hall 1964A, Hall in press) to carbohydrate chemistry give a more thorough discussion of these topics.

GAS-LIQUID CHROMATOGRAPHY

Since the first report (McInnes *et al.* 1958) of the application of gas-liquid chromatography to carbohydrate derivatives in 1958, this technique has found increasing importance as a method for the separation and analysis of these materials. The sensitivity and versatility of gas-liquid chromatography has facilitated the analysis of minute quantities of carbohydrates in glycopeptides, polysaccharide hydrolyzates, and biological samples. The major requirement for gas-liquid chromatography of a material is that it be volatile or that a volatile derivative be

---

[1] Published with approval of the Director as Paper No. 2391, Journal Series, Nebraska Agricultural Experiment Station.

easily prepared. In general, compounds that can be distilled or sublimed, even under reduced pressure, are amenable to separation by gas-liquid chromatography. The application of gas-liquid chromatography to carbohydrates lagged behind the development of this technique with other classes of compounds because of the difficulty of preparing volatile derivatives by rapid and general techniques. Much of the early work with carbohydrates was done with methyl ethers or acetates (Bishop 1964) which were suitable derivatives since they were sufficiently volatile.

Caution must be exercised in the application of gas-liquid chromatography to separation of carbohydrate derivatives. The carbohydrate derivative should be stable under the conditions used for gas-liquid chromatography. This is rather easily tested by recovery of the carbohydrate derivative unchanged from the effluent gas stream. An example of this occurs with fully methylated glycosides which can be recovered quantitatively and unchanged from the effluent gas stream, showing that no anomerization, hydrolysis or decomposition has occurred during the separation. It has been shown that carbohydrate derivatives can undergo four general types of modification during gas-liquid chromatography (Bishop *et al.* 1963). These reactions were deamidation, change in size of the sugar ring, rearrangement of acetal or ketal groups and degradative rearrangement of acetylated amino sugars.

Trimethylsilyl Ether Derivatives

The facile preparation of volatile derivatives of carbohydrates can be accomplished by synthesis of the O-trimethylsilyl ether derivatives (Hedgley and Overend 1960; Sweeley *et al.* 1963). These derivatives are prepared by treating the carbohydrate in pyridine with an excess of trimethylchlorosilane and hexamethyldisilazane. Trimethylsilylation occurs completely for most carbohydrates within a few minutes at room temperature, although some oligosaccharides react more slowly. The sample can be injected directly onto the column of the gas-liquid chromatograph without further separation or purification. The use of pyridine as the reaction solvent has the advantage of reducing the extent of mutarotation during the reaction. Thus, pure, single anomers of hexoses and pentoses are observed to generally show only a single peak. Sweeley and co-workers studied the trimethylsilyl ethers of nearly 100 carbohydrates and derivatives and found them amenable to separation by gas-liquid chromatography using appropriate conditions. The compositions of aqueous equilibrium mixtures of hexoses and pentoses were determined by Sweeley and co-workers (1963) and the percentages of the α-aldopyranose forms present are shown in Table 17. Third components, labeled γ-forms, were observed in the aqueous equilibrium mixtures of arabinose, xylose, and galactose. The relative amounts of these components increased upon refluxing the sugar with pyridine and were then observed for lyxose and glucose also. These components were attributed to furanose isomers. A derivative of the third isomer of galactose was isolated by preparative gas-liquid chromatography and identified as trimethylsilyl 2,3,5,6-tetra-O-(trimethyl-silyl)-β-D-galactofuranoside (Shallenberger and Acree 1966). A derivative of a fourth isomer has also been isolated in the same manner from

TABLE 17

PERCENTAGE OF THE α-ALDOPYRANOSE FORM AT EQUILIBRIUM

| | NMR at | | | | | |
	60 Mcps[1]	100 Mcps[2]	Rotation[3]	Bromine Oxidation[3]	Calculated[4]	GLC[5]
Galactose	35	27	29.6	31.4	36	31.9
Glucose	36	36	36.2	37.4	36	39.8
Mannose	64	67	68.8	68.9	68	72.0
Xylose	29	33	34.8	32.1	36	41.3
Lyxose	69	71	76.0	79.7	73	74.1
Ribose	18*	20	—	—	11	—
Arabinose	63	63	73.5	67.6	61	50.8

[1] Rudrum and Shaw (1965) (30° C).
[2] Lemieux and Stevens (1966) (35° C).
[3] Isbell (1962).
[4] Eliel et al. (1965).
[5] Sweeley et al. (1963).
* 70° C.

the pyridine solution and identified as trimethylsilyl 2,3,5,6-tetra-O-(tri-methylsilyl)-α-D-galactofuranoside (Acree *et al.* 1968). The isomer distribution obtained from α-D-galactopyranose in pyridine (8 hr at 80°), as determined by gas-liquid chromatography of the trimethylsilyl ether derivatives, contained 31.7% α-D-galactopyranose, 31.2% β-D-galacto-pyranose, 13.7% α-D-galactofuranose, and 23.4% β-D-galactofuranose. After several weeks in pyridine at 25°, the final equilibrium contained 33.8%, 49.0%, 5.1%, and 12.1%, respectively.

Gas-liquid chromatography of trimethylsilyl ether derivatives has been extensively applied to the quantitative analysis of monosaccharides. Mixtures of fucose, mannose, galactose, and glucose were quantitatively analyzed by gas-liquid chromatography of the trimethylsilyl ether derivatives on polyethylene glycol succinate columns using an instrument equipped with an argon ionization detector (Richey *et al.* 1964). Sawardeker and Sloneker (1965) reported an improved procedure for the

analysis of mixtures of monosaccharides as trimethylsilyl ether derivatives using a Carbowax 20M liquid phase and an instrument equipped
with hydrogen flame ionization detector. The separation of the trimethylsilyl ethers obtained by Sawardeker and Sloneker is shown in
Fig. 64. Analysis of synthetic mixtures of these sugars gave quantitative
results except when mannose and arabinose were present, since an
anomer of arabinose overlapped the α-D-mannose peak. Xylose also
could not be determined in the presence of glucose. The columns were
operated isothermally for these separations. Mixtures of arabinose,

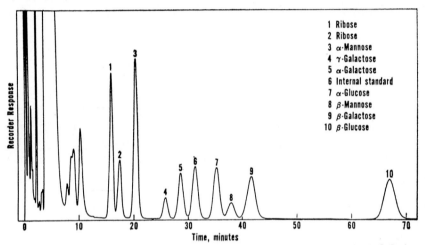

1 Ribose
2 Ribose
3 α-Mannose
4 γ-Galactose
5 α-Galactose
6 Internal standard
7 α-Glucose
8 β-Mannose
9 β-Galactose
10 β-Glucose

From Sawardeker and Sloneker (1965). Copyright 1964 by the American Chemical Society.
Reprinted by permission of the copyright owner

FIG. 64. GAS-LIQUID CHROMATOGRAPHIC SEPARATION OF TRIMETHYLSILYL ETHER
DERIVATIVES OF MONOSACCHARIDES

Conditions: 12 ft Carbowax 20M, 15% on Chromosorb W (HMDS Treated, 80-100
Mesh). Temp 170°, Helium 100 ml/min.

xylose, mannose, galactose, and glucose present in wood product hydrolyzates have been quantitatively determined as trimethylsilyl ether
derivatives using temperature programming to eliminate the problem
that the tailing of pyridine presented to the determination of arabinose
(Brower *et al.* 1966). Gas-liquid chromatography of trimethylsilyl
ether derivatives has been applied to the quantitative analysis of glucose
in commercial corn sugar and corn syrups (Alexander and Garbutt
1965).

The major mono- through tetrasaccharide components of corn syrup
have been determined as their trimethylsilyl ether derivatives (Brobst
and Lott 1966A). A method was developed for the derivatization of

these higher saccharides using trifluoroacetic acid and a large excess of hexamethyldisilazane. The excess hexamethyldisilazane allowed preparation of the derivatives in the presence of water. This avoided the necessity of drying the sample and was directly applicable to commercial corn syrups. The amount of sucrose in hard candy was determined in this manner, indicating the applicability of this method to

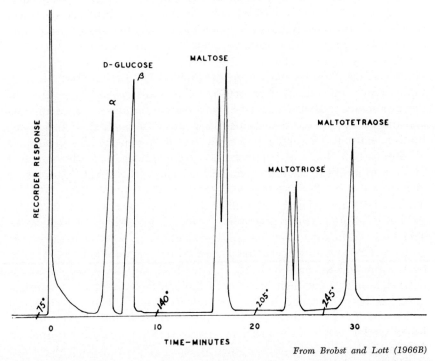

From Brobst and Lott (1966B)

FIG. 65. PROGRAMMED-TEMPERATURE GAS-LIQUID CHROMATOGRAPHIC SEPARATION OF TRIMETHYLSILYL ETHER DERIVATIVES OF MALTO-OLIGOSACCHARIDES

Conditions: 2 ft SE-52, 0.25% on Glassport M (60–80 Mesh). Temp. 75°–245° at 6.4°/min, Helium 80 ml/min.

the analysis of corn syrup-sucrose mixtures. An improved method, utilizing glass beads as a support for the liquid phase in a programmed temperature separation of the mono- through tetrasaccharide components, has been reported and applied to the determination of these carbohydrates in corn syrup and wort (Brobst and Lott 1966B). The separations achieved with this procedure are shown in Fig. 65.

A major problem encountered in the utilization of trimethylsilyl ether derivatives of carbohydrates, particularly monosaccharides, for quantitative analysis by gas-liquid chromatography is the formation of multiple peaks for each sugar. This is especially true when investigating biological samples since these are normally isolated from aqueous media and thus represent the equilibrium mixture for the sugar. As many as four derivatives can be formed from each monosaccharide as the result of anomerization and ring isomerization, each of which will produce a detector response. As a consequence of this, a complex mixture containing several monosaccharides may produce a multiplicity of peaks which will prevent complete separation of all the peaks from one another and accurate quantitative analysis cannot then be achieved.

Glucose, galactose, galactosamine, and sialic acid in neutral glycolipids and gangliosides have been determined by gas-liquid chromatography (Sweeley and Walker 1964). The oligosaccharide portion of the glycolipids was first converted to monosaccharides by methanolysis and the products were then trimethylsilylated for qualitative and quantitative analysis. A similar procedure has been developed for analysis of the carbohydrate components of glycopeptides and glycoproteins (Bolton et al. 1965; Clamp et al. 1967). The carbohydrate portions of glycopeptides or glycoproteins were converted to methyl glycosides by methanolysis and trimethylsilyl ether derivatives then formed for analysis by gas-liquid chromatography. Methyl glycosides of 6-deoxy-L-galactose (L-fucose), D-mannose, D-galactose, 2-acetamido-2-deoxy-D-glucose (N-acetyl-D-glucosamine) and N-acetylneuraminic acid have been estimated quantitatively by this method and the optimum conditions determined for their analysis in biological materials. This procedure has been extended to the determination of D-glucose, D-galactose and D-mannose in a fungal glucoamylase (Lineback and Horner unpublished work). One of the problems involved in the analysis of products from a methanolysis reaction is that as many as four methyl glycosides can be formed for each monosaccharide present and each glycoside will appear in the chromatogram. Since the amount of any sugar is equal to the sum of its isomers present, it is necessary to separate these isomers for quantitative analysis. This can become a problem of considerable importance when several carbohydrate components are present. In the case of the sugars mentioned above, conditions were obtained which resulted in the separation of all of the component peaks.

Alditol Acetates

Reduction of monosaccharides to the corresponding alditols with sodium borohydride and separation of the alditols formed circumvents the appearance of multiple peaks for each monosaccharide since alditols cannot isomerize. In 1961 fully acetylated alditols were first separated by gas-liquid chromatography (Gunner *et al.* 1961), but the conditions used did not allow sufficient separation of ribitol from L-arabinitol, glucitol from galactitol, or fucitol from rhamnitol for quantitative analysis. Conditions necessary for complete separation of alditol acetates derived from ten commonly occurring monosaccharides on a single liquid phase have been used for quantitative analysis of monosaccharide mixtures (Sawardeker *et al.* 1965).

Gas-liquid chromatography of alditol acetates has been applied to the quantitative analysis of the neutral sugars of several glycoproteins including thyrotropic hormone, thyroglobulin, and ovalbumin (Kim *et al.* 1967). The carbohydrate residues were removed from the glycoprotein by acid hydrolysis, reduced with sodium borohydride to the alditols, and acetylated for analysis by gas-liquid chromatography using the organosilicone polyester liquid phase (ECNSS-M) reported by Sawardeker and co-workers (1965). This same liquid phase has been used to separate some partially methylated alditols (arabinitol, xylitol, glucitol, galactitol, and mannitol) as their acetates, and has been used for the analysis of a dextran and a fungal β-glucan (Björndal *et al.* 1967A). The methylated polysaccharides were hydrolyzed, reduced with sodium borohydride, and then the partially methylated alditols acetylated to obtain the derivatives desired for gas-liquid chromatography. This procedure should be of value in methylation analysis of polysaccharides as a complement to analysis of methyl glycosides by gas-liquid chromatography.

The sugars in plant cell-wall polysaccharides, following hydrolysis with trifluoroacetic acid and subsequent reduction with sodium borohydride, have been analyzed by gas-liquid chromatography of their alditol acetates (Nevins *et al.* 1967; Albersheim *et al.* 1967). A mixed liquid phase containing poly(ethylene glycol succinate), poly(ethylene glycol adipate), and silicone XF1150 was used for the analyses. This procedure enabled the determination of as little as 0.25 μg of rhamnose, fucose, arabinose, xylose, mannose, galactose, or glucose (Fig. 66). *myo*-Inositol was used as an internal standard. Integration of the peak areas was done electronically and duplicate analyses were sufficient to yield data with an accuracy of ± 5%.

Other Alditol Derivatives

The major carbohydrate components in starch syrups have also been determined as their alditol trimethylsilyl ether derivatives (Cayle *et al.* 1968). The formation of anomeric forms was avoided by first reducing

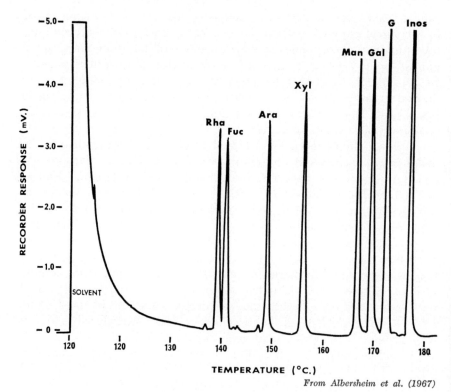

From *Albersheim et al.* (1967)

FIG. 66. PROGRAMMED-TEMPERATURE GAS-LIQUID CHROMATOGRAPHIC SEPARATION
OF ALDITOL ACETATES

Conditions: 4 ft 0.2% poly(ethylene glycol adipate), 0.2% poly(ethylene glycol
succinate) and 0.4% silicone XF1150 on Gas-Chrom P (100-120 Mesh). Temp
maintained at 120° for 10 min after injection, then raised 1°/min.

the carbohydrates to the corresponding alditols with sodium borohydride. Single peaks of glucitol, maltitol, sucrose, maltotriitol, and ribitol
were obtained from glucose, maltose, sucrose, maltotriose, and ribose
(internal standard), respectively.

NUCLEAR MAGNETIC RESONANCE SPECTROSCOPY

In the decade since Lemieux and co-workers (1958) reported the
first application of nmr spectroscopy to carbohydrates and their derivatives, high-resolution nmr spectroscopy has become a firmly established
technique for the investigation of these materials. Proton magnetic
resonance spectroscopy has offered the carbohydrate chemist a method

for: following the progress of a reaction or determining the components of a mixture by observing the appearance or disappearance of nmr signals which are unique to individual components of the mixture, determining the configuration of carbohydrates and their derivatives both at the anomeric center and at other centers in the molecule and investigating the conformations of carbohydrates in solution. This latter facet has assumed particular significance because of the importance of conformational control in chemical and biochemical reactions.

No attempt will be made in this presentation to discuss the nuclear magnetic resonance technique beyond a brief introduction to the parameters involved in interpretation of an nmr spectrum. A number of books are available dealing with theoretical aspects (Pople *et al.* 1959) and applications to organic chemistry (Jackman 1959; Roberts 1959; Roberts 1961; Bhacca and Williams 1964). This discussion will also be limited to proton magnetic resonance.

Interpretation of Spectra

The utilization of nmr spectroscopy has been facilitated by the fact that first-order analysis of nmr spectra may be accomplished with a relatively simple nonmathematical understanding of the nmr phenomenon. Spectra may be interpreted by a consideration of three parameters whose values are measured directly from the spectrum in first-order analyses. These parameters are the chemical shift, the multiplicity of a signal with the attendant spacings (approximate coupling constants), and the area beneath a multiplet.

Chemical Shift

When an organic molecule containing covalently bonded hydrogen atoms is placed in a strong magnetic field (H_o) and irradiated with radiofrequency radiation, absorption of the radiation occurs at a frequency (ν) which is characteristic of the magnetic environment of the hydrogen nuclei according to the expression

$$h\nu = \frac{\mu H_o(1 - \sigma)}{I}$$

where h is Planck's constant, μ is the nuclear magnetic moment, σ is the screening constant for the particular nucleus, and I is the nuclear-spin quantum number. It is impractical to measure the frequency and magnetic field at which these absorptions occur with sufficient accuracy to calibrate the spectrum in absolute units. A reference substance (standard) is used, internally or externally, to provide a reference point for the calibration of the spectrum. The most commonly used internal standard for solutions of carbohydrates in organic solvents is tetra-methylsilane and for aqueous solutions is sodium 2,2-dimethyl-2-silapentane-5-sulfonate. The absorptions of the methyl protons of these substances yield single, sharply defined peaks conveniently located in the spectrum to serve as a reference for the proton resonances of other organic molecules.

The chemical shift (resonance position) of a proton in the nmr spectrum is the frequency separation between the reference signal and the signal for that proton. The magnitude of the chemical shift is directly proportional to the applied magnetic field or to what is equivalent, the rf oscillator frequency. It is more convenient to express the chemical shift as a nondimensional unit which is independent of the field strength rather than using the above equation which necessitates specification of either the magnetic field or the resonance frequency used in the measurement. This is accomplished by the expression

$$\delta = \frac{H_s - H_r}{H_r}$$

where δ is the chemical shift and H_s and H_r are the resonance field strengths of the sample and reference, respectively. However, since spectra are usually calibrated in cycles per second (cps), rather than in units of magnetic field strength, this equation may be written

$$\delta = \frac{\Delta \times 10^6}{\text{oscillator frequency (cps)}}$$

where Δ is the separation (in cps) of the resonance of the sample and the reference. The oscillator frequencies used in nmr spectrometers are in Mcps (60 Mcps) and the factor 10^6 is used to give convenient numbers for proton spectroscopy. The chemical shift, δ, is then expressed in parts per million (ppm).

One of the most significant advances in instrumentation in recent years has been the development of spectrometers operating at increased magnetic fields, such as in the 100 Mcps spectrometer. Currently, the most common nmr spectrometers are those operating at 60 Mcps. A spectrometer operating at 220 Mcps has been used (Bhacca and Horton 1967; Holland et al. 1967) to obtain a spectrum of 1-thio-α-L-arabino-pyranose tetraacetate (Fig. 67A) which was amenable to first-order analysis. At 100 Mcps, the low-field nmr signals of this compound (Fig. 67B) were insufficiently resolved for interpretation. The availability of spectrometers operating at high applied magnetic fields have greatly increased the amount of information that can be obtained from the nmr spectrum of many carbohydrates and their derivatives, i.e., the separation between reference and sample signal is 3.7 times greater at 220 Mcps than at 60 Mcps.

The chemical shift of a proton in a nmr spectrum is dependent on

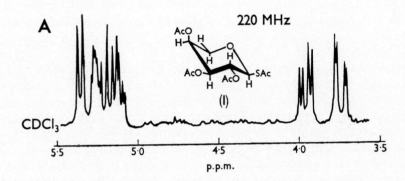

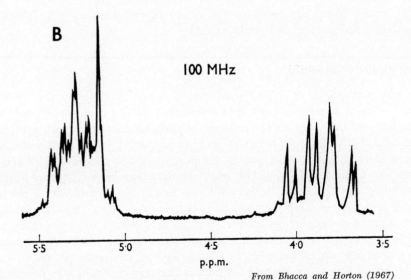

From Bhacca and Horton (1967)

FIG. 67. THE LOW-FIELD PORTION OF THE NUCLEAR MAGNETIC RESO-
NANCE SPECTRUM OF 1-THIO-α-L-ARABINOPYRANOSE TETRAACETATE IN
DEUTEROCHLOROFORM

(A) 220 Mcps spectrum (B) 100 Mcps spectrum.

the electronic and geometrical environment of that proton. The dependence of chemical shift on molecular geometry was observed by Lemieux and co-workers (1958) when they found that equatorial acetoxy groups produce their signal at a higher applied magnetic field than axial acetoxy groups. In contrast, equatorial ring protons produce their signal at a lower field than chemically similar but axially oriented protons. The dependence of the chemical shift on the electronic environment of the proton is indicated by the observation that the chemical shift of a proton is, in general, a function of the electron density around the nucleus. The most common example of this behavior in the spectra of carbohydrates is the characteristic low-field shift of the anomeric proton due to the deshielding effect of the ring-oxygen atom which decreases the electron density around the anomeric hydrogen nucleus. Another example is the shift to lower field of the ring protons in acetylated carbohydrates compared to the chemical shifts of the corresponding ring proton when the hydroxyl group is free. It has been shown that the chemical shifts of the ring protons of acetylated sugars are strongly dependent on configurational changes at positions other than the neighboring positions (Lemieux and Stevens 1965). The dependency of the chemical shift on the electronic and geometrical environment of the proton forms the basis for the utility of this technique in investigating configuration and conformation.

Multiplicity of Signal

The multiplicity of a proton resonance signal depends on the extent of the electron-coupled spin-spin interaction of the proton with neighboring protons. The most common coupling interaction is with protons at adjacent positions (between protons separated by 2 or 3 bonds). However, instances of long-range coupling over several bonds (between protons separated by four or more bonds) are known. Long-range coupling is dependent on the conformation of the carbohydrate molecule and the resulting interactions that may come into play as a result of suitable molecular geometry. The most common example of spin-spin interactions in carbohydrates resulting in a multiplicity of the nmr signal is the appearance of the signal for the anomeric proton of an aldohexopyranose as a doublet which is the result of spin-spin coupling to the adjacent proton at C-2. The doublet at 5.35 ppm in Fig. 67A is the signal from the proton at C-1 (anomeric proton) of 1-thio-α-L-arabinopyranose tetraacetate. This proton is spin-spin coupled to the proton at C-2 with a spacing (coupling constant) of 7.0 cps (Bhacca and Horton 1967).

The multiplicity of a nmr signal is, in general, given by the expression

$$\text{multiplicity} = 2nI + 1$$

where n is the number of neighboring equivalent nuclei of spin I. For the proton, this becomes $n + 1$ since the spin of the proton is $\frac{1}{2}$. The multiplicity of an nmr signal thus becomes a useful index to the number of adjacent interacting protons.

The electron-coupled spin-spin interactions are independent of the strength of the applied magnetic field. Thus, the coupling constant does not alter with changes in field strength of the spectrometer, although the chemical shift does. In first-order analysis, the coupling constant is measured directly from the spectrum as the apparent spacings within multiplets. However, this may not be the true coupling constant due to intervention of second-order effects.

The coupling constant shows an angular dependence on the relative orientations of the protons involved in the spin-spin coupling. This was first observed for carbohydrates by Lemieux and co-workers (1958). Karplus (1959) rationalized this angular dependence with an equation of the form

$$J = J_o \cos^2 \phi + K$$

where J is the coupling constant between hydrogen atoms attached to adjacent carbon atoms and separated by a projected valency angle (dihedral angle) of ϕ. J_o and K are parameters whose values were evaluated as $K = -0.28$, and $J_o = 8.5$ for $0° < \phi < 90°$, and $J_o = 9.5$ for $90° < \phi < 180°$. The general form of this relationship is shown in Fig. 68 and has become known as the Karplus curve. This relationship provides a method for calculating the angle between adjacent hydrogen atoms utilizing the spin-spin coupling constants determined from the nmr spectrum. When these hydrogen atoms are part of a cyclic system, as often occurs with carbohydrates, the calculated angles define the shape of the ring. This method has been widely used in investigating the conformations of carbohydrates in solution.

However, calculations obtained from the Karplus relationship can only be considered to be approximate as it is now known that the coupling constant is affected by parameters other than dihedral angle. The Karplus curve appears to be subject to displacement depending on the electronegativities of the substituent groups and the hybridization

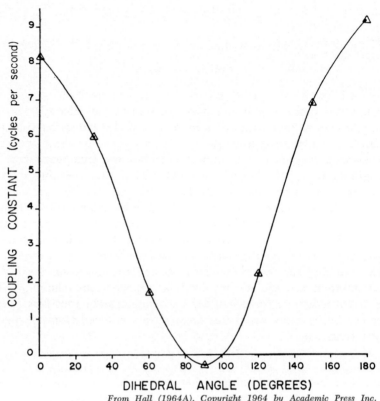

DIHEDRAL   ANGLE   (DEGREES)

*From Hall (1964A). Copyright 1964 by Academic Press Inc.
Reproduced by permission of the copyright owner.*

FIG. 68.   THE KARPLUS CURVE, CALCULATED FROM THE
ORIGINAL PARAMETERS

of the bridging carbon atoms (Lemieux *et al.* 1962; see Hall 1964A for a discussion of the application of the Karplus relationship to carbohydrate conformations).   Calculations from the relationship are still extremely useful since one can distinguish between hydrogens on adjacent carbon atoms having a projected bond angle of 180° and those having a projected bond angle of 60°.   The coupling constant of the former is observed to be 2 to 3 times greater than the latter (see Fig. 68).

Area Beneath Multiplet

The intensity of a signal in the nmr spectrum of a compound is directly proportional to the number of protons resonating at that position.   Thus, determination of the areas under the multiplets of a spectrum can be used to determine the relative ratios of the different types of hydrogen present in the organic molecule. Most nmr spectrometers are equipped with an integrator which is used to directly measure the relative intensity of spectral peaks.

Sample Size

One major problem involved in the utilization of nmr spectroscopy for the investigation of carbohydrates and their derivatives has been the relatively large amount of sample required. Until recently, the minimum sample required for a 60 Mcps spectrometer was approximately 10 mg and the most commonly used sample involved 50 mg or more. The availability of a practical microcell has reduced the sample size by a factor of 4 to 5 allowing usable spectra to be obtained from samples of 1–5 mg. The introduction of the time averaging computer resulted in a significant reduction in the amount of sample required. The time averaging computer electronically averages the random noise from successive traces of the spectrum while accumulating the resonance signal. The signal-to-noise ratio is increased approximately equal to the square root of the number of scans. Thus, by scanning the spectrum a sufficient number of times, usable spectra can be obtained from microgram quantities of sample with a 60 Mcps spectrometer. The sensitivity of a 100 Mcps spectrometer is several times that of a 60 Mcps spectrometer and even smaller sample sizes can then be used, especially when used with a time averaging computer.

Application of NMR Spectroscopy to Carbohydrates

Applications to Monosaccharides.—The chemical shifts of the anomeric protons of solutions of pentoses and hexoses in deuterium oxide have been determined in several laboratories (Lenz and Heeschen 1961; Hall 1964B; Rudrum and Shaw 1965). At 60 Mcps, the anomeric protons gave the only resolved multiplets and appeared as doublets at a lower field than the remaining protons in the spectrum. The assignment of the anomeric proton peaks in the spectrum of an equilibrium solution was made by comparing the spectrum with one made immediately after solution of the sugar, when only one species was present, and by utilizing the observation of Lemieux and co-workers (1958) that axial ring protons produce their signal at higher fields than chemically similar but equatorially oriented ring protons. The relative amounts of each form present in the equilibrium solution could then be determined by integration of the area under the multiplet assigned to the

anomeric proton of that form. A more detailed study, using both double
irradiation and specific deuteration techniques, of the nmr spectra of
hexoses and pentoses at 100 Mcps has been reported (Lemieux and
Stevens 1966) in which the chemical shifts of a number of protons at
positions other than the anomeric center have been determined for
deuterium oxide solutions. The 100 Mcps spectra of mannose, glucose,
and galactose are shown in Fig. 69, and in each spectrum the doublets
arising from the anomeric protons of the two anomeric forms are clearly
present at lower fields than the remaining protons. The percentage of

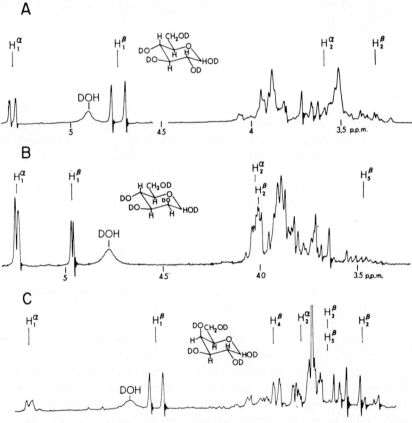

From Lemieux and Stevens (1966). Reproduced by permission of the Natl. Res. Council of
Canada from the Canadian J. Chem. 44, 249 (1966)

FIG. 69.   100 MCPS NUCLEAR MAGNETIC RESONANCE SPECTRA OF ALDOHEXOSES AT
TAUTOMERIC EQUILIBRIUM IN DEUTERIUM OXIDE AT 35°

(A) D-Glucose. (B) D-Mannose. (C) D-Galactose.

the α-aldopyranose form present in the equilibrium solutions of several sugars, as determined by several methods, is shown in Table 17. It can be seen that the values obtained by nmr spectroscopy agree reasonably well with those found by other methods. The differences noted for galactose, lyxose, and arabinose might be attributed to the fact that these sugars exist, to a small extent, in the furanose form. However, the 100 Mcps spectrum of the equilibrium mixture of each of these sugars does not show signals attributed to the furanose isomers (Lemieux and Stevens 1966). Signals attributed to furanose isomers were observed only in the nmr spectra of D-ribose, 2-deoxy-D-*erythro*-pentose (2-deoxy-D-ribose) and D-altrose. Angyal and Pickles (1967) detected the presence of furanose forms in the nmr spectra of aqueous solutions of D-allose, D-altrose, D-gulose, D-talose, D-arabinose and 3-deoxy-D-*ribo*-hexose and determined the percentage of the furanose and pyranose isomers present. No signals corresponding to furanoses were detected in the spectra of equilibrium solutions of D-glucose, D-mannose, D-galactose, D-xylose, D-lyxose, and 2-deoxy-D-*arabino*-hexose.

The detailed study of the 100 Mcps spectra of the hexoses and pentoses in deuterium oxide solution provided considerable information concerning the conformations of sugars in aqueous solution (Lemieux and Stevens 1966). Using the chemical shifts of the ring protons of β-D-xylopyranose and β-D-glucopyranose, and a set of empirical rules correlating the chemical shift with the configuration of the sugar, it was possible to calculate the chemical shifts of protons in other pyranose structures within a useful degree of accuracy. A similar set of empirical rules for correlating the chemical shift with the configurational and conformational properties of acetylated sugars has been reported (Lemieux and Stevens 1965).

One of the problems encountered in the study of the nmr spectra of aqueous solutions of unsubstituted carbohydrates is that of mutarotation. This makes it extremely difficult to study pure anomeric forms of free sugars, since mutarotation starts upon dissolution of the sugar and the nmr spectrum contains the signals of both anomers. However, advantage has been taken of the fact that sugars are soluble in dimethyl sulfoxide (DMSO) and do not mutarotate to any appreciable extent in this solvent. Casu et al. (1964) have determined the spectra of free sugars in DMSO. In these spectra, the resonance positions of the hydroxyl groups are also observed. The strong hydrogen bonding of the hydroxyl protons to the solvent shifts the hydroxyl resonances downfield and reduces the rate of proton exchange sufficiently to permit observation of hydroxyl proton splitting (Chapman and King 1964).

The signals due to the hydroxyl protons may then be removed by adding a small amount of deuterium oxide to the DMSO solution. In the spectral range accessible with DMSO as solvent (up to 6 τ), the signals of all the hydroxyl protons and the anomeric proton are observed. For the ring protons which give their signal above this position and are obscured by the methyl resonances of DMSO, it is necessary to use DMSO-d_6 as the solvent. The hydroxyl protons are coupled to adjacent protons in the same manner as are ring protons. As a result, the C-1 hydroxyl appears as a doublet by virtue of spin-spin coupling with the anomeric proton while the C-6 hydroxyl in hexopyranoses appears as a triplet due to spin-spin coupling with the 2 protons of the hydroxymethyl function. Hydroxyl resonances do not appear to be useful in the determination of conformation of the pyranose ring. Their chemical shifts do not appear to depend only on the axial or equatorial orientation of the hydroxyl bond with respect to the ring, but are also related to the extent to which they can hydrogen bond with the DMSO solvent (Casu et al. 1965). Conformational studies have been carried out in DMSO solutions using a modified Karplus equation to calculate the dihedral angle defined by the ring protons at C-1 and C-2 from the coupling constants determined from the nmr spectrum (Casu et al. 1965). These studies indicated that the hexoses and pentoses studied had the same conformations in DMSO solutions as in deuterium oxide solutions.

Nmr spectroscopy has been extensively applied to structural determinations in the field of carbohydrate chemistry by making comparisons of the chemical shifts and coupling constants of the material under investigation with those of known standards. The literature in this area is extensive and will not be discussed. Rather extensive investigations have also been made of the conformations of carbohydrates and their derivatives in solution by nmr spectroscopy using dihedral angles calculated with the Karplus relationship from the coupling constants measured from the spectra. These investigations have given insight into the preferred conformations of many carbohydrates in solution. This knowledge is of particular significance in studying reaction mechanisms of chemical and biochemical reactions under conformational control. With instruments of increased sensitivity and wider dispersal of signals in the nmr spectrum, and as the parameters governing the Karplus relationship become more fully understood and calculated, the utilization of nmr spectroscopy will continue to play an ever more important role in investigating the configuration and conformation of carbohydrates and their derivatives.

Applications to Oligosaccharides and Polysaccharides.—Nmr spectroscopy has found only limited application to the areas of oligosaccharides and polysaccharides due to the increased complexity of the spectra of these materials and, in some cases, to problems of solubility. The presence of many nearly equivalent protons leads to overlapping signals in the spectra of these substances. These signals are often too poorly resolved to be of use in obtaining parameters which can be related to the composition, configuration, or conformation of the compound under investigation. However, the increased sensitivity of spectrometers operating at higher magnetic fields and the increased dispersal of the signals in the spectra obtained with these spectrometers indicates the feasibility of applying nmr spectroscopy to the study of structure and conformation of oligosaccharides and polysaccharides.

Pasika and Cragg (1963A) utilized nmr spectroscopy to determine the amount of branching in a dextran. The spectra were determined in deuterium oxide using a 60 Mcps spectrometer. Isomaltotriose was used as a model compound to aid in peak assignments. In the spectrum of this compound (Fig. 70A), the anomeric protons of the 2 glucose residues involved in the α-D-$(1 \rightarrow 6)$ linkage gave rise to the multiplet at 5.05 ppm. The glucose unit at the reducing end produced signals at 5.29 ppm and 4.67 ppm assigned to the anomeric protons of the α- and β- anomers, respectively. These latter protons disappear in the spectra of polysaccharides due to their vanishing ratio relative to the anomeric protons of the nonreducing units. The spectrum of a linear dextran (Fig. 70B) revealed signals assigned to the anomeric protons of the α-D-$(1 \rightarrow 6)$ linkages at the same chemical shift (5.05 ppm) as the corresponding protons in isomaltotriose. The spectrum of a branched dextran (Fig. 70C) contained signals assigned to the anomeric protons involved in the α-D-$(1 \rightarrow 6)$ linkages (5.05 ppm) and an additional multiplet at 5.40 ppm assigned to the anomeric protons of non-$(1 \rightarrow 6)$ linkages [$(1 \rightarrow 3)$ linkages in the dextran]. The ratio of the intensity of the 2 sets of signals from the anomeric protons [$(1 \rightarrow 6)$ linkages/non-$(1 \rightarrow 6)$ linkages] was used to determine the relative amount of branching in the dextran. This ratio (71 : 29) agreed very closely with the ratios (70 : 30) obtained from periodate oxidation and methylation studies. High-resolution nmr spectroscopy may provide a rapid means for determining the extent of branching in polysaccharides.

Van der Veen (1963) determined the chemical shift and coupling constants of the anomeric protons of a number of glycosides and oligosaccharides of glucose and galactose in deuterium oxide. The chemical shifts and observed splittings were correlated with the configuration of

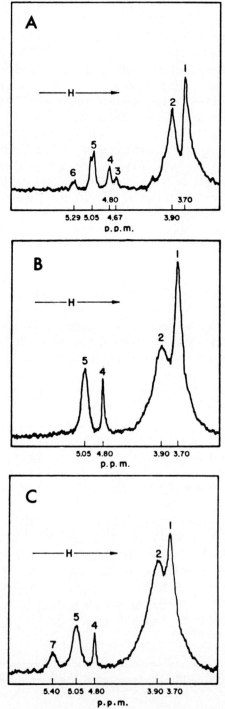

From Pasika and Cragg (1963A). Reproduced by permission of the Natl. Res. Council of Canada from the Canadian J. Chem. 41, 293 (1963)

the glycoside. The chemical shift of the anomeric protons also appeared to be useful in determining the conformation of the carbohydrate by indicating the equatorial or axial orientation of the anomeric proton. Further studies of the nmr spectra of glucoses and polyglucoses in deuterium oxide solution led to the proposal that the glucopyranose ring exists in the C1 conformation in both the linear polymers of glucose and the cyclodextrins (Rao and Foster 1963; Glass 1965). Investigation of the nmr spectra of glucoses and polyglucoses in DMSO solution have confirmed that the glucopyranose ring also exists in the C1 conformation in these solutions (Casu et al. 1966). The resonances of the hydroxyl protons, observed in the DMSO solutions, added significant information since it was observed that two hydroxyl signals in the spectra of maltose, maltooligosaccharides, amylose and the cyclodextrins were shifted to significantly lower field than the signals of the remaining hydroxyls. This was attributed to the existence of an intramolecular hydrogen bond between the hydroxyls at C-2 and C-3' of contiguous glucose units. Models indicated that C1 chair conformations were consistent with strong intramolecular hydrogen bonding between these hydroxyl groups and were even stabilized by this internal association. The chemical shifts and apparent coupling constants for the anomeric protons of a number of mono-, oligo-, and polysaccharides are shown in Table 18.

Nmr spectroscopy has been used to confirm the α-linkages of a series of mannose oligosaccharides derived from a mannan of *Saccharomyces cerevisiae* and to indicate that the mannopyranose units appear to be present in C1 conformations (Lee and Ballou 1965A). Analysis of the signals due to anomeric protons in the nmr spectra of the *myo*-inositol di-, tri-, tetra-, and penta- mannosides obtained from the *myo*-inositol phospholipids from *Mycobacterium tuberculosis* and *Mycobacterium phlei* confirmed the structure proposed for these mannosides and the presence of α-mannosidic linkages (Lee and Ballou 1965B). The configuration of the anomeric linkages in the serotype 2 capsular polysaccharide produced by *Aerobacter aerogenes* NCTC 243 were established by an investigation of the nmr spectra of oligosaccharides derived from the polysaccharide (Gahan et al. 1967). The extent of degradation by chain scission or branch hydrolysis during sulfation of dextrans was found to be negligible from a study of the nmr spectra of

FIG. 70. 60 MCPS NUCLEAR MAGNETIC RESONANCE SPECTRA OF ISOMALTOTRIOSE (A), NRRL B-512 LINEAR DEXTRAN (B) AND NRRL B-742 BRANCHED DEXTRAN (C) IN DEUTERIUM OXIDE

TABLE 18

CHEMICAL SHIFTS AND COUPLING CONSTANTS OF SOME MONO-, OLIGO-,
AND POLYSACCHARIDES

| | Chemical Shift[1] Anomeric Protons | | | | | |
| | Reducing[2] | | Nonreducing | | Sol- | Refer- |
Carbohydrate	H_e	H_a	H_e	H_a	vent	ence
Glucose	4.68 (3.5)[3]	5.26 (7.5)			D_2O	4
Glucose	5.08 (3.0)	5.70 (6.5)			DMSO	5
Mannose	4.75 (1.7)	5.03 (1.0)			D_2O	4
Galactose	4.66 (2.8)	5.32 (7.1)			D_2O	4
Galactose (α)	5.05 (<3.0)				DMSO	5
Xylose	4.74 (3.1)	5.35 (7.4)			D_2O	4
Xylose (α)	5.15 (3.5)				DMSO	5
Lyxose	4.92 (4.2)	5.06 (1.5)			D_2O	4
Arabinose	4.66 (2.7)	5.40 (7.2)			D_2O	4
Ribose	5.09 (2.1)	5.01 (6.4)			D_2O	4
Maltose	4.80 (3.3)	5.35 (7.3)	4.62 (3.0)		D_2O	6
Maltose (β)		5.70 (6.5)	5.01 (3.0)		DMSO	7
Cellobiose	4.78 (3.3)	5.35 (7.4)		5.50 (7.0)	D_2O	8
Cellobiose (β)		5.60 (7.0)		5.71 (6.5)	DMSO	7
Gentiobiose	4.78 (3.5)	5.34 (7.6)		5.50 (7.7)	D_2O	8
Lactose	4.78 (3.7)	5.36 (7.3)		5.58 (7.1)	D_2O	8
Sucrose			4.59 (3.2)		D_2O	8
α,α-Trehalose			4.81 (3.2)		D_2O	8
α,α-Trehalose			5.15 (3.0)		DMSO	7
Melibiose	4.79 (2.9)	5.33 (7.7)	5.04 (2.9)		D_2O	8
Maltotriose	4.80 (3.7)	5.37 (7.3)	4.63 (3.5)		D_2O	6
Maltotriose			5.00		DMSO	7
Amylose			4.90 (3.0)		DMSO	7
Laminarin				5.25 (6.7)	D_2O	6
(β-1 → 3)						
Laminarin				5.56 (7.0)	DMSO	7
Linear dextran			5.09 (2.3)		D_2O	6
(α-1 → 6)						
Linear dextran			5.40 (2.5)		DMSO	7
Branched dextran			5.07, 4.73		D_2O	6
70% α-1 → 6						
30% α-1 → 3						
Branched dextran			5.10, 5.35		DMSO	7
Crown gall poly-				5.12 (6.2)	D_2O	6
saccharide						
(β-1 → 2)						
α-Cyclodextrin			4.95 (2.7)		D_2O	6
α-Cyclodextrin			5.19 (3.0)		DMSO	7
β-Cyclodextrin			5.17 (3.0)		DMSO	7
γ-Cyclodextrin			4.93 (3.5)		D_2O	6

[1] Chemical shift in τ values ($\tau = 10 - \delta$).
[2] H_e represents an equatorial anomeric proton while H_a represents the equivalent axial anomeric proton of the reducing glycose or nonreducing glycosyl units as indicated.
[3] Number in parentheses is the observed spacing (approximate coupling constant, $J_{1,2}$) in cps.
[4] Lemieux and Stevens (1966).
[5] Casu et al. (1965).
[6] Glass (1965).
[7] Casu et al. (1966).
[8] Van der Veen (1963).

the dextrans before and after sulfation (Pasika and Cragg 1963B), and evidence was obtained for the preferential substitution of sulfate at C-2 in the anhydroglucose unit. An nmr study of sulfated mucopolysaccharides in deuterium oxide has revealed information concerning the glycosidic linkages, the position of the sulfate group, the nature of the repeating unit and branching (Inoue and Inoue 1966).

Much work remains to be done before nmr spectroscopy can be applied to oligosaccharides and polysaccharides to the extent that it has been to monosaccharides or their derivatives. One of the major requirements at the present time is the systematic investigation of model compounds and series of homologous oligosaccharides to establish the chemical shift and coupling constant parameters that can then be applied to investigations of polysaccharides. With the recent improvements in instrumentation, it appears probable that nmr spectroscopy will make major contributions to an understanding of the configuration and conformation of polysaccharides and, perhaps, to analyses for these materials.

MASS SPECTROMETRY

In 1958 the first application of mass spectrometry to carbohydrate derivatives was reported (Finan *et al.* 1958) and since that time has found increasing application to problems in this area. As in the case of gas-liquid chromatography, application of mass spectrometry to carbohydrates and their derivatives was initially limited by the relatively low volatility of free carbohydrates and also by the complex spectra obtained from some derivatives. These problems have been reduced by the introduction of new inlet techniques for substances that have very low volatility and by systematic studies of certain carbohydrate derivatives to determine the characteristic fragmentation patterns and rearrangements of the molecular ions. The sensitivity of the mass spectral technique offers significant advantages since mass spectra can be obtained with about 0.1 mg of sample and sometimes on amounts as low as 0.1–1.0 µg (Kochetkov and Chizhov 1966). One of the limitations is the problem of manipulating such small samples.

Mass spectrometry promises to have significant potential for applications to the field of carbohydrate chemistry. The technique may be applied to: determining molecular weights, assigning the position of substituents such as methyl, deoxy, or amino functions, determining the size of a ring, obtaining information concerning stereochemistry, identifying partially methylated monosaccharides and for structural analysis of di- and oligosaccharides. The application of mass spectrometry to the analysis of oligosaccharides obtained by controlled acid hydrolysis of polysaccharides may prove useful in the study of structural problems in polysaccharide chemistry. The molecular weight determination of a carbohydrate is often difficult due to the absence of its molecular ion peak. The molecular ions of carbohydrates are often unstable and undergo extensive fragmentation.

Due to their volatility, the carbohydrate derivatives which have been most extensively investigated by mass spectrometry are the acetates, O-methyl ethers, and acetals (Kochetkov and Chizhov 1966). The mass spectra of these derivatives generally do not exhibit molecular ion peaks, while the spectra of monosaccharide dialkyl dithioacetals or their acetates contain molecular ion peaks. These latter derivatives have been found to be closest to the ideal for structural analysis of monosaccharides but yield almost no information concerning the stereochemistry of the monosaccharide. Methyl ethers and acetates may be used for determination of ring size while alkylidene acetals are preferable for study of stereochemical features. Mass spectra of amino sugars (DeJongh and Hanessian 1965; Heyns and Müller 1965; Heyns and Müller 1966A; Heyns and Müller 1966B; Kochetkov et al. 1966), anhydro sugars (Heyns and Scharmann 1966), and deoxy sugars (Biemann et al. 1963; DeJongh and Hanessian 1966) have been studied including derivatives in furanose, pyranose, and acyclic forms. The mass spectra of the trimethylsilyl ether derivatives of several monosaccharides, their methyl glycosides and disaccharides have been reported (Chizhov et al. 1967). These derivatives may prove suitable for the use of mass spectrometry in the investigation of the structures of higher saccharides.

The use of mass spectrometry to identify partially methylated monosaccharides has been extended to the structural analysis of polysaccharides or carbohydrate-containing compounds (Kochetkov and Chizhov 1966). The polysaccharide is methylated and then subjected to methanolysis to produce methyl glycosides. These are methylated with deuteriomethyl iodide and the position of the trideuteriomethyl groups is determined by mass spectrometry. The position of the trideuteriomethyl groups corresponds to free hydroxyl groups in the original glycoside. In the case of polymers containing different linkages or different carbohydrate components, the problem of separation of the partially methylated glycosides prior to deuteriomethylation is encountered.

The versatility of mass spectrometry is greatly extended when the mass spectrometer is combined with the gas chromatograph so that the effluent from the gas chromatograph is introduced directly into the inlet system of the mass spectrometer. Mass spectrometry is compatible with gas-liquid chromatography because of its sensitivity and since most compounds purified by gas-liquid chromatography are also sufficiently volatile to be analyzed by mass spectrometry. Commercial gas chromatograph-mass spectrometer combinations are available. This technique promises to be of considerable utility in analysis of polysaccharides, glycoproteins, glycolipids, and other carbohydrate-containing compounds when derivatives can be obtained that are separable by gas-liquid chromatography and that also yield information concerning the structure of the material. The mass spectra of methyl glycosides of partially methylated sugars have been obtained following acetylation (DeJongh and Biemann 1963). These derivatives are readily obtained from methylated polysaccharides following glycosidation and acetylation and can be separated by gas-liquid chromatography. However, the problem of multiple peaks from each monosaccharide is again encountered making separation more difficult and often incomplete. Separation of partially methylated alditol acetates by gas-liquid chromatography has been reported (Björndal et al. 1967A) and the mass spectra of these derivatives have been determined (Björndal et al. 1967B). The mass spectra were determined in a combined gas chromatograph-mass spectrometer unit using the partially methylated alditol acetates as samples. This combined gas chromatograph-mass spectrometer method should be useful for characterization of the different methylated sugars obtained from methylation analysis of polysaccharides and other carbohydrate-containing materials. The sensitivity of this method should make it particularly applicable to samples obtained from biological systems.

SUMMARY

Gas-liquid chromatography is an extremely sensitive technique which can be applied to the separation and quantitative analysis of carbohydrates from a wide variety of sources. It can aid in the identification of carbohydrates when used with appropriate standards. Nuclear magnetic resonance spectroscopy finds its greatest use in investigations of configuration and conformation. It appears to be a rapid means for determining the amount of branching in polysaccharides and should find increasing utilization as an analytical technique. Mass spectrometry promises to be a particularly sensitive technique for structural investigations.

It can be seen that nuclear magnetic resonance spectroscopy, gas-liquid chromatography, and mass spectrometry have found widespread application to carbohydrate chemistry and have placed sensitive analytical techniques at the disposal of workers in this field. With the advances in instrumentation and methodology, these techniques promise to be of even wider utility and value.

DISCUSSION

T. J. Siek.—Is IR of any use in initial analysis of a complex carbohydrate as a means of getting some idea of what might be present in this complex material?

D. R. Lineback.—IR, used with the proper precautions, can yield a great deal of information (Spedding 1964). However, there are conflicting reports in the literature concerning the application of IR to the identification of anomeric configuration of carbohydrates.

T. J. Schoch.—Do you think either NMR, IR, or UV spectroscopy could solve the problem of the nature and location of the binding of hydrate water by carbohydrates. Let me give you an example. Cornstarch binds moisture. At a given relative humidity it will bind one mole of water which we presume is on carbon six. On the other hand potato starch binds 1½ moles of water as moisture which we presume is a shared water between the moistures on carbon six. Could the location of these be established and the nature of the bonding be identified? Could there be two types of peaks corresponding to the two types of water?

D. R. Lineback.—I would hesitate to say that NMR could give information on something as complex as starch. However, I would point out that wide line NMR has been used to quite an extent in measuring water content. In the slide with the spectra of the two glucose anomers in dimethylsulfoxide (DMSO), it was shown that the C-2, 3, and 4 hydroxyl signals of the beta-D-anomer, which would be expected to be doublets because of the proton coupling, appeared as a single doublet. In the alpha anomer, however, each of these is a distinct doublet signal at a different chemical shift, indicating that each of the hydroxyls in the alpha anomer is involved in a different type of hydrogen bonding or solvation with the DMSO solvent. In the beta anomer they all appear to be solvated in the same manner. Thus, on a simple molecule we can obtain a great deal of information, but on a complex molecule such as starch we could not identify positional types.

D. French.—In partial discussion of Dr. Schoch's question, in the Jeffrey and Chu (1967) paper on the structure of β-methyl maltoside, they showed that the 1 water of crystallization contained 4 hydrogen bonds. The water furnished 2 H's as H-bond donors, also an acceptor for 2 H bonds. They further pointed out that each of the hydroxyl groups in this crystal structure was fully occupied in hydrogen bonding.

I would suggest that the answer to your question might very well come from a fuller study of the crystal organization of the starch in which you see how the water has to fit in to the structure in order to satisfy the H-bonding requirement. It will probably turn out that the water is not identified with any single hydroxyl group but probably forms bridges between two molecular chains or glucose units.

R. A. Pieringer.—Many times in GLC, the solvent peak interferes or could interfere with the early components coming off the column. Have you used the solid injection technique? If you have, can you evaluate it?

D. R. Lineback.—No. We have not used this technique.

BIBLIOGRAPHY

ACREE, T. E., SHALLENBERGER, R. S., and MATTICK, L. R. 1968. Mutarotation of D-galactose. Tautomeric composition of an equilibrium solution in pyridine. Carbohydrate Res. 6, 498–502.

ALBERSHEIM, P., NEVINS, D. J., ENGLISH, P. D., and KARR, A. 1967. A method for the analysis of sugars in plant cell-wall polysaccharides by gas-liquid chromatography. Carbohydrate Res. 5, 340–345.

ALEXANDER, R. J., and GARBUTT, J. T. 1965. Use of sorbitol as internal standard in determination of D-glucose by gas-liquid chromatography. Anal. Chem. 37, 303–305.

ANGYAL, S. J., and PICKLES, V. A. 1967. Equilibria between furanoses and pyranoses. Carbohydrate Res. 4, 269–270.

BHACCA, N. S., and HORTON, D. 1967. Application of 220 MHz nuclear magnetic resonance to the solution of stereochemical problems: the anomeric configuration and favored conformation of 1-thio-α-L-arabinopyranose tetraacetate. Chem. Commun., 867–869.

BHACCA, N. S., and WILLIAMS, D. H. 1964. Applications of NMR Spectroscopy in Organic Chemistry. Illustrations from the Steroid Field. Holden-Day, San Francisco.

BIEMANN, K., DEJONGH, D. C., and SCHNOES, H. K. 1963. Application of mass spectrometry to structure problems. XIII. Acetates of pentoses and hexoses. J. Am. Chem. Soc. 85, 1763–1771.

BISHOP, C. T. 1964. Gas-liquid chromatography of carbohydrate derivatives. Advan. Carbohydrate Chem. 19, 95–147.

BISHOP, C. T., COOPER, F. P., and MURRAY, R. K. 1963. Reactions of carbohydrate derivatives during gas-liquid chromatography. Can. J. Chem. 41, 2245–2250.

BJÖRNDAL, H., LINDBERG, B., and SVENSSON, S. 1967A. Gas-liquid chromatography of partially methylated alditols as their acetates. Acta Chem. Scand. 21, 1801–1804.

BJÖRNDAL, H., LINDBERG, B., and SVENSSON, S. 1967B. Mass spectrometry of partially methylated alditol acetates. Carbohydrate Res. 5, 433–440.

BOLTON, C. H., CLAMP, J. R., DAWSON, G., and HOUGH, L. 1965. The quantitative analysis of glycopeptides and glycoproteins by gas-liquid chromatography. Carbohydrate Res. 1, 333–335.

BROBST, K. M., and LOTT, C. E., JR. 1966A. Determination of some components in corn syrup by gas-liquid chromatography of the trimethylsilyl derivatives. Cereal Chem. 43, 35–43.

BROBST, K. M., and LOTT, C. E., JR. 1966B. Analysis of carbohydrate mixtures by gas-liquid chromatography. Proc. Am. Soc. Brewing Chemists, 71–75.

BROWER, H. E., JEFFERY, J. E., and FOLSOM, M. W. 1966. Gas chromatographic sugar analysis in hydrolyzates of wood constituents. Anal. Chem. 38, 362–364.

BUDZIKIEWICZ, H., DJERASSI, C., and WILLIAMS, D. H. 1964. Structure Elucidation of Natural Products by Mass Spectrometry, Vol. II. Holden-Day, San Francisco.

CASU, B., REGGIANI, M., GALLO, G. G., and VIGEVANI, A. 1964. Hydroxyl proton resonances of sugars in dimethylsulphoxide solution. Tetrahedron Letters, 2839–2843.

CASU, B., REGGIANI, M., GALLO, G. G., and VIGEVANI, A. 1965. NMR spectra and conformation of glucose and some related carbohydrates in dimethylsulphoxide solution. Tetrahedron Letters, 2253–2259.

CASU, B., REGGIANI, M., GALLO, G. G., and VIGEVANI, A. 1966. Hydrogen bonding and conformation of glucose and polyglucoses in dimethylsulphoxide solution. Tetrahedron 22, 3061–3083.

CAYLE, T., VIERBROCK, F., and SCHIAFFINO, J. 1968. Gas chromatography of carbohydrates. Cereal Chem. 45, 154–161.

CHAPMAN, O. L., and KING, R. W. 1964. Classification of alcohols by nuclear magnetic resonance spectroscopy. J. Am. Chem. Soc. 86, 1256–1258.

CHIZHOV, O. S., MOLODTSOV, N. V., and KOCHETKOV, N. K. 1967. Mass spectrometry of trimethylsilyl ethers of carbohydrates. Carbohydrate Res. 4, 273–276.

CLAMP, J. R., DAWSON, G., and HOUGH, L. 1967. The simultaneous estimation of 6-deoxy-L-galactose (L-fucose), D-mannose, D-galactose, 2-acetamido-2-deoxy-D-glucose (N-acetyl-D-glucosamine) and N-acetylneuraminic acid (sialic acid) in glycopeptides and glycoproteins. Biochem. Biophys. Acta 148, 342–349.

DEJONGH, D. C. In press. In The Carbohydrates, 3rd Edition, W. Pigman, and D. Horton (Editors). Academic Press, New York.

DEJONGH, D. C., and BIEMANN, K. 1963. Application of mass spectrometry to structure problems. XIV. Acetates of partially methylated pentoses and hexoses. J. Am. Chem. Soc. 85, 2289–2294.

DEJONGH, D. C., and HANESSIAN, S. 1965. Characterization of amino sugars by mass spectrometry. J. Am. Chem. Soc. 87, 3744–3751.

DEJONGH, D. C., and HANESSIAN, S. 1966. Characterization of deoxy sugars by mass spectrometry. J. Am. Chem. Soc. 88, 3114–3119.

DEJONGH, D. C., HRIBAR, J. D., and HANESSIAN, S. In press. Mass spectrometry in carbohydrate chemistry. Glycosides and O-isopropylidene ketals of deoxy sugars. Advan. Chem. Ser.

ELIEL, E. L., ALLINGER, N. L., ANGYAL, S. J., and MORRISON, G. A. 1965. Conformational Analysis. John Wiley & Sons, New York.

FINAN, P. A., REED, R. I., and SNEDDEN, W. 1958. The application of the mass spectrometer to carbohydrate chemistry. Chem. Ind. (London), 1172.

GAHAN, L. C., SANDFORD, P. A., and CONRAD, H. E. 1967. The structure of the serotype 2 capsular polysaccharide of *Aerobacter aerogenes*. Biochemistry *6*, 2755–2767.

GLASS, C. A. 1965. Proton magnetic resonance spectra of D-glucopyranose polymers. Can. J. Chem. *43*, 2652–2659.

GUNNER, S. W., JONES, J. K. N., and PERRY, M. B. 1961. The gas-liquid partition chromatography of carbohydrate derivatives. Part I. The separation of glycitol and glycose acetates. Can. J. Chem. *39*, 1892–1899.

HALL, L. D. 1964A. Nuclear magnetic resonance. Advan. Carbohydrate Chem. *19*, 51–93.

HALL, L. D. 1964B. Chemical shifts of carbohydrates. Tetrahedron Letters, 1457–1460.

HALL, L. D. In press. In The Carbohydrates, 3rd Edition, W. Pigman, and D. Horton (Editors). Academic Press, New York.

HANESSIAN, S. 1964. Deoxy sugars. Advan. Carbohydrate Chem. *21*, 143–207.

HEDGLEY, E. J., and OVEREND, W. G. 1960. Trimethylsilyl derivatives of carbohydrates. Chem. Ind. (London), 378–380.

HEYNS, K., GRÜTZMACHER, H. F., SCHARMANN, H., and MÜLLER, D. 1966. Carbohydrate structural analysis by mass spectroscopy. Fortschr. Chem. Forsch. *5*, 448–490.

HEYNS, K., and MÜLLER, D. 1965. Mass spectroscopic investigations. VIII. The mass spectra of permethylated N-acetyl-aminosugars. Tetrahedron *21*, 3151–3169.

HEYNS, K., and MÜLLER, D. 1966A. Mass spectroscopic investigations. X. Mass spectra of additional permethylated N-acetylamino sugars; isomeric 2-methyl-2-acetamidodeoxyhexopyranosides, determination of ring size. Tetrahedron Letters, 449–458.

HEYNS, K., and MULLER, D. 1966B. Mass spectroscopic investigations. XI. Mass spectra of permethylated hexosamine hydrochlorides. Tetrahedron Letters, 617–621.

HEYNS, K., and SCHARMANN, H. 1966. Mass spectroscopic investigations. IX. Mass spectroscopic investigations of permethylated anhydro sugars. Carbohydrate Res. *1*, 371–392.

HOLLAND, C. V., HORTON, D., MILLER, M. J., and BHACCA, N. S. 1967. Nuclear magnetic resonance studies on acetylated 1-thioaldopyranose derivatives. J. Org. Chem. *32*, 3077–3086.

INOUE, S., and INOUE, Y. 1966. Nuclear magnetic resonance study of sulfated mucopolysaccharides. Biochem. Biophys. Res. Commun *23*, 513–517.

ISBELL, H. S. 1962. Oxidation of aldoses with bromine. J. Res. Natl. Bur. Std. *66A*, 233–238.

JACKMAN, L. M. 1959. Applications of Nuclear Magnetic Resonance Spectroscopy in Organic Chemistry. Pergamon Press, New York.

JEFFREY, G. A., and CHU, S. S. C. 1967. The crystal structure of methyl β-maltopyranoside. Acta Cryst. *23*, 1038–1040.

KARPLUS, M. 1959. Contact electron-spin coupling of nuclear magnetic moments. J. Chem. Phys. *30*, 11–15.

KIRCHER, H. W. 1962. Gas-liquid partition chromatography of sugar derivatives. Methods Carbohydrate Chem. *1*, 13–20.

KIM, J. H., SHOME, B., LIAO, T. H., and PIERCE, J. G. 1967. Analysis of neutral sugars by gas-liquid chromatography of alditol acetates. Anal. Biochem. 20, 258–274.

KOCHETKOV, N. K., and CHIZHOV, O. S. 1966. Mass spectrometry of carbohydrate derivatives. Advan. Carbohydrate Chem. 21, 39–93.

KOCHETKOV, N. K., CHIZHOV, O. S., and ZOLOTAREV, B. M. 1966. Mass spectrometry of hexosamines. Carbohydrate Res. 2, 89–92.

LEE, Y. C., and BALLOU, C. E. 1965A. Preparation of mannobiose, mannotriose, and a new mannotetraose from Saccharomyces cerevisiae mannan. Biochemistry 4, 257–264.

LEE, Y. C., and BALLOU, C. E. 1965B. Complete structure of the glycophospholipids of mycobacteria. Biochemistry 4, 1395–1404.

LEMIEUX, R. U., KULLNIG, R. K., BERNSTEIN, H. J., and SCHNEIDER, W. O. 1958. Configurational effects on the proton magnetic resonance spectra of six-membered ring compounds. J. Am. Chem. Soc. 80, 6098–6105.

LEMIEUX, R. U., and STEVENS, J. D. 1965. Substitutional and configurational effects on chemical shift in pyranoid carbohydrate derivatives. Can. J. Chem. 43, 2059–2070.

LEMIEUX, R. U., and STEVENS, J. D. 1966. The proton magnetic resonance spectra and tautomeric equilibria of aldoses in deuterium oxide. Can. J. Chem. 44, 249–262.

LEMIEUX, R. U., STEVENS, J. D., and FRASER, R. R. 1962. Observations on the Karplus curve in relation to the conformation of the 1,3-dioxolane ring. Can. J. Chem. 40, 1955–1959.

LENZ, R. W., and HEESCHEN, J. P. 1961. The application of nuclear magnetic resonance to structural studies of carbohydrates in aqueous solution. J. Polymer Sci. 51, 247–261.

LINEBACK, D. R., and HORNER, R. L. Unpublished data. University of Nebraska, Lincoln.

McINNES, A. G., BALL, D. H., COOPER, F. P., and BISHOP, C. T. 1958. Separation of carbohydrate derivatives by gas-liquid partition chromatography. J. Chromatog. 1, 556–557.

NEVINS, D. J., ENGLISH, P. D., and ALBERSHEIM, P. 1967. The specific nature of plant cell-wall polysaccharides. Plant Physiol. 42, 900–906.

PASIKA, W. M., and CRAGG, L. H. 1963A. The detection and estimation of branching in dextran by proton magnetic resonance spectroscopy. Can. J. Chem. 41, 293–299.

PASIKA, W. M., and CRAGG, L. H. 1963B. Proton magnetic resonance spectroscopy and the sulphation of linear and branched dextran. Can. J. Chem. 41, 777–782.

POPLE, J. A., SCHNEIDER, W. G., and BERNSTEIN, H. J. 1959. High-resolution Nuclear Magnetic Resonance. McGraw-Hill, New York.

RAO, V. S. R., and FOSTER, J. F. 1963. On the conformation of the D-glucopyranose ring in maltose and in higher polymers of D-glucose. J. Phys. Chem. 67, 951–952.

RICHEY, J. M., RICHEY, H. G., JR., and SCHRAER, R. 1964. Quantitative analysis of carbohydrates using gas-liquid chromatography. Anal. Biochem. 9, 272–280.

ROBERTS, J. D. 1959. Nuclear Magnetic Resonance. Applications to Organic Chemistry. McGraw-Hill, New York.

ROBERTS, J. D. 1961. An Introduction to the Analysis of Spin-Spin Splitting in High-Resolution Nuclear Magnetic Resonance Spectra. W. A. Benjamin, New York.

RUDRUM, M., and SHAW, D. F. 1965. The structure and conformation of some monosaccharides in solution. J. Chem. Soc. (London), 52–57.

SAWARDEKER, J. S., and SLONEKER, J. H. 1965. Quantitative determination of monosaccharides by gas liquid chromatography. Anal. Chem. 37, 945–947.

SAWARDEKER, J. S., SLONEKER, J. H., and JEANES, A. 1965. Quantitative determination of monosaccharides as their alditol acetates by gas liquid chromatography. Anal. Chem. 37, 1602–1604.

SHALLENBERGER, R. S., and ACREE, T. E. 1966. The identity of γ-D-galactose. Carbohydrate Res. 1, 495–497.

SPEDDING, H. 1964. Infrared spectroscopy and carbohydrate chemistry. Advan. Carbohydrate Chem. 19, 23–49.

SWEELEY, C. C., BENTLEY, R., MAKITA, M., and WELLS, W. W. 1963. Gas-liquid chromatography of trimethylsilyl derivatives of sugars and related substances. J. Am. Chem. Soc. 85, 2497–2507.

SWEELEY, C. C., and WALKER, B. 1964. Determination of carbohydrates in glycolipides and gangliosides by gas chromatography. Anal. Chem. 36, 1461–1466.

VAN DER VEEN, J. M. 1963. An n.m.r. study of the glycoside link in glycosides of glucose and galactose. J. Org. Chem. 28, 564–566.

CHAPTER 10

J. N. BeMiller | Biochemical Methods
of Carbohydrate Analysis

INTRODUCTION

At this time, any discussion of biochemical methods of carbohydrate analysis must be more of a catalog of possibilities than of present uses. For example, no enzymic methods are listed with the current "official" methods of sugar analysis (DeWhalley 1964), although some "official" AOAC methods do use enzymes. This is not to say, however, that biochemical methods of carbohydrate analysis are not available, but only that they are not being used.

A number of enzymic methods have been proposed and collected together by Bergmeyer (1965). These methods have been used to determine certain carbohydrates in blood and urine, especially; few, with the exception of the determination of D-glucose with glucose oxidase, have been used with foods. Certainly, the others have potential applicability in this field. For this reason, methods which might be adapted to foods are presented without regard to current use in the hope that this presentation may stimulate those concerned with food analysis to investigate the use of enzymic methods to determine carbohydrates and carbohydrate derivatives.

The discussions are not intended to detail preparation and properties of the enzymes, to present methods in detail, nor to be a bibliography of uses. Rather they are intended to suggest enzyme systems which may be used and to point out, where possible, problems which may arise. Most of the enzymes needed are commercially available and may be obtained from such places as C. F. Boehringer and Soehne GmbH, Mannheim, Germany; British Drug Houses Ltd., Poole, England; Calbiochem, Los Angeles, Calif.; Farmochimica Cutolo Calosi, Naples, Italy; Gallard-Schlesinger Chemical Mfg. Corp., Carle Place, L.I., N.Y.; General Biochemicals Inc., Chagrin Falls, Ohio; Koch-Light Laboratories Ltd., Colnbrook, Bucks, England; Mycofarm-Delft, Delft, Netherlands; Nutritional Biochemicals Corp., Cleveland, Ohio; Schwarz Bioresearch, Inc., Orangeburg, N.Y.; Seravac Laboratories Ltd., Maidenhead, England; Sigma Chemical Co., St. Louis, Mo.; and Worthington Biochemical Corp., Freehold, N.J. Several of these companies are excellent sources of bibliographies on the methods and enzymes used in them.

As stated above, methods are not given in detail. One reason for this is that applications to specific food products have not been made.

In some cases extractions of either the carbohydrates or interfering substances may be necessary; in some removal of proteins and/or lipids may be a necessary pretreatment before carbohydrate determination, and so on.

Special mention should be made of the very thorough work of Friedemann *et al.* (1967) in developing methods for the determination of total carbohydrate available for metabolism. Their methods which involve the use of Rhozyme-S, a product of Rohm and Haas Co., Philadelphia, Pa., that quantitatively catalyzes the hydrolysis of starch, glycogen, sucrose, maltose, cellobiose and lactose, are not described here because the remainder of this paper will deal with methods for the determination of specific carbohydrates. The reader is referred to their paper for details and a review of methods which attempt to determine metabolizable carbohydrates.

Finally, it should be pointed out that many enzymes are known which have potential application to analytical methods although they have not as yet been used for that purpose. One example is the enzyme that oxidizes lactose to lactobionic acid (Nishizuka and Hayaishi 1962); many others are also known and await someone interested in methods development to apply them.

At the end of the chapter, brief reports of some nonenzymic biochemical methods are given.

MONOSACCHARIDES

Many of the enzymic methods for the determination of monosaccharides depend on oxidation (dehydrogenation) of the substrate or one of its reaction products followed by measurement of the NADH (reduced nicotinamide adenine dinucleotide; DPNH) or NADPH (reduced nicotinamide adenine dinucleotide phosphate; TPNH) formed in the reaction. In each of these determinations, suitable blanks must be run to determine that neither the sample nor the enzyme preparation contains other dehydrogenases, NADH or NADPH oxidases, nucleotidases, or redox compounds which will alter the results. Erroneous results can also arise from the action of carbohydrases either during extraction or analysis. In food analysis, this should be watched carefully as carbohydrases are often used in the processing of foods (Reed 1966). Isomerizations which are either enzyme-catalyzed and may occur during extraction or analysis or base-catalyzed and may occur during extraction or deproteinization (Johnson and Fusaro 1965) will interfere; and it is possible that specific products would contain enzyme inhibitors, perhaps as preservatives. Finally, since NADH and NADPH are most commonly determined by measurement of the 340 nm (or 366 nm) absorbance, any absorbance of the sample in this region will interfere.

PENTOSES

L-Arabinose and D-Xylose

L-Arabinose isomerase (L-arabinose ketol-isomerase; 5.3.1.4) catalyzes the reaction

$$\text{L-Arabinose} \rightleftharpoons \text{L-}erythro\text{-pentulose (L-ribulose)}$$

and can be used to determine L-arabinose (Horecker 1965A). The equilibrium of the reaction usually favors L-arabinose but, if the reaction is effected in pH 8.2 borate buffer (0.1M), the equilibrium favors L-ribulose which can be determined with the cysteine-carbazole reaction (Dische and Borenfreund 1951; Axelrod and Jang 1954). The enzyme can be obtained from *Lactobacillus plantarum* (Horecker 1965A).

Another enzyme, D-xylose isomerase (D-xylose ketol-isomerase; 5.3.1.5), from the same organism catalyzes the reaction

$$\text{D-Xylose} \rightleftharpoons \text{D-}threo\text{-pentulose (D-xylulose)}$$

and can be used to determine D-xylose (Mitsuhashi and Lampen 1953; Horecker 1965B). Again, the equilibrium of the reaction is normally in favor of the aldopentose, but if the reaction is effected in pH 8.2 borate buffer (0.1M) the pentulose is favored and can be determined with the cysteine-carbazole reaction (Dische and Borenfreund 1951; Axelrod and Jang 1954). With L-ribulose, full color is developed in 10 min while with D-xylulose 100 min is required (Ashwell and Hickman 1954).

D-Xylose isomerase preparations always contain L-arabinose isomerase activity. Therefore, the sample must first be treated with L-arabinose isomerase and then with D-xylose isomerase when the first reaction is complete. By removing an aliquot after the first enzyme treatment, the same reaction mixture can be used for successive determinations of L-arabinose and D-xylose.

HEXOSES

D-Glucose

Determination with Glucose Oxidase and Peroxidase.—Glucose oxidase (D-glucose:O_2 oxidoreductase; 1.1.3.4) specifically oxidizes β-D-glucopyranose (Franke and Deffner 1939; Keilin and Hartree 1945, 1948A,B, 1952A) to D-glucono-1,5-lactone which is spontaneously hydrolyzed to D-gluconic acid (Bentley and Neuberger 1949). At 20° β-D-glucopyranose is oxidized 150 times more rapidly than α-D-glucopyranose (Keilin and Hartree 1952A). Even so, glucose oxidase can be used to oxidize completely an equilibrium mixture of the anomers because even highly purified enzyme preparations contain mutarotase (Keilin and Hartree 1952B) and the reaction time can be adjusted so that all D-glucose is oxidized.

$$\beta\text{-D-Glucopyranose} + H_2O + O_2 \rightarrow \text{D-glucono-1,5-lactone} + H_2O_2$$

D-Glucose was first determined with glucose oxidase manometrically (Keilin and Hartree 1948B) and then spectrophotometrically in a coupled enzyme reaction (Keston 1956; Froesch and Renold 1956; Teller 1956).

$$H_2O_2 + DH_2 \rightarrow 2H_2O + D$$

Hydrogen peroxide in the presence of peroxidase oxidizes a chromogenic hydrogen donor (DH_2) to a colored derivative, the amount of which is a measure of the D-glucose oxidized. o-Dianisidine is commonly used as the hydrogen donor. The absorption spectrum of the dye formed from it has a broad maximum around 460 nm; measurements are frequently made at 400 to 450 nm. o-Toluidine (Salomon and Johnson 1959) and 2,6-dichlorophenolindophenol (2,6-dichlorobenzenone-indophenol) (Dobrick 1958) have also been used as hydrogen donors.

Specific directions for the determination of D-glucose can be found in publications of Hugget and Nixon (1957), Bergmeyer and Bernt (1965), the Worthington Biochemical Corp., Freehold, N. J., and Sigma Chemical Co., St. Louis, Mo. Reagents are commercially available which contain glucose oxidase, peroxidase, a chromogenic hydrogen donor, and buffer, and these are quite popular for D-glucose determinations. An automated method has been developed (Malmstedt and Hicks 1960).

Glucose oxidase is found in several fungal species. Its isolation from *Pencillium notatum,* purification, and properties were reviewed by Bentley (1955). However, most commercial enzyme preparations in the United States are from *Aspergillus niger* (Underkofler 1957). This enzyme has a molecular weight of 150,000, a broad pH range from 4 to 7 with the optimum at pH 5.5, and an isoelectric point at pH 4.2. It is a glycoprotein containing two molecules of flavin adenine dinucleotide.

The enzymic determination of D-glucose has most often been used with blood and urine, but the method has also been widely used with corn syrup (Whistler *et al.* 1953) and ought to be applicable for use with other food products. Since the enzyme is specific for β-D-glucopyranose, the enzymic determination of D-glucose anomers is possible (Jørgensen and Jørgensen 1966).

Interference with the analysis could occur if the glucose oxidase preparation is contaminated with enzymes such as maltase, invertase, lactase, trehalase, amylase, or other carbohydrases. Reports of such contaminations have been made (Adams *et al.* 1960; Hlaing *et al.* 1961; Sols and de la Fuente 1961; Crowne and Mansford 1962; Scharlach *et al.* 1962; Johnson and Fusaro 1965). Tris [tris(hydroxymethyl)aminomethane] buffer has been used to lower the maltase (Dahlqvist 1961; Sols and de la Fuente 1961; White and Subers 1961) and invertase (Blecher and Glassman 1962) activities of glucose oxidase preparations, but commercial preparations are now available in which these other activities are either absent or negligible without employing Tris buffer. However, it is well to check for other activities by use of purified substrates such as maltose, lactose, sucrose, trehalose, starch, etc.

Other interfering substances are the enzyme catalase and large amounts of reducing substances, such as ascorbic acid, which can compete with the oxidation of the dye by peroxide and peroxidase, hydrogen peroxide (Kleppe 1966) and bisulfite (Swoboda and Massey 1966) which inactivate the enzyme, and oxidants which oxidize the dye.

Glucose oxidase is specific for β-D-glucose. It has been reported that the following give 1% or less as much reaction: D-arabinose, L-arabinose, D-xylose, D-allose, D-altrose, D-fructose, D-galactose, L-glucose, L-gulose, D-idose, D-mannose, L-sorbose, D-talose, 3-deoxy-D-*ribo*-hexose, 5-deoxy-D-*xylo*-hexose, 5-thio-D-glucose, 3-*O*-methyl-D-glucose, 6-*O*-methyl-D-glucose, 4,6-*O*-benzylidene-D-glucose, methyl α-D-glucopyranoside, 1,5-anhydro-D-glucitol, inositol, phosphorylated sugars, lactose, maltose, melibiose, raffinose, and sucrose (Bernt 1965; Keilin and Hartree 1952A; McComb *et al.* 1957; McComb and Yushok 1958; Sols and de la Fuente 1957; Adams *et al.* 1960; Hlaing *et al.* 1961; Gibson *et al.* 1964; Pazur and Kleppe 1964). The same authors report that 4,6-di-*O*-methyl-D-glucose and 4-deoxy-D-*xylo*-hexose give 2% or less as much reaction, 6-deoxy-6-fluoro-D-glucose less than 4%, 6-deoxy-D-glucose 10%, 4-*O*-methyl-D-glucose 15%, and 2-deoxy-D-*arabino*-hexose 12–25%. In our laboratory, we have found that small amounts of 4-*O*-methyl-D-glucose inhibits the determination of D-glucose, perhaps competitively. The same may be true of other sugars, as it has been reported that glucose oxidase is inhibited by D-arabinose (Adams *et al.* 1960). In addition, substrate inhibition has been noted (Nicol and Duke 1966). Since inhibition by peroxide has already been found (Kleppe 1966), inhibition by the other product, D-glucono-1,5-lactone, should be investigated.

In spite of these potential interferences, glucose oxidase has been and will continue to be used extensively for D-glucose determinations, although it should perhaps be used with more care than it has been.

Recently, glucose oxidase has been made insoluble for column packing by covalent attachment to polymeric matrices, and automated methods for D-glucose determination have been worked out using these materials (Hicks and Updike 1966; Updike and Hicks 1967). Glucose oxidase on carboxymethylcellulose is available commercially from Seravac Laboratories Ltd.

Other information on glucose oxidase and its use can be found in reviews by Voigt (1961), Underkofler (1961), Primo Yufera (1961), Bentley (1963), and Reed (1966).

Determination with Hexokinase and Glucose 6-Phosphate Dehydrogenase.—D-Glucose can also be determined by the following reactions (Slein 1965):

D-Hexose + ATP → D-hexose 6-phosphate + ADP

D-Glucose 6-phosphate + NADP$^+$ ⇌ 6-phospho-D-glucono-

$$1,5\text{-lactone} + NADPH + H^+$$

The NADPH so formed is determined spectrophotometrically.

Both reactions proceed stoichiometrically and quantitatively under proper conditions. The first reaction is catalyzed by hexokinase (ATP:D-hexose 6-phosphotransferase; 2.7.1.1), and the second by glucose 6-phosphate dehydrogenase (D-glucose 6-phosphate:NADP oxidoreductase; 1.1.1.49).

Hexokinases (Kunitz and McDonald 1946; Brown 1951; Crane 1962) are not specific for D-glucose but also catalyze C-6 phosphorylation of 2-amino-2-deoxy-D-glucose, D-mannose, and D-fructose. Commercial preparations are obtained from yeast. This enzyme has a pH optimum of 8–9 and is most stable at about pH 5. It is inhibited by heavy metal ions; Mg^{++} is required. The enzyme can be purified by the method of Darrow and Colowick (1962) and should be free of D-glucose which is sometimes added as a stabilizing agent during purification (Berger *et al.* 1946; Kunitz and McDonald 1946).

Glucose 6-phosphate dehydrogenase has high substrate specificity (Glaser and Brown 1955; Lowry *et al.* 1961; Brown and Clarke 1963) which gives specificity to the overall reaction. It has a pH optimum at 7.4; the overall reaction is run at pH 7.4–8.0 in 0.05M Tris [tris(hydroxymethyl)aminomethane] buffer since this enzyme is inhibited by phosphate. It can be purified by the methods of Kornberg (1950) or Glaser and Brown (1955).

Interferences in this assay system can come from the presence of 6-phosphogluconic dehydrogenase, enzymes which oxidize or destroy NADPH, phosphoglucose isomerase, phosphomannose isomerase, phosphoglucomutase, and carbohydrases which release D-glucose. Hence, the possible presence of these enzymes should be checked.

A reagent is available commercially which contains all the components needed for D-glucose determinations by this method.

Determination with Hexokinase, Pyruvic Kinase, and Lactic Dehydrogenase.—Because commercial glucose 6-phosphate dehydrogenase usually contains 6-phosphogluconic dehydrogenase which interferes with the determination of D-glucose, Pfleiderer and Grein (1957) and Pfleiderer (1965) recommend the following reactions:

$$\text{D-Glucose} + \text{ATP} \rightarrow \text{D-glucose 6-phosphate} + \text{ADP}$$
$$\text{ADP} + \text{Phosphoenolpyruvate} \rightarrow \text{ATP} + \text{pyruvate}$$
$$\text{Pyruvate} + \text{NADH} + \text{H}^+ \rightleftharpoons \text{L-lactate} + \text{NAD}^+$$

The NAD$^+$ so formed is determined spectrophotometrically.

All reactions proceed stoichiometrically and quantitatively under proper conditions. D-Glucose is phosphorylated with ATP, a reaction catalyzed by hexokinase and discussed in the preceding section. The ADP formed in the first reaction is converted to ATP in the second reaction catalyzed by pyruvic kinase (ATP:pyruvate phosphotransferase; 2.7.1.40) (Kubowitz and Ott 1944; Kornberg and Pricer 1951; Strominger 1955; Tietz and Ochoa 1958), an enzyme requiring Mg^{++} and K$^+$. The third reaction is the indicator reaction and is catalyzed by lactic dehydrogenase (L-lactate:NAD oxidoreductase; 1.1.1.27) which is discussed elsewhere.

Since this determination depends on a stoichiometric formation of ADP in the first reaction, anything which alters the stoichiometry will interfere. Possibilities are 2-amino-2-deoxy-D-glucose, D-mannose, and D-fructose which are phosphorylated by hexokinase (Kunitz and McDonald 1946; Brown 1951; Crane 1962). Therefore, this method cannot be used in the presence of other hexoses. High D-glucose values would result if the sample contains much ADP, but the presence of ADP can be offset by adding all components except hexokinase so that the second and third reactions can take place before the D-glucose determination. Following determination of this blank value, the reaction sequence could be started with hexokinase. The system must be free from adenylate kinase (ATP:AMP phosphotransferase; 2.7.4.3) (Thorn *et al.* 1959).

D-Galactose

Galactose oxidase (D-galactose:O_2 oxidoreductase; 1.1.3.c) oxidizes D-galactopyranose and many of its derivatives at C-6, converting the primary hydroxymethyl group to an aldehyde group (Avigad et al. 1961, 1962). The reaction can be exemplified by the oxidation of methyl α-D-galactopyranoside:

Methyl α-D-galactopyranoside + O_2 → methyl α-D-*galacto*-hexodialdo-
1,5-pyranoside + H_2O_2

D-Galactopyranosyl residues of oligo- and polysaccharides are oxidized in addition to free D-galactose. Avigad et al. (1962) found that the tetrasaccharide stachyose was oxidized six times faster than D-galactose and the polysaccharide guaran twice as fast. Schlegel et al. (1968) in a detailed study of the substrate specificity of galactose oxidase found that derivatives of D-galactose with substituents on the hydroxyl group at C-4 are not oxidized, and neither are 2-amino-2-deoxy-D-galactose residues with glycosyl substituents at C-3.

The enzyme is obtained from *Dactylium dendroides* and is commercially available. It has been used to determine D-galactopyranosyl units and their derivatives (Avigad et al. 1962; DeVerdier and Hjelm 1962; Barker et al. 1963; Bradley and Kanfer 1964; Fischer and Zapf 1964A,B; Sempere et al. 1965; Hjelm 1967).

Commonly, the reaction is followed by coupling the reaction to one involving the oxidation of a chromogen with hydrogen peroxide catalyzed by peroxidase as is done with glucose oxidase. o-Toluidine (Avigad et al. 1961; Rorem and Lewis 1962), o-dianisidine (Fischer and Zapf 1964A) and benzidine (Roth et al. 1965) have been used. Reagents containing galactose oxidase, peroxidase, reduced chromogen and buffer are commercially available. However, it has been noted that D-galactose is not completely oxidized (Avigad et al. 1962; Avigad 1967), unusual kinetics are sometimes observed (Avigad et al. 1962; Morell et al. 1966), and analytical data are not consistent (Avigad et al. 1962). Hence, further applications of this enzyme must be carefully scrutinized. It is claimed that the Worthington Biochemical Corp. reagent which uses a different chromogen and glycine buffer gives a satisfactory linear standard curve.

Inhibition by chloride, bisulfite, cynanide, formate, hydrazine, hydroxylamine, azide, and diethyldithiocarbamate has been noted.

Gancedo et al. (1966) have recently reported that a crude preparation of galactose oxidase from *D. dendroides* oxidized D-galactose and D-galactosides (better) such as melibiose and raffinose to uronic acids.

D-Fructose

D-Fructose can be determined by the same method employed for the determination of D-glucose with hexokinase and glucose 6-phosphate dehydrogenase as discussed previously with the addition of phosphoglucoisomerase (D-glucose 6-phosphate ketol-isomerase; 5.3.1.1). After D-glucose has been determined, phosphoglucoisomerase is added to catalyze the reaction (Klotzsch and Bergmeyer 1965):

$$\text{D-Fructose 6-phosphate} \rightleftharpoons \text{D-glucose 6-phosphate}$$

D-Fructose 6-phosphate is formed from D-fructose in the hexokinase-catalyzed reaction. The overall reaction is stoichiometric with respect to the formation of NADPH.

GLYCOSIDES

β-D-Glucosides

β-Glucosidase (β-D-glucoside glucohydrolase; 3.2.1.21) (Veibel 1950; Larner 1960) can be obtained from sweet almonds by the method of Hestrin et al. (1955). It is commercially available in a purified form which contains traces of β-galactosidase, α-galactosidase, and α-mannosidase activities. Emulsin, the crude product obtained by grinding and defatting almonds, contains the contaminating enzymes in much higher amounts. Fungi and bacteria which produce cellulolytic enzymes also make β-glucosidase (see Cellulose).

It should be possible to determine β-D-glucosides in foods using β-glucosidase coupled with a specific method for the determination of D-glucose. However, in our laboratory it was found that the absorbance of saligenin (maximum at 515 nm), the product of the hydrolysis of salicin, interferes with the determination of D-glucose with the coupled enzyme system containing glucose oxidase, peroxidase, and o-dianisidine (measured at 400 nm).

The pH optimum of β-glucosidase varies with the substrate; for salicin it is pH 5.0 (Baruah and Swain 1957). The enzyme is most stable at pH 7.1; it is inactivated relatively rapidly above pH 8.2 and below pH 4.1 (Helferich and Schniedmuller, 1931). The rate of hydrolysis is affected by several cations and oxidants (Helferich et al. 1934; Helferich and Petersen 1935; Helferich and Schmitz-Hillebrecht 1935). It is powerfully, specifically, and competitively inhibited by D-glucono-1,5-lactone (Conchie and Levvy 1957; Reese and Mandels 1957, 1960; Norkrans and Wahlstrom 1961) so that β-glucosidase and glucose oxidase cannot be used concurrently but must be used successively.

β-ᴅ-Glucuronosides

β-Glucuronidase (β-ᴅ-glucuronide glucuronohydrolase; 3.2.1.31) can be obtained from a variety of sources. That from the marine mollusc *Patella vulgata* is commercially available. It has a pH optimum of pH 3.5–4.0 (Cox 1959). Also commercially available is that from bovine liver. It has a pH optimum of 4.5 which is usually used (Levvy and Marsh 1959). Both enzymes are stable over a wide pH range from <5.0 to >8.0 (Levvy *et al.* 1957). The commercial enzymes have no other appreciable carbohydrase activity associated with them.

It should be possible to determine simple β-ᴅ-glucuronosides in foods using β-glucuronidase and a reducing sugar method or a specific method for ᴅ-glucuronate discussed elsewhere. However, the enzyme is powerfully, specifically, and competitively inhibited by ᴅ-glucaro-1,4-lactone (Smith 1944; Marsh 1962; Harigaya 1964). It is also inhibited by Cu^{++}, Hg^{++}, and Ag^+ ions.

β-Glucuronidase is most often used for the determination of steroid conjugates (Cohen 1951; Fishman 1957; Wakabayashi and Fishman 1961; Kirk *et al.* 1962; Cohen 1964; Becker 1965).

OLIGO- AND POLYSACCHARIDES

Oligo- and polysaccharides can be determined by the use of enzymes which are specific for their hydrolysis, coupled with or followed by enzymic or other reactions which allow measurement of a product of the hydrolysis. If a coupled enzyme system is used, concentrations of enzymes must be chosen so that the hydrolytic reaction is the rate-limiting reaction. Large amounts of a product of the reaction present as free sugar in the sample can interfere with the determination, while smaller amounts will cause high blanks. It may be possible in some cases to remove monosaccharides from di- and trisaccharides by adsorption of the latter on carbon, washing with water, and elution of the di- and trisaccharides with 10% ethanol.

Alternative methods for determining the extent of hydrolysis by a specific carbohydrase include measurement of the change in optical rotation and measurement of the increase in reducing power. However, the accuracy of polarimetric methods is low if the sample contains relatively large amounts of optically active compounds, and the accuracy of reducing sugar methods is good only if the sample contains very small amounts of free sugars.

Of course, the enzyme preparation used must be free of other carbohydrases which release the same sugar(s). In addition, carbohydrases usually have transferase activity also, so reactions must be effected in dilute solutions. And the problems associated with the determination of monosaccharides also apply to the determination of the hydrolysis products.

Lactose

β-Galactosidase (β-D-galactoside galactohydrolase; 3.2.1.23) from *Escherichia coli* (Hu *et al.* 1959) catalyzes the hydrolysis of lactose:

$$\text{Lactose} + H_2O \rightarrow \text{D-galactose} + \text{D-glucose}$$

This reaction coupled with a determination of D-glucose (Reithel 1965) or D-galactose discussed elsewhere allows the determination of lactose. Other oligosaccharides containing D-glucose attached to D-galactose by a β-D-(1 → 4) linkage will also be hydrolyzed but normally are present in very small amounts.

Crude β-galactosidase preparations frequently have significant absorption at 340 nm and, in addition, contain a nucleotidase, both of which can interfere with assays which depend on the measurement of NADPH.

The enzyme-catalyzed oxidation of lactose to lactobionic acid has been reported (Nishizuka and Hayaishi 1962).

Sucrose

Invertase (β-fructofuranosidase) (β-D-fructofuranoside fructohydrolase; 3.2.1.26) from yeast catalyzes the hydrolysis of sucrose:

$$\text{Sucrose} + H_2O \rightarrow \text{D-glucose} + \text{D-fructose}$$

This reaction coupled with a determination of D-glucose (Bergmeyer and Klotzsch 1965) or D-fructose discussed elsewhere allows the determination of sucrose.

Commercial invertase preparations are frequently stabilized with hexoses and must be dialyzed against distilled water before use. Checks must be made for other carbohydrase activity.

Invertase catalyzes the hydrolysis of saccharides which contain unsubstituted β-D-fructofuranosyl units, including raffinose and gentianose and their homologs and inulin. Only sucrose yields equimolar amounts of D-glucose and D-fructose; deviations will tell something about the types of saccharides containing β-D-fructofuranosyl units in the sample.

Treatment with invertase followed by determination of the D-glucose formed in this reaction with glucose oxidase has been used to determine sucrose in unclarified beet molasses (Taeufel *et al.* 1965), without correction for raffinose (see below).

Raffinose

Invertase (β-fructofuranosidase) (β-D-fructofuranoside fructohydrolase; 3.2.1.26) from yeast catalyzes the hydrolysis of raffinose to D-fructose and melibiose (6-O-α-D-galactopyranosyl-β-D-glucopyranose). Melibiase, an α-galactosidase (α-D-galactoside galactohydrolase; 3.2.1.22), catalyzes the hydrolysis of melibiose:

$$\text{Melibiose} + H_2O \rightarrow \text{D-galactose} + \text{D-glucose}$$

Hence, treatment with invertase followed by treatment with α-galactosidases ought to allow the successive determination of sucrose and raffinose. However, it is difficult to separate the two activities and, indeed, melibiase is sold commercially as an invertase-melibiase mixture (Siegfried de Whalley 1965). Therefore, treatment with the mixture followed by a specific determination of D-galactose, D-glucose, and D-fructose might be a useful alternative. Such coupled reactions need to be investigated as does purification of the enzymes themselves to separate the carbohydrase activities.

Cellulose

"Cellulase" is a collective term for the enzymes which catalyze the reaction:

$$\text{Cellulose (insoluble)} \rightarrow \text{soluble carbohydrate}$$

It would seem then that it could be used for the determination of cellulose (Halliwell 1965A). However, problems are involved.

"Cellulase" actually refers to a group of enzymes: an enzyme(s) responsible for a prehydrolytic step wherein β-D-(1 → 4)-glucan chains are swollen or hydrated (C_1 activity), a β-D-1,4-glucan glucanohydrolase (3.2.1.4) (C_x activity), and a β-glucosidase (β-D-glucoside glucohydrolase; 3.2.1.21). The latter two are an endo- and an exoenzyme, respectively. "Cellulases," usually obtained commercially from *Aspergillus niger* or *Trichoderma viride*, will also catalyze the hydrolysis of carboxymethylcellulose and frequently hemicelluloses which are β-D-(1 → 4)-xylans (Bishop and Whitaker 1955; Thomas 1956; Kooiman 1957; Sørensen 1957; BeMiller *et al.* 1968). The state of the cellulose is important, as regenerated cellulose or cellulose swollen with mineral acids is more reactive (Walseth 1952; Whitaker 1953; Sison *et al.* 1958; BeMiller *et al.* 1968).

Halliwell (1965A) has described a method for the determination of cellulose which measures the weight loss after enzyme-catalyzed hydrolysis by "cellulase." Impurities and solubles must be removed first; then contaminating carbohydrases give much less interference with the determination.

Alternatively, D-glucose released can be determined with glucose oxidase as discussed previously (Meyers *et al.* 1960; BeMiller *et al.* 1968).

Chitin

A partially purified chitinase (chitin glycanohydrolase; 3.2.1.14) from *Streptomyces griseus* is now commercially available and may be considered for use in the determination of chitin if it can be coupled with a determination of D-glucosamine (Brown 1965).

Pectin and Alginates

Hydrolysis of pectin and pectate is catalyzed by a variety of enzymes (Reed 1966). None of these yield more than 30% D-galacturonic acid. The transeliminases (lyases) yield oligosaccharides terminated at the nonreducing end with a 4,5-unsaturated derivative with a specific ultraviolet absorbance and, perhaps, offer the best potential for development of a method.

Enzymes which catalyze the hydrolysis of alginic acid to both D-mannuronic acid and unsaturated oligosaccharides are also known. These reactions, which parallel those of pectin, offer a possibility for the enzymic determination of alginates if they can be coupled with a determination of the products. Eller and Payne (1963) isolated a bacterium, *Alginomonas alginica,* which can oxidize D-mannuronic acid and which could yield an enzyme that would be useful in such an analysis.

Starch

Starch can be determined by hydrolysis with glucoamylase (α-1,4-glucan glycoanhydrolase; 3.2.1.3) to yield D-glucose followed by determination of D-glucose (Thivend *et al.* 1965A, 1965B). The latter determination can be done in various ways, including oxidation with glucose oxidase.

Glucoamylase catalyzes the hydrolysis of both the α-D-$(1 \rightarrow 4)$ and the α-D-$(1 \rightarrow 6)$ bonds of starch, the former about 30 times more rapidly than the latter. Hence, this method will not quantitatively distinguish between starch and glycogen, as may be necessary in the analysis of sweet corn (Peat *et al.* 1956). Dextrins and the malto-oligosaccharides of corn syrup will also be hydrolyzed, so a separation based on solubilities will be necessary to distinguish between these compounds (Friedemann *et al.* 1967).

Older methods used malt extract for the incomplete hydrolysis of starch or Taka-Diastase for the complete hydrolysis of starch under carefully controlled conditions and are reviewed by Friedemann *et al.* (1967).

Gums

Rohm and Haas Co., Philadelphia, Pa., markets a "Pentosanase-hexosanase" preparation which catalyzes the hydrolysis of guar gum, locust bean gum, gum arabic, gum tragacanth, and related gums. Coupled with a reducing sugar determination, such a preparation could be used for the collective determination of such compounds. Some might be determined with galactose oxidase as discussed previously. The determination of hemicelluloses has been discussed by Halliwell (1965B), but much more work needs to be done in this area because of the apparent lack of specificity of the commercial enzyme preparation.

ALDITOLS AND CYCLITOLS

Glycerol

Glycerol can be determined by the following reactions (Wieland 1965A):

Glycerol + ATP $\rightarrow$ L-glycerol 1-phosphate + ADP

L-Glycerol 1-phosphate + NAD$^+$ $\rightleftharpoons$ dihydroxyacetone phosphate
$$+ \text{NADH} + \text{H}^+$$

The first reaction is catalyzed by glycerol kinase (ATP : glycerol phosphotransferase; 2.7.1.30) and the second by glycerol phosphate dehydrogenase (L-glycerol 3-phosphate : NAD oxidoreductase; 1.1.1.8). The equilibrium of the indicator reaction lies far to the left; hence, to determine glycerol, dihydroxyacetone phosphate must be removed from the reaction mixture. To do this, dihydroxyacetone phosphate is converted to the hydrazide by adding hydrazine, by buffering the reaction mixture at pH 9.8 with glycine (0.2M), and by using large amounts of the two enzymes. The enzymes must be free of other dehydrogenase activities.

Glycerol dehydrogenase (glycerol : NAD oxidoreductase; 1.1.1.6), an induced enzyme from *Aerobacter aerogenes*, catalyzes the following reaction:

$$\text{Glycerol} + \text{NAD}^+ \rightleftharpoons \text{dihydroxyacetone} + \text{NADH} + \text{H}^+$$

This enzyme can be used for the direct determination of glycerol (Lin and Magasanik 1960; Hagen and Hagen 1962). Ammonium ion greatly increases the affinity of the enzyme for its substrate while zinc ion is inhibitory (Lin and Magasanik 1960). In both methods, increase in NADH is used to follow the oxidation.

Sorbitol (D-Glucitol)

Sorbitol (D-glucitol) is oxidized by a rather nonspecific alditol dehydrogenase (L-iditol : NAD oxidoreductase; 1.1.1.14) to yield D-fructose:

$$\text{D-Glucitol} + \text{NAD}^+ \rightleftharpoons \text{D-fructose} + \text{NADH} + \text{H}^+$$

This reaction can be used to determine sorbitol by measuring the increase in NADH (Williams-Ashman 1965). The reaction is stoichiometric if an excess of NAD$^+$ is used at pH 9.5. However, other alditols, e.g., L-threitol, ribitol, xylitol, allitol, L-iditol and D-*glycero*-D-*gluco*-heptitol are also oxidized. The equilibrium constants for the oxidation of these compounds vary (Hollmann and Touster 1957; Wolfson and Williams-Ashman 1958). Of course, the free ketoses, i.e., the reaction products [D-fructose, L-*glycero*-tetrulose (L-erythrulose), D-*erythro*-pentulose (D-ribulose), D-*threo*-pentulose (D-xylulose), D-*ribo*-hexulose (D-allulose), L-sorbose, and D-*altro*-heptulose (D-sedoheptulose)], if present, can reverse the reaction and reoxidize the NADH.

myo-Inositol

Inositol dehydrogenase (*myo*-inositol : NAD oxidoreductase; 1.1.1.18) can be induced in *Aerobacter aerogenes* and used for the determination of *myo*-inositol according to the reaction (Weissbach 1965):

$$\textit{myo}\text{-Inositol} + \text{NAD}^+ \rightleftharpoons \textit{scyllo}\text{-inosose} + \text{NADH} + \text{H}^+$$

The reaction does not go to completion but can be used with a standard curve. Of the cyclitols and related compounds investigated, only sequoyitol (5-*O*-methyl-*myo*-inositol) gave appreciable oxidation, being oxidized like *myo*-inositol itself.

Charalampous and Abrahams (1956) coupled the reaction with reduction of 2,6-dichlorophenolindophenol (2,6-dichlorobenzenone-indophenol) catalyzed by NADH$_2$: lipoamide oxidoreductase (1.6.4.3).

A *myo*-inositol oxygenase (1.99.2.6) from kidney catalyzes the reaction:

$$myo\text{-Inositol} + O_2 \rightarrow \text{D-glucuronate}$$

which, coupled with a determination of D-glucuronate described elsewhere, can be used to determine *myo*-inositol (Kean and Charalampous 1959).

CARBOHYDRATE ACIDS

Lactate

Muscle lactic dehydrogenase (L-lactate:NAD oxidoreductase; 1.1.1.27) catalyzes the reaction:

$$\text{L-Lactate} + \text{NAD}^+ \rightleftharpoons \text{pyruvate} + \text{NADH} + \text{H}^+$$

The equilibrium of the reaction lies far to the left; hence, to determine lactate by oxidation, pyruvate must be removed from the reaction mixture. To do this, pyruvate is converted to the hydrazide or semicarbazide by adding hydrazine or semicarbazine, respectively (Hohorst 1965; Rosenberg and Rush 1966; Olson 1962). Using relatively high concentrations of NAD$^+$ and enzyme and one of these reagents in pH 9.5 glycine buffer, a quantitative and rapid oxidation of lactate can be obtained. The course of the reaction is followed by determining the reduction of NAD$^+$. The several sources of error have been discussed by Hohorst (1965).

Commercial lactate dehydrogenase preparations usually come from beef heart or rabbit muscle. That from beef heart is more active and is specific for L-(+)-lactate; it has no detectable activity with D-(—)-lactate (Hakala *et al.* 1950). Lactate dehydrogenases from a variety of sources have been extensively studied, and there is extensive literature on the constitution and use of these enzyme preparations. Two others are discussed briefly below.

Yeast lactate dehydrogenase is not NAD-linked; rather it is a flavocytochrome which will oxidize lactate in the presence of ferricyanide by hydrogen transfer (L-lactate : cytochrome c oxidoreductase; 1.1.2.3) (Wieland 1965B; Schon 1965). The decrease in color caused by reduction of the ferricyanide ion is followed at 405 nm.

An advantage of the use of this enzyme is that, with it, lactate can be determined in the presence of a large excess of pyruvate (Wallenfels and Hofmann 1959. A disadvantage is that reducing substances, such as L-ascorbic acid, glutathione, L-cysteine, etc., interfere with the determination. Therefore, blanks for both substrate and enzyme must be determined separately. The enzyme will also reduce other acids such as 2-hydroxybutyric acid which will interfere with the determination if present; the same is true of the animal preparations but to a lesser extent. Only L-(+)-lactate reacts, so with both muscle and yeast lactate dehydrogenase DL-lactate will react to only 50% of the theoretical value.

Acetone extracts of *E. coli B*, however, contain an enzyme system which oxidizes D-lactate to pyruvate (Haugaard 1950; van den Hamer 1965). The hydrogen is transferred to methylene blue which is spontaneously reoxidized by atmospheric oxygen affording a manometric technique. L-Lactate and D-glucono-1,5-lactone are oxidized at a rate less than 3% of that of D-lactate.

Among the other methods presented for the determination of L-lactate are those of Horn and Bruns (1956), Hess (1956), Pfleiderer and Dose (1955), and Holzer and Soling (1965). Labeyrie *et al.* (1959) have published an alternative method for the determination of D-lactate.

D-Gluconate

D-Gluconate can be determined by the following coupled reactions (Leder 1965):

D-Gluconate + ATP → 6-phospho-D-gluconate + ADP

6-Phospho-D-gluconate + NADP$^+$ → CO_2 + D-*erythro*-
pentulose (D-ribulose) 5-phosphate + NADPH + H$^+$

The first reaction is catalyzed by gluconokinase (ATP : D-gluconate 6-phosphotransferase; 2.7.1.12) and the second by 6-phosphogluconate dehydrogenase (6-phospho-D-gluconate : NAD(P) oxidoreductase; 1.1.1.43). Both enzyme preparations should be free of other NADP-linked dehydrogenases.

D-Glucuronate and D-Galacturonate

A uronic acid dehydrogenase (Kilgore and Starr 1959) which catalyzes the oxidation of D-glucuronate and D-galacturonate with the concomitant reduction of NAD$^+$ has been purified from *Pseudomonas syringae* grown on D-galacturonate (Bateman *et al.* 1967). No interfering compounds were found. One possible use of this enzyme is the determination of pectin by coupling the oxidation with exopolygalacturonase.

A NADP-linked D-glucuronate dehydrogenase has been used by Kean and Charalampous (1959).

NONENZYMIC METHODS

Use of Concanavalin A to Determine D-Mannose

There is no specific chemical or enzymic method for the determination of D-mannose. However, Poretz and Goldstein (1967) have developed a procedure for its determination based on a measurement of the degree to which the turbidity formed on interaction of concanavalin A from jack-bean meal (Agrawal and Goldstein 1965) with glycogen is inhibited by it. The reaction is rather specific for D-mannose although D-glucose interferes somewhat and a correction must be made for it. 2-Amino-2-deoxy-D-glucose also causes somewhat higher results but is much less effective as an inhibitor than D-mannose or D-glucose. It is unlikely that appreciable free D-glucosamine would be present in food products but, if it is suspected, it can be measured (Brown 1965). This method could be quite useful both for the determination of both free D-mannose and the D-mannose formed by hydrolysis of polysaccharides.

Quantitative (Differential) Fermentations

Quantitative fermentation methods have been described for the determination of L-arabinose, D-xylose, and D-galactose. These methods, which are little, if ever, applied, are discussed by Wise (1962).

DISCUSSION

R. O. Sinnhuber.—Could you comment on the determination of D-glucosamine? We're interested in chitin and chitinase.

J. N. BeMiller.—A method for the determination of D-glucosamine has been proposed by Brown (1965). I have had no personal experience with it. Commercial chitinases are now available and could be used to determine chitin if they could be coupled with a determination of D-glucosamine. This information was not presented in the talk for lack of time but is included in the published chapter.

M. L. Wolfrom.—There is a chemical method for mannose. The phenylhydrazone method was worked out by Wise and co-workers some time ago. It goes back to Emil Fischer and is an excellent method.

J. N. BeMiller.—I stand corrected.

F. Shipe.—You described cellulases as enzymes which produce water-soluble fragments from cellulose. Would you comment on the existence of enzymes that produce only water-insoluble fragments from cellulose?

J. N. BeMiller.—That was not my own definition, and you know as well as I do that almost everyone has his own definition. I think this definition would be ok when one is talking about commercial "cellulase" preparations, for in that case you're talking about enzymes that do convert cellulose from an insoluble compound to a soluble product. There really is no such thing as "cellulase." You're talking about a collective group of enzymes each of which have to be described specifically. I usually get around all this by calling them cellulolytic enzymes rather than "cellulases." From a single organism you usually have a C_1 enzyme (also called a "hydrogen bondase"), which does not produce water-soluble fragments but is necessary for the action of other enzymes, and various combinations of an endoglucanase, an exoglucanase, and a β-glucosidase which do produce water-soluble fragments as, at least, part of the products.

F. Shipe.—Would enzymes that would change viscosity of cellulose preparations necessarily involve producing water-soluble fragments?

J. N. BeMiller.—I don't think so. I refer the question to a rheologist. I'm really not sure, but it seems possible to me that viscosity could be increased by increasing the amount of more highly hydrated, more swollen, lower molecular weight products without introducing true solubility.

BIBLIOGRAPHY

ADAMS, E. C., JR., MAST, R. L., and FREE, A. H. 1960. Specificity of glucose oxidase. Arch. Biochem. Biophys. *91*, 230–234.

AGRAWAL, B. B. L., and GOLDSTEIN, I. J. 1965. Specific binding of concanavalin A to cross-linked dextran gels. Biochem. J. *96*, 23C–25C.

ASHWELL, G., and HICKMAN, J. 1954. Formation of xylulose phosphate from ribose phosphate in spleen extracts. J. Am. Chem. Soc. *76*, 5889.

AVIGAD, G. 1967. Synthesis of D-galactose-6-*t* and D-galactosides-6-*t*. Carbohydrate Res. *3*, 430–434.

AVIGAD, G., AMARAL, D., ASENSIO, C., and HORECKER, B. L. 1961. Galactodialdolase production with an enzyme from the mold *Polyporus circinatus*. Biochem. Biophys. Res. Commun. *4*, 474–477.

AVIGAD, G., AMARAL, D., ASENSIO, C., and HORECKER, B. L. 1962. The D-galactose oxidase of *Polyporus circinatus*. J. Biol. Chem. *237*, 2736–2743.

AXELROD, B., and JANG, R. 1954. Purification and properties of phosphoriboisomerase from alfalfa. J. Biol. Chem. *209*, 847–855.

BARKER, S. A., PARDOE, G., STACEY, M., and HOPTON, J. W. 1963. Sequential enzyme induction: a new approach to the structure of complex mucoproteins. Nature *197*, 231–233.

BARUAH, P., and SWAIN, T. 1957. β-Glycosidase of potato. Biochem. J. *66*, 321–323.

BATEMAN, D. F., KOSUGE, T., and DeVAY, J. E. 1967. Specific enzymic assay of monogalacturonate and monoglucuronate with uronic acid dehydrogenase. (Abstr.) Phytopath. 57, 1004.

BECKER, J. F. 1965. Steroid glucuronides. II. The hydrolysis of β-D-glucosiduronic acids by β-glucuronidases. Biochim. Biophys. Acta 100, 582–590.

BeMILLER, J. N., TEGTMEIER, D. O., and PAPPELIS, A. J. 1968. Cellulolytic enzymes of Diplodia zeae. Phytopath. 58, 1336–1339.

BENTLEY, R. 1955. Glucose aerodehydrogenase (glucose oxidase). In Methods in Enzymology, Vol. 1, S. P. Colowick, and N. O. Kaplan (Editors). Academic Press, New York.

BENTLEY, R. 1963. Glucose oxidase. In The Enzymes, 2nd Edition, Vol. 7, P. D. Boyer, H. Lardy, and K. Myrback (Editors). Academic Press, New York.

BENTLEY, R., and NEUBERGER, A. 1949. The mechanism of the action of notatin. Biochem. J. 45, 584–590.

BERGER, L., SLEIN, M. W., COLOWICK, S. P., and CORI, C. F. 1946. Isolation of hexokinase from baker's yeast. J. Gen. Physiol. 29, 379–391.

BERGMEYER, H.-U. (Editor). 1965. Methods of Enzymatic Analysis. Verlag Chemie GMBH, Weinheim, Germany, and Academic Press, New York.

BERGMEYER, H.-U., and BERNT, E. 1965. D-Glucose: determination with glucose oxidase and peroxidase.

BERGMEYER, H.-U., and KLOTZSCH, H. 1965. Sucrose.

BERNT, E. 1965. Unpublished results cited by H.-U. Bergmeyer, and E. Bernt 1965. D-Glucose: determination with glucose oxidase and peroxidase.

BROWN, D. H. 1965. D-Glucosamine.

HALLIWELL, G. 1965A. Cellulose.

HALLIWELL, G. 1965B. Hemicelluloses.

HOHORST, H.-J. 1965. L-(+)-Lactate.

HOLZER, H., and SOLING, H.-D. 1965. L-(+)-Lactate: determination with lactic dehydrogenase and the β-acetylpyridine analogue of DPN (AP-DPN).

HORECKER, B. L. 1965A. L-Ribulose and L-arabinose.

HORECKER, B. L. 1965B. D-Xylulose and D-xylose: Determination with D-xylose isomerase.

KLOTZSCH, H., and BERGMEYER, H.-U. 1965. D-Fructose.

LEDER, I. G. 1965. D-Gluconate.

PFLEIDERER, G. 1965. Glycogen: determination as D-glucose with hexokinase, pyruvic kinase and lactic dehydrogenase.

REITHEL, F. J. 1965. Lactose.

SIEGFRIED DE WHALLEY, H. C. 1965. Raffinose.

SLEIN, M. W. 1965. D-Glucose: Determination with hexokinase and glucose-6-phosphate dehydrogenase.

VAN DEN HAMER, C. J. A. 1965. D-Lactate.

WEISSBACH, A. 1965. myo-Inositol.

WIELAND, O. 1965A. Glycerol.

WIELAND, O. 1965B. L-(+)-Lactate: determination with lactic dehydrogenase from yeast.

WILLIAMS-ASHMAN, H. G. 1965. D-Sorbitol. *All in* Methods of Enzymatic Analysis, H.-U. Bergmeyer (Editor). Verlag Chemie GMBH, Weinheim, Germany, and Academic Press, New York.

BISHOP, C. T., and WHITAKER, D. R. 1955. Mixed arabinose-xylose oligosaccharides from wheat straw xylan. Chem. Ind., 119.

BLECHER, M., and GLASSMAN, A. B. 1962. Determination of glucose in the presence of sucrose using glucose oxidase; effect of pH on absorption spectrum of oxidized o-dianisidine. Anal. Biochem. 3, 343–352.

BRADLEY, R. M., and KANFER, J. N. 1964. The action of galactose oxidase on several sphingoglycolipids. Biochim. Biophys. Acta 84, 210–212.

BROWN, D. H. 1951. The phosphorylation of D-(+)-glucosamine by crystalline yeast hexokinase. Biochim. Biophys. Acta 7, 487–493.

BROWN, E., and CLARKE, D. L. 1963. Simplified method for the determination of reduced triphosphopyridine nucleotide by means of enzymic cycling. J. Lab. Clin. Med. 61, 889–892.

CHARALAMPOUS, F. C., and ABRAHAMS, P. 1956. Biochemical studies on inositol. I. Isolation of myo-inositol from yeast and its quantitative enzymatic estimation. J. Biol. Chem. 225, 575–583.

COHEN, S. L. 1951. The hydrolysis of steroid glucuronides with calf spleen glucuronidase. J. Biol. Chem. 192, 147–160.

COHEN, S. L. 1964. Potentiation of glucuronidase hydrolysis of sodium pregnanediol glucuronidate. Can. J. Biochem. 42, 127–138.

CONCHIE, J., and LEVVY, G. A. 1957. Inhibition of glycosidases by aldonolactones of corresponding configuration. Biochem. J. 65, 389–395.

COX, R. I. 1959. The interaction of some factors affecting the kinetics of a molluscan β-glucuronidase. Biochem. J. 71, 763–768.

CRANE, R. K. 1962. Hexokinases and pentokinases. *In* The Enzymes, 2nd Edition, Vol. 6, P. D. Boyer, H. Lardy, and K. Myrback (Editors). Academic Press, New York.

CROWNE, R. S., and MANSFORD, K. R. L. 1962. Studies on the specificity of commercial preparations of glucose oxidase. Analyst 87, 294–296.

DAHLQVIST, A. 1961. Determination of maltase and isomaltase activities with a glucose-oxidase reagent. Biochem. J. 80, 547–551.

DARROW, R. A., and COLOWICK, S. P. 1962. Hexokinase from baker's yeast. *In* Methods in Enzymology, Vol. 5, S. P. Colowick and N. O. Kaplan (Editors). Academic Press, New York.

DEVERDIER, C. H., and HJELM, M. 1962. A galactose-oxidase method for the determination of galactose in blood plasma. Clin. Chim. Acta 7, 742–744.

DEWHALLEY, H. C. S. 1964. ICUMSA Methods of Sugar Analysis. Elsevier Publishing Co., New York.

DISCHE, Z., and BORENFREUND, E. 1951. A new spectrophotometric method for the detection and determination of keto sugars and trioses. J. Biol. Chem. 192, 583–587.

DOBRICK, L. A. 1958. Screening method for glucose of blood serum utilizing glucose oxidase and an indophenol indicator. J. Biol. Chem. 231, 403–409.

ELLER, J., and PAYNE, W. J. 1963. Studies on bacterial utilization of uronic acids. IV. Alginolytic and mannuronic acid oxidizing isolates. J. Bacteriol. 80, 193–199.

FISCHER, W., and ZAPF, J. 1964A. Quantitative determination of galactose by *Dactylium dendroides* galactose oxidase. I. Z. Physiol. Chem. *337*, 186–195. (German)

FISCHER, W., and ZAPF, J. 1964B. Determination of galactose by *Dactylium dendroides* galactose oxidase. II. Determination of blood and urine galactose. Z. Physiol. Chem. *339*, 54–63. (German)

FISHMAN, W. H. 1957. Preparation and assay of substrates for β-glucuronidase. *In* Methods in Enzymology, Vol. 3. S. P. Colowick, and N. O. Kaplan (Editors). Academic Press, New York.

FRANKE, W., and DEFFNER, M. 1939. Glucose oxidase. II. Ann. *541*, 117–150. (German)

FRIEDEMANN, T. E., WITT, N. F., NEIGHBORS, B. W., and WEBER, C. W. 1967. Determination of available carbohydrates in plant and animal foods. J. Nutr. Suppl. 2, Part II, *91*, No. 3, 1–40.

FROESCH, E. R., and RENOLD, A. E. 1956. Specific enzymatic determination of glucose in blood and urine using glucose and oxidase. Diabetes *5*, 1–6.

GANCEDO, J. M., GANCEDO, C., and ASENSIO, C. 1966. Uronic acid formation by enzymic oxidation of galactosides. Biochem. Zeit. *346*, 264–268. (German)

GIBSON, Q. H., SWOBODA, B. E. P., and MASSEY, V. 1964. Kinetics and mechanism of action of glucose oxidase. J. Biol. Chem. *239*, 3927–3934.

GLASER, L., and BROWN, D. H. 1955. Purification and properties of D-glucose-6-phosphate dehydrogenase. J. Biol. Chem. *216*, 67–79.

HAGEN, J. H., and HAGEN, P. B. 1962. An enzymic method for the estimation of glycerol in blood and its use to determine the effect of noradrenaline on the concentration of glycerol in blood. Can. J. Biochem. Physiol. *40*, 1129–1139.

HAKALA, M. T., GLAID, A. J., and SCHWERT, G. W. 1950. Lactic dehydrogenase. II. Variation of kinetic and equilibrium constants with temperature. J. Biol. Chem. *221*, 191–209.

HARIGAYA, S. 1964. Studies on glucosaccharolactones. II. Glucosaccharo-1,4 : 3,6-dilactone and its inhibitory effect on β-glucuronidase. J. Biochem. *56*, 392–399.

HAUGAARD, N. 1950. Lactic acid oxidizing system of *Escherichia coli.* Federation Proc. *9*, 182–183.

HELFERICH, B., and PETERSEN, S. R. 1935. Emulsin. XIX. The action of ozone on almond emulsin. Z. Physiol. Chem. *233*, 75–80. (German)

HELFERICH, B., and SCHMITZ-HILLEBRECHT, E. 1935. Emulsin. XX. The influence of neutral salts on the activity of almond emulsin. Z. Physiol. Chem. *234*, 54–62. (German)

HELFERICH, B., and SCHNIEDMULLER, A. 1931. Emulsin. IV. Z. Physiol. Chem. *198*, 100–104. (German)

HELFERICH, B., WINKLER, S., SCHMITZ-HILLEBRECHT, E., and BACH, M. 1934. Emulsin. XVII. Action of ozone on almond emulsin. Z. Physiol. Chem. *229*, 112–116. (German)

HESS, B. 1956. Kinetic-enzymic determination of L-(+)-lactic acid in human serum and other biological fluids. Biochem. Z. *328*, 110–116. (German)

HESTRIN, S., FEINGOLD, D. S., and SCHRAMM, M. 1955. Hexoside hydrolases.

In Methods in Enzymology, Vol. 1, S. P. Colowick, and N. O. Kaplan (Editors). Academic Press, New York.

HICKS, G. P., and UPDIKE, S. J. 1966. The preparation and characterization of lyophilized polyacrylamide enzyme gels for chemical analysis. Anal. Chem. *38*, 726–730.

HJELM, M. 1967. A methodological study of the enzymatic determination of galactose in human whole blood, plasma and erythrocytes with galactose oxidase. Clin. Chim. Acta *15*, 87–96.

HLAING, T. T., HUMMEL, J. P., and MONTGOMERY, R. 1961. Some studies of glucose oxidase. Arch. Biochem. Biophys. *93*, 321–327.

HOLLMANN, S., and TOUSTER, O. 1957. The L-xylulose-xylitol enzyme and other polyol dehydrogenases of guinea pig liver mitochondria. J. Biol. Chem. *225*, 87–102.

HORN, H. D., and BRUNS, F. H. 1956. Quantitative estimation of L-(+)-lactic acid with lactic acid dehydrogenase. Biochim. Biophys. Acta *21*, 378–380. (German)

HU, A. S. L., WOLFE, R. G., and REITHEL, F. J. 1959. The preparation and purification of β-galactosidase from *Escherichia coli*, ML308. Arch. Biochem. Biophys. *81*, 500–507.

HUGGET, A. ST. G., and NIXON, D. A. 1957. Enzymatic determination of blood glucose. Biochem. J. *66*, 12P.

JOHNSON, J. A., and FUSARO, R. M. 1965. Base-catalyzed isomerization of glucose and its effect on glucose oxidase analysis. Anal. Biochem. *13*, 412–416.

JØRGENSEN, B. B., and JØRGENSEN, O. B. 1966. Enzymatic determination of anomers of glucose released by glucosidases. Acta Chem. Scand. *20*, 1437–1438.

KEAN, E. L., and CHARALAMPOUS, F. C. 1959. New methods for the quantitative estimation of *myo*-inositol. Biochim. Biophys. Acta *36*, 1–3.

KEILIN, D., and HARTREE, E. F. 1945. Properties of catalase. Catalysis of coupled oxidation of alcohols. Biochem. J. *39*, 293–301.

KEILIN, D., and HARTREE, E. F. 1948A. Properties of glucose oxidase (notatin). Biochem. J. *42*, 221–229.

KEILIN, D., and HARTREE, E. F. 1948B. The use of glucose oxidase (notatin) for the determination of glucose in biological material and for the study of glucose-producing systems by manometric methods. Biochem. J. *42*, 230–237.

KEILIN, D., and HARTREE, E. F. 1952A. Specificity of glucose oxidase (notatin). Biochem. J. *50*, 331–341.

KEILIN, D., and HARTREE, E. F. 1952B. Biological catalysis of mutarotation of glucose. Biochem. J. *50*, 341–348.

KESTON, A. S. 1956. Specific colorimetric enzymatic analytical reagents for glucose. Abstr. Papers, Am. Chem. Soc. *129*, 31C.

KILGORE, W. W., and STARR, M. P. 1959. Uronate oxidation by phytopathogenic Pseudomonads. Nature *183*, 1412–1413.

KIRK, B. K., FELGER, C. B., and KATZMAN, P. A. 1962. Hydrolysis of urinary estrogen and 17-ketosteroid glucosiduronides by bacterial and mammalian β-glucuronidases. Arch. Biochem. Biophys. *98*, 206–213.

KLEPPE, K. 1966. The effect of hydrogen peroxide on glucose oxidase from *Aspergillus niger*. Biochem. 5, 139–143.

KOOIMAN, P. 1957. Some properties of cellulase of *Myrothecium verrucaria* and some other fungi. II. Enzymologia 18, 371–384.

KORNBERG, A. 1950. Enzymatic synthesis of triphosphopyridine nucleotide. J. Biol. Chem. 182, 805–813.

KORNBERG, A., and PRICER, W. E. 1951. Enzymatic phosphorylation of adenosine and 2,6-diaminopurine riboside. J. Biol. Chem. 193, 481–495.

KUBOWITZ, F., and OTT, P. 1944. Isolation of fermentation enzymes from human muscle. Biochem. Z. 317, 193–203. (German)

KUNITZ, M., and McDONALD, M. R. 1946. Crystalline hexokinase (heterophosphatase). J. Gen. Physiol. 29, 393–412.

LABEYRIE, F., SLONIMSKI, P. P., and NASLIN, L. 1959. Difference in stereospecificity between lactic dehydrogenases extracted from anaerobically and aerobically grown yeast. Biochim. Biophys. Acta 34, 262–265. (French)

LARNER, J. 1960. Other glucosidases. *In* The Enzymes, 2nd Edition, Vol. 4, P. D. Boyer, H. Lardy, and K. Myrback (Editors). Academic Press, New York.

LEVVY, G. A., HAY, A. J., and MARSH, C. A. 1957. Properties of limpet β-glucuronidase. Biochem. J. 65, 203–208.

LEVVY, G. A., and MARSH, C. A. 1959. Preparation and properties of β-glucuronidase. Advan. Carbohydrate Chem. 14, 381–428.

LIN, E. C. C., and MAGASANIK, B. 1960. The activation of glycerol dehydrogenase from *Aerobacter aerogenes* by monovalent cations. J. Biol. Chem. 235, 1820–1823.

LOWRY, O. H., PASSONNEAU, J. V., SCHULZ, D. W., and ROCK, M. K. 1961. The measurement of pyridine nucleotides by enzymatic cycling. J. Biol. Chem. 236, 2746–2755.

MALMSTEDT, H. V., and HICKS, G. P. 1960. Determination of glucose in blood serum by a new rapid and specific automatic system. Anal. Chem. 32, 394–398.

MARSH, C. A. 1962. Inhibition of β-glucuronidase by endogenous saccharate during hydrolysis of urinary conjugates. Nature 194, 974–975.

McCOMB, R. B., and YUSHOK, W. D. 1958. Colorimetric estimation of D-glucose and 2-deoxy-D-glucose with glucose oxidase. J. Franklin Inst. 265, 417–422.

McCOMB, R. B., YUSHOK, W. D., and BATT, W. G. 1957. 2-Deoxy-D-glucose, a new substrate for glucose oxidase (glucose aerodehydrogenase). J. Franklin Inst. 263, 161–165.

MEYERS, S. P., PRINDLE, B., and REYNOLDS, E. S. 1960. Cellulolytic activity of marine fungi. Degradation of ligno-cellulose material. Tappi 43, 534–538.

MITSUHASHI, S., and LAMPEN, J. O. 1953. Conversion of D-xylose to D-xylulose in extracts of *Lactobacillus pentosus*. J. Biol. Chem. 204, 1011–1018.

MORELL, A. G., VAN DEN HAMER, C. J. A., SCHEINBERG, I. H., and ASHWELL, G. 1966. Physical and chemical studies on ceruloplasmin. IV. Preparation of radioactive, sialic acid-free ceruloplasmin labeled with tritium on terminal D-galactose residues. J. Biol. Chem. 241, 3745–3749.

NICOL, M. J., and DUKE, F. R.   1966.   Substrate inhibition with glucose oxidase.   J. Biol. Chem. *241*, 4292–4293.

NISHIZUKA, Y., and HAYAISHI, O.   1962.   Enzymic formation of lactobionic acid from lactose.   J. Biol. Chem. *237*, 2721–2728.

NORKRANS, B., and WAHLSTROM, L.   1961.   Studies on inhibition of cellulolytic enzymes.   Physiol. Plantarum *14*, 851–860.

OLSON, G. F.   1962.   Optimal conditions for the enzymic determination of L-lactic acid.   Clin. Chem. *8*, 1–10.

PAZUR, J. H., and KLEPPE, K.   1964.   The oxidation of glucose and related compounds by glucose oxidase from *Aspergillus niger*.   Biochem. *3*, 578–583.

PEAT, S., WHELAN, W. J., and TURVEY, J. R.   1956.   The soluble polyglucose of sweet corn (*Zea mays*).   J. Chem. Soc., 2317–2322.

PFLEIDERER, G., and DOSE, K.   1955.   Enzymic determination of L(+)-lactic acid with lactic acid dehydrogenase.   Biochem. Z. *326*, 436–441. (German)

PFLEIDERER, G., and GREIN, L.   1957.   Enzymic determination of D(+)-glucose in blood.   Biochem. Z. *328*, 499–506.   (German)

PORETZ, R. D., and GOLDSTEIN, I. J.   1967.   Protein-carbohydrate interaction. VIII.   A turbidimetric method for the analysis of D-mannose.   Carbohydrate Res. *4*, 471–477.

PRIMO YUFERA, E.   1961.   Glucose oxidase and catalase as additives in the preservation of foods.   Rev. Agroquim. Tecnol. Alimentos *1*, 51–56. (Spanish)

REED, G.   1966.   Enzymes in Food Processing.   Academic Press, New York.

REESE, E. T., and MANDELS, M.   1957.   Chemical inhibition of cellulases and β-glucosidases.   Res. Rept., Pioneering Res. Div., Quartermaster Res. and Develop. Center, Natick Mass., Microbiol. Ser. *17*.

REESE, E. T., and MANDELS, M.   1960.   Chemical inhibition of enzymes. II. Gluconolactone as inhibitor of carbohydrases.   Ind. Microbiol. *1*, 171–180.

ROREM, E., and LEWIS, J. C.   1962.   A test paper for the detection of galactose and certain galactose-containing sugars.   Anal. Biochem. *3*, 230–235.

ROSENBERG, J. C., and RUSH, B. F.   1966.   Enzymic-spectrophotometric determination of pyruvic acid and lactic acid in blood.   Methodologic aspects.   Clin. Chem. *12*, 299–307.

ROTH, H., SEGAL, S., and BERTOLI, D.   1965.   The quantitative determination of galactose—An enzymic method using galactose oxidase, with applications to blood and other biological fluids.   Anal. Biochem. *10*, 32–52.

SALOMON, L. L., and JOHNSON, J. E.   1959.   Enzymatic microdetermination of glucose in blood and urine.   Anal. Chem. *31*, 453–456.

SCHARLACH, G., ROBBINS, J., and WOLFF, J.   1962.   Contamination of commercial preparations of glucose oxidase with beta-fructofuranosidase.   Nature *194*, 1249–1250.

SCHLEGEL, R. A., GERBECK, C. M., and MONTGOMERY, R.   1968.   Substrate specificity of D-galactose oxidase.   Carbohydrate Res. *7*, 193–199.

SCHON, R.   1965.   A simple and sensitive enzymic method for determination of L(+)-lactic acid.   Anal. Biochem. *12*, 413–420.

SEMPERE, J. M., GANCEDO, C., and ASENSIO, C.   1965.   Determination of galactosamine and N-acetylgalactosamine in the presence of other hexosamines with galactose oxidase.   Anal. Biochem. *12*, 509–515.

Sison, B. C., Jr., Schubert, W. J., and Nord, F. F. 1958. On the mechanism of enzyme action. LXV. A cellulolytic enzyme from the mold *Poria vaillantii.* Arch. Biochem. Biophys. 75, 260–272.

Smith, F. 1944. Lactones of glucosaccharic acid. Part III. Mono- and di-lactones. J. Chem. Soc., 633–636.

Sols, A., and de la Fuente, G. 1957. On the substrate specificity of glucose oxidase. Biochim. Biophys. Acta 24, 206–207.

Sols, A., and de la Fuente, G. 1961. Hexokinase and other enzymes of sugar metabolism in the intestine. *In* Methods in Medical Research, Vol. 9, J. H. Quastel (Editor). Year Book Medical Publishers, Chicago.

Sørensen, H. 1957. Microbial decomposition of xylan. Acta Agr. Scand., Suppl. 1.

Strominger, J. L. 1955. Enzymic synthesis of guanosine and cytidine triphosphates: A note on the nucleotide specificity of the pyruvate phosphokinase reaction. Biochim. Biophys. Acta 16, 616–617.

Swoboda, B. E. P., and Massey, V. 1966. On the reaction of the glucose oxidase from *Aspergillus niger* with bisulfite. J. Biol. Chem. 241, 3409–3416.

Taeufel, K., Behnke, U., and Wersuhn, H. 1965. Determination of sucrose in beet molasses by means of glucose oxidase. Z. Zuckerind. 15, 462–467. (German)

Teller, J. D. 1956. Direct quantitative colorimetric determination of serum or plasma glucose. Abstr. Papers, Am. Chem. Soc. 130, 69C.

Thivend, P., Mercier, C., and Guilbot, A. 1965A. Application of glucoamylase to starch determination. Staerke 17, 276–281. (German)

Thivend, P., Mercier, C., and Guilbot, A. 1965B. Estimation of starch in complex media. Ann. Biol. Animale Biochim. Biophys. 5, 513–526. (French)

Thomas, R. 1956. Fungal cellulases. VII. *Stachybotrys atra:* Production and properties of cellulolytic enzyme. Austral. J. Biol. Sci. 9, 159–183.

Thorn, W., Isselhard, W., and Muldener, B. 1959. Glycogen, glucose, and lactic acid contents of warm-blooded animals under different experimental conditions and under anoxia determined by optical enzyme tests. Biochem. Z. 331, 545–562. (German)

Tietz, A., and Ochoa, S. 1958. "Fluorokinase" and pyruvic kinase. Arch. Biochem. Biophys. 78, 477–493.

Underkofler, L. A. 1957. Properties and applications of the fungal enzyme glucose oxidase. Proc. Intern. Symp. Enzyme Chem. (Tokyo and Kyoto), 486–490.

Underkofler, L. A. 1961. Glucose oxidase: production, properties, and present and potential applications. Soc. Chem. Ind. (London), Monograph 11, 72–86.

Updike, S. J., and Hicks, G. P. 1967. Reagentless substrate analysis with immobilized enzymes. Science 158, 270–272.

Veibel, S. 1950. β-Glucosidase. *In* The Enzymes, Vol. 1, Part 1, J. B. Sumner, and K. Myrback (Editors). Academic Press, New York.

Voigt, J. 1961. Glucose oxidase extraction, properties, and applications. Ernaehrungsforschung 6, 563–578. (German)

Wakabayashi, M., and Fishman, W. H. 1961. The comparative ability of β-glucuronidase preparations (liver, *Escherichia coli, Helix pomatia,* and

Patella vulgata) to hydrolyze certain steroid glucosiduronic acids. J. Biol. Chem. *236*, 996–1001.

WALLENFELS, K., and HOFMANN, D. 1959. Mechanism of hydrogen transfer with pyridine nucleotides. XVIII. Nonenzymic reduction of pyruvic acid with a reduced diphosphopyridine nucleotide model system. Tetrahedron Letters *15*, 10–13. (German)

WALSETH, C. S. 1952. Occurrence of cellulases in enzyme preparations from microorganisms. Tappi *35*, 228–233.

WHISTLER, R. L., HOUGH, L., and HYLIN, J. W. 1953. Determination of D-glucose in corn sirups. Anal. Chem. *25*, 1215–1216.

WHITAKER, D. R. 1953. Purification of *Myrothecium verrucaria* cellulase. Arch. Biochem. Biophys. *43*, 253–268.

WHITE, J. W., JR., and SUBERS, M. H. 1961. A glucose oxidase reagent for maltase assay. Anal. Biochem. *2*, 380–384.

WISE, L. E. 1962. Fermentation methods for quantitative determination of sugars. *In* Methods in Carbohydrate Chemistry, Vol. 1, R. L. Whistler, M. L. Wolfrom, J. N. BeMiller, and F. Shafizadeh (Editors). Academic Press, New York.

WOLFSON, S. K., JR., and WILLIAMS-ASHMAN, H. G. 1958. Enzymatic determination of sorbitol in animal tissues. Proc. Soc. Exp. Biol. Med. *99*, 761–765.

Reactions and Interactions of Carbohydrates

CHAPTER 11

A. S. Perlin | Carbohydrate Oxidations

The subject of oxidation touches on many aspects of the chemistry and biology of the carbohydrates. With the background of this Symposium in mind one naturally thinks of oxidations involved in the metabolism of carbohydrates. One can point also to important physiological oxidations, such as that of the ascorbic acid-dehydroascorbic acid system, or to the classic alkaline copper oxidation associated with sugar analysis of biological fluids. The long history of technical concern with carbohydrate oxidations originates in such processes as deterioration of textiles, aging of viscose, and manufacture of oxidized starches or of ascorbic acid.

There surely are very few known oxidants that have not been tested on some form of carbohydrate in one context or another, and in several specific areas the literature is very extensive. In the present paper, therefore, only a few aspects of the subject can be mentioned and emphasis is focused on those areas that have seen substantial recent activity.

We can begin with some comments on oxidation of the hemiacetal group of sugars. Reference has been made already to the important alkaline copper reaction, in which the sugar is oxidized to a carboxylic acid, and many other oxidants can effect this transformation, often in a highly selective and quantitative manner. Because of its great significance for the analysis of carbohydrates, this type of oxidation alone has spawned an enormous chemical and biochemical literature.

Also outstanding in this category is the bromine oxidation of aldoses to aldonolactones. Isbell's early studies with Hudson (1932) and Pigman (1933) of this reaction played a key role in the elucidation of structure of the sugars, and stimulated the development of conformational analysis. Recently Isbell has reappraised the mechanism of this reaction in the light of current concepts (Isbell 1962, 1964). He has shown that for aldoses having high stability in only one of the chair conformations, the equatorial anomer is much more rapidly oxidized than the axial anomer (e.g, β-D-glucose : α-D-glucose $= 40 : 1$). As the relative stabilities of the two chair conformations become more nearly equal the difference in oxidation rates between the anomers decreases (e.g., β-D-gulose : α-D-gulose $= 6 : 1$). A plausible rationalization of these and other properties of the reactions has been provided by depicting a transition state for the bromine-aldose complex in which O-1,C-1,C-2,C-5, and the ring O approach coplanarity (i.e., a conformation intermediate between that of the aldose and the derived lactone, and also one that should be stabilized by resonance involving the ring oxygen atom). It is likely, then, that the free energy change required to pass from the ground state to this transition state should be smaller when OH-1 is equatorial than when axial, because the former arrangement results in a comparatively small overall conformational realignment. There is relatively little direct reaction between a conformationally-stable axial anomer (e.g., α-D-glucose) and bromine, because isomerization to the reactive equatorial anomer provides a lower energy pathway (see also Barker *et al.* 1960)—this has been elegantly demonstrated by using tritium isotope effects. However, as the energy barrier to conformational change becomes smaller, differences in ΔF (and hence in rates) decrease, and both anomers (e.g., of D-lyxose or D-gulose) are oxidized directly because they readily assume a conformation in which OH-1 is equatorial.

Glucose oxidase ("notatin") undoubtedly is the most important example of *enzymic* means for converting an aldose to an aldonic acid. It is specific for β-D-glucose (Keilin and Hartree 1952), and hence is sterically analogous to bromine in its selectivity for the equatorial anomer. Similarly also, its reaction is characterized by a large isotope effect, which shows that the C-1 to hydrogen bond of the sugar is broken in the rate determining step (Isbell and Sniegoski 1964).

Turning to the other end of a sugar molecule—i.e.—to the primary carbinol group, one again finds a number of selective oxidants available. Of considerable industrial importance in this context is nitrogen dioxide (N_2O_4), which has been utilized for the manufacture of polyuronide oxycelluloses. That is, either as a gas or in an inert solvent such as carbon tetrachloride, this reagent promotes oxidation of the hydroxymethyl group of a glycosyl residue to a carboxyl group with relatively minor attack on secondary hydroxyl groups (McGee *et al.* 1947). There are also many examples of this reaction having been achieved catalytically with oxygen or air in the presence of platinum and in a weakly alkaline medium (Heyns and Paulsen 1962). Probably even more selective than nitrogen dioxide, catalytic oxidation is highly valuable for synthesis of uronic acid derivatives (Mehltretter 1953), and also finds use for selective degradation of some polysaccharides by oxidation followed by partial acid hydrolysis (Aspinall and Nicholson 1960).

Catalytic oxidation can be utilized as well for preparing aldoses, because in neutral solution oxidation of the hydroxymethyl group is largely arrested at the aldehyde stage (Heyns and Paulsen 1962). Dimethylsulfoxide in the presence of a suitable electrophile (see below) also converts a primary carbinol group to an aldehyde (Pfitzner and Moffatt 1963), although other free hydroxyl groups may be affected simultaneously. Dicyclohexylcarbodiimide is an effective promoter of this reaction (Pfitzner and Moffatt 1963), but not acetic anhydride (Godman and Horton 1968), and sulfur trioxide-pyridine is perhaps most satisfactory for the purpose (Cree *et al.* 1969).

Of considerable recent interest in this area, although of much more restricted application, is D-galactose oxidase (Avigad *et al.* 1962). This enzyme, isolated initially from *Polyporus circinatus,* converts the primary carbinol group of D-galactose to an aldehyde group. Since D-galactopyranosyl residues in the form of glycosides or as nonreducing end units of oligo- or polysaccharides also are susceptible to attack at this position, the enzyme promises to be of considerable value for selectively modifying D-galactosides. For example, the use of galactose oxidase followed by chemical oxidation of the aldehyde groups formed, has been used to convert D-galactosyl units of several complex carbohydrates to uronic acid units (see also Rogers and Thompson 1968).

Regarding the oxidation of secondary hydroxyl groups of sugars, enzymic methods again can be cited as the most selective examples in this category. Oxidation by *Acetobacter suboxydans,* discovered over 70 yr ago, continues to provide a ready source of 2-ketones from polyols containing the L-*erythro* configuration (Bertrand 1896; Hann *et al.* 1938). The specificity of this dehydrogenation is not modified by a variety of substituents—e.g., acetamido, diethyl dithioacetal—so long as they are 3 or 4 carbon atoms removed from the site of oxidation (Jones and co-workers 1961, 1965). Recently it has been found that A. *suboxydans* can promote other oxidations as well, i.e., conversion of the L-threo configuration to a 1,3-dihydroxy-2-keto compound (Arcus and Edson 1956), and oxidation of OH-5 of D-galactose (but not of D-xylose- or D-lyxose-) diethyl dithioacetal to a keto group (Williams and Jones

1967). Recently also, the list of microbiological oxidations developed has been extended to include position −3 of D-glucopyranosyl or D-galactopyranosyl residues, the pertinent dehydrogenase being found in *Agrobacterium tumifaciens* (Grebner and Feingold 1965). Examples of the action of this enzyme are the oxidation of sucrose to β-D-fructo-furanosyl-α-D-*ribo*-hexopyranosid-3-ulose, and of lactose to 4-*O*-β-D-*xylo*-hexopyranosyl-3-ulose-D-glucose.

The past several years have seen a great deal of renewed interest in the chemical oxidation of isolated secondary hydroxyl groups of sugar derivatives (see Theander 1962). This situation has been stimulated on the one hand by a desire to utilize carbonyl compounds for various synthetic purposes and, on the other, by the discovery of efficient oxidation methods.

In the category of synthetic uses may be mentioned various specifically labeled sugars. For example, by oxidation of 1,2-*O*-isopropylidene-D-glucurone to the 5-oxo derivative, followed by borodeuteride reduction one may obtain the 5-deutero or 5,6,6'-trideutero D-glucose derivatives (Mackie and Perlin 1965B). By analogous procedures, using boro-hydride-*t* as the reducing agent, various tritiated sugars are obtained (e.g. Barnett and Corina 1966; Hogenkamp 1966).

In several instances, this sequence of oxidation-reduction constitutes a very effective way to obtain a less common epimer by starting with a readily-available one. A good example is the conversion of 1,2 : 5,6-di-*O*-isopropylidene-D-glucose via the 3-oxo compound to the D-allose derivative (Theander 1964). Other examples are the conversion of compounds having the *galacto* or *manno* configuration to ones having the *talo* configuration (Collins and Overend 1965; Horton and Jewell 1967), of the D-*gluco* to the L-*ido* configuration (Kuzuhara *et al.* 1967), and of the D-*arabino* to the D-*ribo* configuration (Tong *et al.* 1967). Simultaneously, at least three laboratories (McDonald 1967; James *et al.* 1967; Cree and Perlin 1968) developed the same route to the relatively rare ketose, D-allulose (D-psicose), which involves an overall high yield inversion by oxidation-reduction at position −3 of 1,2 : 4,5-di-*O*-isopropylidene-D-fructose.

Successful inversion of configuration in this way depends, of course, on the direction of the reduction step, and the desired control of the latter is not easy to achieve. In the examples cited the major product obtained, fortunately, has the desired configuration. Although reducing agents vary widely in stereoselectivity, one cannot as yet readily predict the outcome when application is made to ketones in the sugar series. In our own experience, the most striking difference found is the D-fructose-to-D-allulose conversion, mentioned above: borohydride reduction yields the *allulo* isomer almost exclusively, whereas sodium amalgam affords only the *fructo* isomer (although in very low yield) (Cree and Perlin 1968). Not only the nature of the reductant, but the structure of the ketone is important. For example, after hydrolysis of the 5,6-ketal group of the 3-oxo derivative in question, borohydride reduction yields a 1 : 1 ratio of the *allulo* and *fructo* isomers.

These various factors may be mentioned also in the context of possible routes to glycosides and oligosaccharides that are not readily attainable in other ways. Theander (1957) has shown, for example, that catalytic reduction of the 2-keto compound obtained by chromate oxidation of methyl β-D-glucoside yields mainly methyl β-D-mannoside. This suggests a possible general entry route to difficultly accessible β-D-mannopyranosides from readily obtainable β-D-glucopyranosides (Fig. 71). Of pertinence also, therefore, is the conversion of some residues of cellulose (as the 6-*O*-trityl derivative) to β-D-mannosyl residues, by dimethylsulfoxide–acetic anhydride oxidation followed by borohydride reduction (Bredereck 1967). On the other hand, α-D-mannopyranosides are relatively easily synthesized (Gorin and Perlin 1961), and oxidation at C-2 followed by reduction could provide a route to α-D-glucopyranosides (Fig. 71). In fact, the direction of the reduction step has already been shown in one instance to lead to the α-D-gluco product (Lemieux *et al.* 1968).

A number of recent fruitful synthetic applications involve the conversion of ketones to amines, *via* reduction of intermediate oximes or hydrazones, or to branched sugars, *via* Grignard and related additions (e.g., see Lindberg and Theander 1959; Overend 1963).

The most widely employed oxidant at present in this area is dimethylsulfoxide. Its use for oxidation of hydroxyl groups was reported initially by Pfitzner and Moffatt (1963, 1965), who utilized dicyclohexylcarbodiimide to promote the reaction. Most likely the reactive species is

CH_3—$\overset{\oplus}{S}$═CH_2 which functions as the hydrogen acceptor (Epstein and Sweat 1966), but also may lead to formation of a thiomethyl ether (R—O—CH_2—S—CH_3) as a by-product. Several other modifications of this oxidant have been described in which dimethylsulfoxide is combined with acetic anhydride (Albright and Goldman 1965), phosphorus pentoxide (Onodera *et al.* 1968), or the SO_3-pyridine complex (Parikh and Doering 1967). There appears to be relatively little to choose between these various modifications although, in our experience (Cree *et al.* 1969), that employing the SO_3-pyridine complex makes for easier workup of the reaction mixture and leads to the formation of relatively little thiomethyl ether. Ruthenium tetroxide in carbon tetrachloride also is a highly effective oxidant (Beynon *et al.* 1966), and a variant of this reagent, using a catalytic proportion of ruthenium dioxide and excess of periodate to generate tetroxide continuously, has been described (Parikh and Jones 1965).

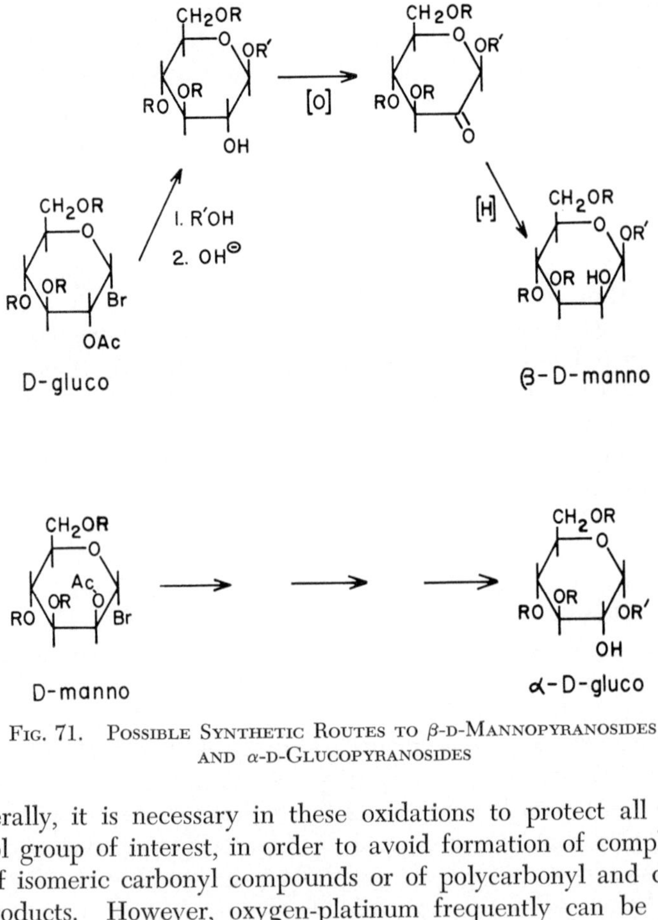

FIG. 71. POSSIBLE SYNTHETIC ROUTES TO β-D-MANNOPYRANOSIDES AND α-D-GLUCOPYRANOSIDES

Generally, it is necessary in these oxidations to protect all but the carbinol group of interest, in order to avoid formation of complex mixtures of isomeric carbonyl compounds or of polycarbonyl and degradation products. However, oxygen-platinum frequently can be used to oxidize at one position in the presence of unsubstituted hydroxyl groups because, almost exclusively, it will attack a carbinol group in which the hydroxyl function is axially oriented (Heyns and Paulsen 1962).

A type of oxidation distinctly different from those just considered is the glycol cleavage reaction of periodate or lead tetraacetate, for many years probably the most widely employed oxidation in the carbohydrate field.

For many purposes periodate and lead tetraacetate are interchangeable reagents, but sometimes their relative behavior is quite different. A striking recent example of this is furnished by the finding that cleavage of the 1,2-diol group of 3-O-methyl D-glucose by periodate yields the 4-formate of 2-O-methyl D-arabinose, whereas lead tetraacetate gives the 3-formate (Perlin 1964; Mackie and Perlin 1965A). That is, periodate degrades the pyranose sugar directly, whereas during the lead tetraacetate reaction the sugar assumes a furanose form, most probably by complexing with the tetravalent lead atom. However, some sugars—notably D-mannose—are oxidized by lead tetraacetate in the pyranose form.

Although the glycol-cleaving oxidants are highly specific, they frequently promote "overoxidation." This nonspecific behavior most commonly involves oxidation products related to tartronaldehyde, containing an active hydrogen atom. The latest evidence (Hudson and Barker 1967) shows convincingly that the mechanism of such overoxidation by periodate involves hydroxylation of an intermediate enolic species. Thus, the trialdehyde produced by oxidation of 1,4-anhydro-D-allitol has been isolated in an enolic form that is oxidized to carbon dioxide, formic acid, and formaldehyde, just as in the uninterrupted reaction. Many attempts to eliminate such overoxidation have been unsuccessful, but it has been reported recently (Szabo and Szabo 1967) that virtually only the cleavage reaction is observed when periodate is used with 0.1 N sulfuric acid as the solvent. Relatively few compounds have been examined under these conditions and, of course, some might be too sensitive in such a medium, but it will be important to determine how widely applicable these modified reaction conditions are.

The dialdehydes produced by cleavage of cyclic derivatives have received a good deal of attention during the past several years. It is recognized that such compounds can exist in a variety of forms, e.g., as a hydrated acyclic dialdehyde, a hemialdal or internal acetal (Guthrie 1961). The dialdehydes obtained from methyl aldopentopyranosides, for example, have been shown by nmr spectroscopy to exist in deuterium oxide as a 1 : 1 mixture of the acyclic hydrate and the hemialdal, and almost exclusively as the latter in dimethylsulfoxide or pyridine (Perlin 1966). The dialdehyde obtained by scission of the 2,3,4-triol group of 1,5-anhydropentitols, although seemingly a very simple compound, shows rather complex behavior. In addition to the acyclic hydrated form, which is stable in deuterium oxide, there are 2 major hemialdals—the 2 hydroxyl groups of 1 are equatorial, whereas the other species contains 1 equatorial and 1 axial hydroxyl group and is depicted by interconverting chairs (Greenberg and Perlin 1966).

Dialdehydes serve as an excellent starting point for the synthesis of 3-amino-3-deoxy cyclic derivatives (Baer and Fischer 1959). Ring closure is achieved by condensing both carbonyl groups of the dialde- hyde with one molecule of nitromethane. The resulting 3-nitro deriva- tive, which can exist in a variety of isomeric forms involving the new asymmetric centres at C-2, C-3, and C-4, is then reduced to the amino compound (Baer 1962).

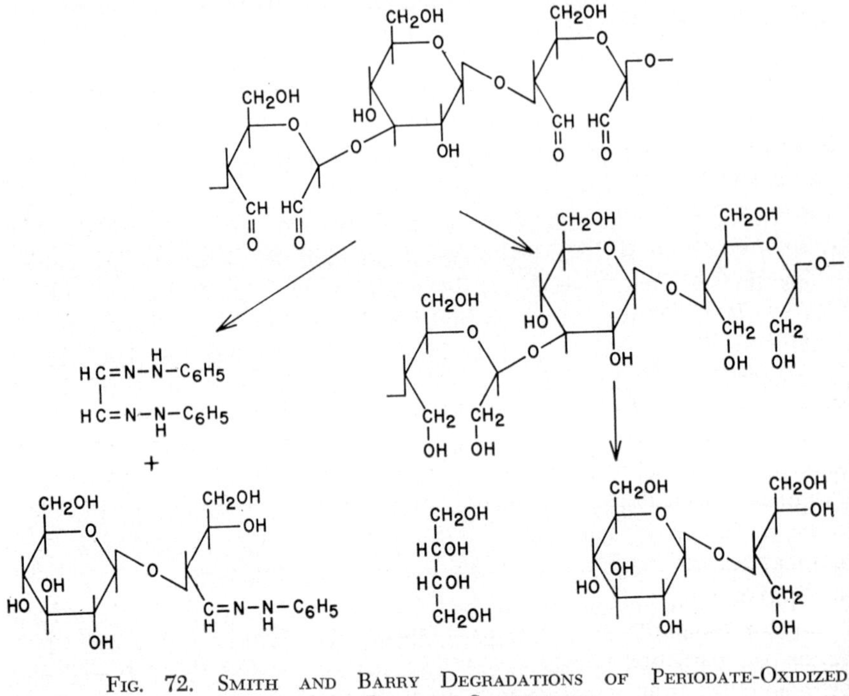

FIG. 72. SMITH AND BARRY DEGRADATIONS OF PERIODATE-OXIDIZED COMPLEX SACCHARIDES

Of very great benefit for the determination of structure of polysac- charides and oligosaccharides has been the development in recent years of the Smith degradation (Goldstein *et al.* 1965). This procedure is similar to the classic Barry degradation (Barry 1943), in that the dialdehyde produced by glycol cleavage oxidation is utilized as the site of a selective fragmentation of the molecule (Fig. 72). In the Barry procedure this step is effected with phenylhydrazine, which forms glyoxal bis-phenylhydrazone. The Smith degradation, which generally is much more satisfactory to apply, involves reduction of the dialdehyde to a polyol acetal which is highly labile in acid. The overall reaction thus enables one to separate those residues of a complex saccharide that are susceptible to oxidation from those that are not attacked, and hence provides an excellent means for examining the detailed structure of the molecule. A typical application is shown by Smith degradation of the $(1 \rightarrow 3)(1 \rightarrow 4)$ β-D-glucans of cereals: a major product is 2-O-β-D-gluco-pyranosyl-D-erythritol, which demonstrates that 3- and 4- linked residues alternate in the polymer (Goldstein et al. 1965). The method is adaptable also for selective removal of nonreducing end units of oligo-saccharides (Parrish et al. 1960).

It is not always easy to recover polysaccharide dialdehydes from a reaction mixture for use in these degradations. A recently reported oxidation procedure, in which lead tetraacetate is used in dimethylsulfox-ide-acetic acid, makes for a more satisfactory recovery of the product (Zitko and Bishop 1966) and should benefit future studies in this area.

Some vic-diols are remarkably resistant to glycol cleavage under the usual conditions, probably because they cannot attain an orientation favorable for complexing with the oxidant (Dimler 1952). Lead tetra-acetate in pyridine generally is effective for oxidation of such diols (Goldschmid and Perlin 1960). An example of this kind of resistant glycol has recently been encountered in a microbiol glycolipid (Tulloch et al. 1967). The sophorosyl disaccharide unit of this compound is partially acylated and at first had appeared to contain only one vic-diol group. By using lead tetraacetate-pyridine, however, a second diol group has been detected which, together with other data, has necessi-tated a change in the original assignment of structure. This example emphasizes the continuing need for caution in working with complex carbohydrates.

In this rather brief review an attempt has been made to survey the kinds of oxidation reactions used by the carbohydrate researcher. Although far from ideal, the methods and techniques available are of great value for detecting carbohydrates, determining their constitution, and for the preparation and synthesis of a wide range of sugars and derivatives.

DISCUSSION

G. N. Bollenback.—I believe you mentioned the biological interest in allulose. Would you mind being a little bit more specific?

A. S. Perlin.—I have to take this for granted from Dr. Simpson who is collaborating with us in work of this kind. There have been references to allulose as a naturally occurring sugar in some plants.   Dr. Simpson is interested in isomerases which might be involved in interconversions relating to allulose.   Dr. Hough, for example, has published in this area from a biological point of view.   Allulose has been found in several plant sources and allulose-6-phosphate is the product that we started out to make with this oxidation.   In fact we did make it, because we obtained diacetone allulose very readily by this method. Putting it into acid converts the pyranose to the furanose form and phosphorylation at the 6 position gives you allulose-6-phosphate.   Dr. Simpson is looking at this from an enzymatic point of view.   I can find the specific references but I don't have them here with me.

   D. French.—Would you comment on the relative rate of periodate (IO_4^-) oxidation on different types of vicinal hydroxyl groups?   As a case in point with D-glucofuranose you indicated there was cleavage between carbons 1 and 2 leaving carbons 5 and 6 uncleaved.   Also with cellotriose you showed cleavage of the nonreducing terminal glucose unit and the reducing terminal glucose unit without touching the central glucose unit.   These different situations would give different rates of attack.   Could you summarize the current picture on this?

   A. S. Perlin.—There is a great variability here and this has been for us a more interesting aspect of glycol cleavage oxidations.   The relative rates vary enormously with $Pb(OAc)_4$ from the hemiacetal type of glycol situation here to one internally.   The examples I used involved lead tetraacetate oxidation.   To some extent it is paralleled by periodate oxidation but I do not think the reaction is usually as clean cut.   Hemiacetal glycols are very easily cleaved, for a variety of steric and electronic reasons, at 50 to 100x the rate of a glycoside.   With a disaccharide you can get almost selective cleavage at the reducing end relative to the glycosidic end.   The same applies to the other case you mentioned.   In D-glucofuranose, itself, you have the 1,2-diol group and the 5,6-diol group.   The former is oxidized far more rapidly than the 5,6-diol group.   So one can certainly get selective oxidation with lead tetraacetate in cases of this kind.   As a matter of fact many years ago we oxidized an octose in which you had the possibility of a triol sticking out at the end.   Yet, that wasn't touched at all.   The oxidation went down two units and the formate ester blocked these lower ends . . . so it is highly selective.   The central units of oligosaccharides are generally reduced in activity, probably through steric factors.   You can oxidize reducing ends first of all, nonreducing ends second, and central units last, given the right situation.   Generally, furanoses are faster than pyranoses although it depends on the configuration.   Mannopyranose is oxidized very rapidly indeed.

   R. Montgomery.—Would you comment on the reactivity of those sugar derivatives in which the primary alcoholic function has been oxidized to an aldehyde? I ask the question with reference again to the consideration of the galactose oxidase reaction where we have noted that if you follow the availability of aldehyde as the oxidation proceeds, that the aldehyde concentration comes up to a maximum and then gradually disappears in some cases nearly to zero, as if there is a secondary reaction taking place. Have you noted this with the synthetic (excuse the nomenclature) 6-aldehyo derivatives?

A. S. Perlin.—These reactions are exceedingly fast in dimethysulfoxide as a rule and so by doing preliminary runs, using say thin-layer or gas-phase chromatography to follow them, we usually try to find out when the oxidation is essentially finished and work it up as quickly as possible. In dimethylsulfoxide oxidations, however, one must get the properly substituted derivative with the OH's you want free. Then you oxidize.

BIBLIOGRAPHY

ALBRIGHT, J. D., and GOLDMAN, L. 1965. Dimethyl sulfoxide–acid anhydride mixtures. New reagents for oxidation of alcohols. J. Am. Chem. Soc. 87, 4214–4216.

ARCUS, A. C., and EDSON, N. L. 1956. Polyol dehydrogenases (II) of *Acetobacter suboxydans* and *Candida utilis*. Biochem. J. 64, 385–394.

ASPINALL, G. O., and NICHOLSON, A. 1960. The catalytic oxidation of European Larch ε-galactan. J. Chem. Soc., 2503–2507.

AVIGAD, G., AMARAL, D., ASENSIO, B. C., and HORECKER, B. L. 1962. D-Galactose oxidase of *Polyporus circinatus*. J. Biol. Chem. 237, 2736–2743.

BAER, H. H. 1962. Cyclizations of dialdehydes with nitromethane. VIII. J. Am. Chem. Soc. 84, 83–89.

BAER, H. H., and FISCHER, H. O. L. 1959. Cyclizations of dialdehydes with nitromethane. II. J. Am. Chem. Soc. 81, 5184–5189.

BARKER, I. R. L., OVEREND, W. G., and REES, C. W. 1960. Oxidation of α- and β-D-glucose with bromine. Chem. and Ind. (London), 1297–1298.

BARNETT, J. E. G., and CORINA, D. L. 1966. A synthesis of D-glucose-5-*t* and D-glucose-5-*t* 6-phosphate. Carbohydrate Res. 3, 134–137.

BARRY, V. C. 1943. Regulated degradation of 1,3-polysaccharides. Nature 152, 537–538.

BERTRAND, G. 1896. Polyol oxidations by *Acetobacter*. Compt. Rend. 126, 762–765.

BEYNON, P. J., COLLINS, P. M., DOGANGES, P. T., and OVEREND, W. G. 1966. Oxidation of carbohydrate derivatives with ruthenium tetroxide. J. Chem. Soc. (C), 1131–1136.

BREDERICK, K. 1967. Synthesis of ketocellulose. Tetrahedron Letters, 695–698.

COLLINS, P. M. and OVEREND, W. G. 1965. A synthesis of 6-deoxy-L-talose. J. Chem. Soc., 1912–1918.

CREE, G. M. and PERLIN, A. S. 1968. Can. J. Biochem. 46, 765–770.

CREE, G. M., MACKIE, D. M., and PERLIN, A. S. 1969. Can. J. Chem. 47, 511–513.
DIMLER, R. J. 1952. 1,6-Anhydrohexofuranoses, a new class of hexosans. Advan. Carbohydrate Chem. 7, 37–52.
EPSTEIN, W. W., and SWEAT, F. W. 1966. Dimethyl sulfoxide oxidations. Chem. Rev. 67, 247–260.
GODMAN, J. L., and HORTON, D. 1968. Reaction of methyl sulfoxide-acetic anhydride with 1,2:3,4-di-O-isopropylidene-α-D-galactopyranose. Carbohydrate Res. 6, 229.
GOLDSCHMID, H. R., and PERLIN, A. S. 1960. Scission of sterically hindered vic-diols. Can. J. Chem. 38, 2280–2284.
GOLDSTEIN, I. J., HAY, G. W., LEWIS, B. A., and SMITH, F. 1965. Controlled degradation of polysaccharides by periodate oxidation, reduction, and hydrolysis. Methods Carbohydrate Chem. 5, 361–370.
GORIN, P. A. J., and PERLIN, A. S. 1961. Configuration of glycosidic linkages in oligosaccharides. IX. Can. J. Chem. 39, 2474–2485.
GREBNER, E. E., and FEINGOLD, D. S. 1965. D-Aldohexopyranoside dehydrogenase of Agrobacterium tumifaciens. Biochem. Biophys. Res. Commun. 19, 37–42.
GREENBERG, H., and PERLIN, A. S. 1966. Unpublished.
GUTHRIE, R. D. 1961. The "dialdehydes" from the periodate oxidation of carbohydrates. Advan. Carbohydrate Chem. 16, 105–158.
HANN, R. M., TILDEN, E. B., and HUDSON, C. S. 1938. The oxidation of sugar alcohols by Acetobacter suboxydans. J. Am. Chem. Soc. 60, 1201–1203.
HEYNS, K., and PAULSEN, H. 1962. Selective catalytic oxidation of carbohydrates employing platinum catalysts. Advan. Carbohydrate Chem. 17, 169–221.
HOGENKAMP, H. P. C. 1966. The synthesis of D-ribose-3-t. Carbohydrate Res. 3, 239–241.
HORTON, D., and JEWELL, J. S. 1967. Synthesis of 1,6-anhydro-β-D-talopyranose from derivatives of D-galactose and D-mannose. Carbohydrate Res. 5, 149–160.
HUDSON, B. G., and BARKER, R. 1967. The overoxidation of carbohydrates with sodium metaperiodate. J. Org. Chem. 32, 2101–2109.
ISBELL, H. S. 1962. Oxidation of aldose with bromine. J. Res. Natl. Bur. Std. 66A, 233–239.
ISBELL, H. S., and HUDSON, C. S. 1932. The course of the oxidation of the aldose sugars by bromine water. B.S. J. Res. 8, 327–338.
ISBELL, H. S., and PIGMAN, W W. 1933. The oxidation of α and β glucose and a study of the isomeric forms of the sugar in solution. B.S. J. Res. 10, 337–356.
ISBELL, H. S., and SNIEGOSKI, L. T. 1964. Tritium-labelled compounds X. Isotope effects in the oxidation of aldoses-1-t with bromine. J. Res. Natl. Bur. Std. 68A, 145–151.
JAMES, K., TATCHELL, A. R., and RAY, P. K. 1967. The synthesis of D-psicose derivatives using recent oxidative procedures. J. Chem. Soc. (C), 2681.

JONES, J. K. N., PERRY, M. B., and Turner, J. C. 1961. The synthesis of acetamido-deoxy ketoses by *Acetobacter suboxydans*. Can. J. Chem. *39*, 965–972.

JONES, J. K. N., *et al.* 1965. The oxidation of sugar acetals and thioacetals by *Acetobacter suboxydans*. Can. J. Chem. *43*, 955–959.

KEILIN, D., and HARTREE, E. F. 1949. Specificity of glucose oxidase (notatin). Biochem. J. *50*, 331–341.

KUZUHARA, H., and FLETCHER, H. G., JR. 1967. Syntheses with partially benzylated sugars. IX. J. Org. Chem. *32*, 2535–2537.

LEMIEUX, R. V., SUEMITSU, R., and GUNNER, S. W. 1968. A stereoselective synthesis of α-D-glucopyranosides. Can. J. Chem. *46*, 1040–1041.

LINDBERG, B., and THEANDER, O. 1959. Amino-deoxy- and deoxy-sugars from methyl 3-oxo-β-D-glucopyranoside. Acta Chem. Scand. *13*, 1226–1230.

MACKIE, W., and PERLIN, A. S. 1965A. Nuclear magnetic resonance spectral observations on the glycol-scission of deuterated D-glucose. Can. J. Chem. *43*, 2645–2651.

MACKIE, W., and PERLIN, A. S. 1965B. 1,2-O-isopropylidene-α-D-glucofuranose-5-*d* and -5,6,6-*d₃*. Can J. Chem. *43*, 2921–2924.

McDONALD, E. J. 1967. A new synthesis of D-psicose (D-*ribo*-hexulose). Carbohydrate Res. *5*, 106–108.

McGEE, P. A., *et al.* 1947. Investigation of the properties of cellulose oxidized by nitrogen dioxide. V. J. Am. Chem. Soc. *69*, 355–361.

MEHLTRETTER, C. L. 1953. The chemical synthesis of D-glucuronic acid. Advan. Carbohydrate Chem. *8*, 231–249.

ONODERA, K., HIRANO, S., and KASHIMURA, N. 1968. Oxidation of carbohydrates with methyl sulfoxide containing phosphorus pentaoxide. I. Synthesis of some aldosuloses and aldosiduloses. Carbohydrate Res. *6*, 276.

OVEREND, W. G. 1963. Recent developments in the chemistry of carbohydrates. Chem. and Ind. (London), 342.

PARIKH, J. R., and DOERING, W. von E. 1967. Sulfur trioxide in the oxidation of alcohols by dimethyl sulfoxide. J. Am. Chem. Soc. *89*, 5505–5507.

PARIKH, V. M., and JONES, J. K. N. 1965. Oxidation of sugars with ruthenium dioxide-sodium periodate. Can J. Chem. *43*, 3452.

PARRISH, F. W., PERLIN, A. S., and REESE, E. T. 1960. Selective enzymolysis of poly-β-D-glucans, and the structure of the polymers. Can. J. Chem. *38*, 2094–2104.

PERLIN, A. S. 1964. Ring contraction during the lead tetraacetate oxidation of reducing sugars. Can. J. Chem. *42*, 2365–2374.

PERLIN, A. S. 1966. Nuclear magnetic resonance spectra of glycol-cleavage oxidation products of methyl aldopentopyranosides. Can. J. Chem. *44*, 1757–1764.

PFITZNER, K. E., and MOFFATT, J. G. 1963. The synthesis of nucleoside-5'-carboxyl aldehydes. J. Am. Chem. Soc. *85*, 3027.

PFITZNER, K. E., and MOFFATT, J. G. 1965. Sulfoxide-carbodiimide reactions. I. A facile oxidation of alcohols. J. Am. Chem. Soc. *87*, 5661–5670.

ROGERS, J. K., and THOMPSON, N. S. 1968. The oxidation of D-galactopyranosyl residues of polysaccharides to D-galactopyranosyluronic acid residues. Carbohydrate Res. *7*, 66–75.

Szabo, P., and Szabo, L.   1967.   The reaction of malonaldehyde with perio-
    date, and its application to the determination of vicinal diols in deoxy
    sugar derivatives.   Carbohydrate Res. *4*, 206–224.
Theander, O.   1957.   Oxidation of glycosides V.   Acta Chem. Scand. *11*,
    1557–1564.
Theander, O.   1962.   Dicarbonyl carbohydrates.   Advan. Carbohydrate Res.
    17, 223–229.
Theander, O.   1964.   1,2:5,6-D-*O*-isopropylidene derivatives of D-gluco-
    hexodialdose and D-*ribo*-hexose-3-ulose.   Acta. Chem. Scand. *18*, 2209.
Tong, G. L., Lee, W. W., and Goodman, L. 1967.   Synthesis of some
    3-*O*-methyl purine ribonucleosides.   J. Org. Chem. *32*, 1984–1986.
Tulloch, A. P., Hill, A., and Spencer, J. F. T.   1967.   A new type of
    macrocyclic lactone from *Torulopsis apicola*.   Chem. Commun. (London),
    584–586.
Williams, D. T., and Jones, J. K. N. 1967.   Further experiments on the
    oxidation of sugar acetals and thioacetals by *Acetobacter suboxydans*.   Can.
    J. Chem. *45*, 741–744.
Zitko, V., and Bishop, C. T.   1966.   Oxidation of polysaccharides by lead
    tetraacetate in dimethylsulfoxide.   Can. J. Chem. *44*, 1749–1756.

CHAPTER 12

Thelma M. Reynolds

Nonenzymic Browning
Sugar-Amine Interactions

INTRODUCTION

The interaction between a reducing sugar and an amine initiates a sequence of reactions whose end-product is a brown pigment. The most important reactions leading to nonenzymic browning in foods are those between aldoses and free amino acids, or the free amino groups of proteins. The rate of the initial reaction increases as the water content of the system decreases. The effects of sugar-amine interactions are therefore most pronounced in dried and concentrated foods. Relatively massive reactions may occur and a high proportion of an important component may be involved; for example, the lysine residues in dried milk may be rendered indigestible by reaction with lactose. In situations where the supply of protein is critical for nutritional welfare, the effect of nonenzymic browning on the biological value of foods is likely to be more important than its effect on palatability.

The deliberate induction of nonenzymic browning in order to produce desirable odors and flavors is rather different. The desirable products are formed in trace amounts, and the flavors are usually spoiled if the browning goes too far. Even so, nutritional values are sometimes endangered.

Flavors derived from nonenzymic browning reactions were discussed by Hodge (1967) at the 4th Symposium on Foods held at Corvallis. This aspect of nonenzymic browning is therefore given second place, although some reactions that yield flavoring compounds are discussed.

Reactants

Reducing sugars that may be involved in nonenzymic browning reactions include aldoses (pentoses, hexoses, or disaccharides), uronic acids, and ketoses. The interaction between an aldose, or uronic acid, and any type of primary or secondary amine (or amino acid) may lead to browning, although there is a considerable variation in the reactivities of both sugars and amines. Ketoses react with aromatic amines, but secondary reactions leading to browning do not occur; however, the reaction between ketoses and other types of amine (and amino acid) leads to browning.

219

Five Steps to Nonenzymic Browning

Five steps can be recognized in the sequence of reactions leading to browning, as follows:— (1) the formation of a glycosylamine, a reversible reaction; (2) the rearrangement of the glycosylamine to a ketoseamine or aldoseamine; (3) the formation of a diketoseamine or of a diamino-sugar; (4) the degradation of the amino-sugar, by the loss of one or more moles of water, to amino or nonamino intermediates, and the reaction of these intermediates with inhibitors of browning (if present); (5) condensation of the amine with intermediates formed in Step 4, followed by polymerization to brown pigments.

This sequence of reactions has been deduced from the preparation and properties of pure compounds, from their identification in foods, and from qualitative and quantitative studies of changes in model systems. Reaction mechanisms have been proposed for Steps 1 to 4, but the mechanism of the formation of brown pigments has not been elucidated.

The nonamino intermediates of Step 4 can also be formed from sugars by the action of acids, alkalies, and heat, but the sugar-amine reactions of Steps 1 to 3 lead to the formation of these intermediates under mild conditions.

Reviews

The chemistry of nonenzymic browning reactions in foods and model systems, including advances over the period 1951 to 1965, has been reviewed by Danehy and Pigman (1951), Hodge (1953), Ellis (1959), Heyns and Paulsen (1960), and Reynolds (1963, 1965). Hodge (1967) and Streuli (1967) reviewed the production of flavors and odors from nonenzymic browning reactions. The following reviews are concerned with reactions involved in, or related to, nonenzymic browning:—Hodge (1955)—the Amadori rearrangement; Ellis and Honeyman (1955)—glycosylamines; Anet (1964)—3-deoxyglycosuloses and the degradation of carbohydrates; Paulsen (1966)—carbohydrates containing nitrogen or sulfur in the "hemiacetal" ring.

GLYCOSYLAMINES AND REARRANGEMENT PRODUCTS

Formation of Glycosylamines

Glycosylamines are commonly prepared by dissolving the sugar in the amine and heating, or by heating methanolic solutions of the reactants (Ellis and Honeyman 1955; Reynolds 1963, 1965). For example,

D-glucose (1) reacts with ethylamine (2) to give D-glucosylethylamine (3) and water. The equilibrium, for any given amine, depends on the

$$(1) \qquad\qquad (2) \qquad\qquad\qquad (3)$$

water content of the system; very little glycosylamine will be formed in dilute aqueous solutions. In general, aldoses react with amines more readily than ketoses, and strong bases react more readily than weak bases. Isbell and Frush (1958) proposed a mechanism for the reaction.

Rearrangement of Aldosylamines

In the presence of an acidic catalyst, aldosylamines undergo the Amadori rearrangement to a 1-amino-1-deoxyketose (or ketoseamine); alternatively, the aldose and amine, in the presence of a catalyst, give a ketoseamine. D-Glucose reacts with glycine to give 1-deoxy-1-glycino-D-fructose (or D-fructoseglycine) (4), the amino acid, itself, acting as

$$(4) \qquad\qquad\qquad (5)$$

the catalyst. Most Amadori products crystallize in the β-D-pyranose form, as shown (4). The furanose ring-form has not been observed in ketoseamines. Compounds like 1-deoxy-1-glycino-D-*threo*-pentulose (5), formed from D-xylose and glycine, crystallize in the open-chain form and show a carbonyl band in the infrared.

Aromatic amines yield Amadori products with ease in the presence of catalytic amounts of hydrochloric or acetic acid; exceptions are aromatic amines whose basicity has been lowered by substituents. On the other hand, alkylamines, piperidine, and other strongly basic amines, yield Amadori products with difficulty and molar quantities of a catalyst, such as an organic acid, are used to obtain good yields (see Hodge 1955; Hodge and Fisher 1963A; Reynolds 1963, 1965).

Reaction Between Uronic Acids and Amines

In a series of papers over the period 1958 to 1962, Heyns and co-workers described the preparation of a range of Amadori products from the reaction between glucuronic acid (6) and amines (see Reynolds 1965). These compounds, which are 6-amino-6-deoxy-D-*lyxo*-5-hex-ulosonic acids, crystallized in either an open-chain form (7) or, unlike the ketoseamines, in a furanose ring-form (8), depending on the type of amine.

```
      CHO                 CH₂NRR'
       |                    |
      HCOH                 CO
       |                    |                         HOOC    O
      HOCH                 HOCH                             \ / \   OH
       |                    |                          HO   X
      HCOH                 HCOH                         |       CH₂NRR'
       |                    |                     HO
      HCOH                 HCOH
       |                    |
      COOH                 COOH

      (6)            R = H, R' = phenyl       R = H, R' = butyl

                     NRR' = N-(amino acid)    NRR' = piperidino
                              (7)                    (8)
```

Galacturonic acid was more reactive than glucuronic acid, but excessive browning prevented the isolation of Amadori products in some cases, including the reaction with amino acids. Glucuronic acid was more reactive than glucose.

Rearrangement of Ketosylamines

Ketosylamines can be rearranged in the presence of an acidic catalyst to 2-amino-2-deoxyaldoses (or aldoseamines). The reaction is analogous to the Amadori rearrangement of aldosylamines and has been named the Heyns rearrangement. This reaction introduces a new asymmetric center at C-2, so that fructose and an amine can give a 2-amino-2-deoxyglucose and a 2-amino-2-deoxymannose. However, the nature of the products of the Heyns rearrangement depends on the ease with which it occurs. When the rearrangement is difficult ketose-amines are formed, or even isomeric ketoses. The products isolated from the reaction between D-fructose and different types of amine were as follows:—(1) ammonia, primary alkylamines, and ω-amino acids gave 2-amino-2-deoxyglucoses; (2) α-amino acids gave mixtures of glucose-amino, mannoseamino, and fructoseamino acids, in proportions varying with the structure of the amino acid and the reaction conditions; (3) piperidine gave mainly fructosepiperidine; (4) other strongly basic secondary amines gave D-psicose. Ketosylamines derived from aromatic amines could not be rearranged (for references see Reynolds 1965).

Preparation of Rearrangement Products

In all cases a system with a low water content was used, and some-times anhydrous conditions were necessary. The reaction conditions required adjustment even for compounds of the same type. Rearrange-ment products have been prepared from liquid amines by (1) dis-solving the sugar in the amine to form the glycosylamine, (2) adding acetic acid, in proportions up to one mole per mole amine, depending on the type of amine, (3) heating, and (4) adding a solvent to crystal-lize the product. Alternatively, the sugar, amine, and catalyst were dissolved in a suitable solvent.

No catalyst was required in the case of amino acids. The sugar and amino acid were heated together in methanolic solution, or in syrups containing about ten per cent of water, with the addition of sodium bisulfite to inhibit browning. The ketoseamino acids were isolated by elution or displacement chromatography on columns of cation exchange resins.

Hodge and Fisher (1963A) described the preparation of Amadori products; for references to the preparation of ketoseamines see Reynolds (1963, 1965), for the preparation of aldoseamines and aminohexulosonic acids see Reynolds (1965).

Mechanism of the Amadori and Heyns Rearrangements

Weygand (1940) proposed a mechanism for the Amadori rearrange-ment that was later elaborated, with one modification, by Isbell and Frush (1958) (Fig. 73). The acidic catalyst adds on to the ring oxygen of the aldosylamine (9). Then a flow of electrons from the nitrogen atom of (10) promotes the formation of a Schiff's base (11); this step is inhibited with very weak bases by the strong attraction of the sub-stituents (RR') for the electrons of the nitrogen atom. The electronic shifts indicated in (11) lead to the formation of an enolic amine (12); this step is difficult with strong bases, because the substituents (RR') have little attraction for the electrons of the nitrogen atom. However,

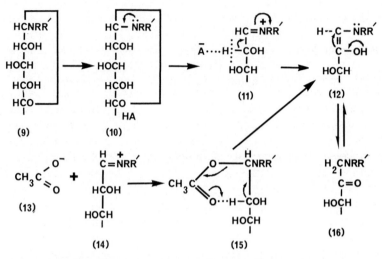

FIG. 73. MECHANISM OF THE AMADORI REARRANGEMENT

good yields of Amadori products were obtained from strong bases in the presence of molar proportions of an organic acid. To explain this exceptionally strong catalytic effect, Isbell and Frush (1958) suggested that the carboxylate ion (13) and the imonium ion (14) combined to form a transient intermediate (15); an intramolecular decomposition, as shown, would liberate the carboxylic acid and produce the enolic amine (12). By a tautomeric shift, as indicated, the enolic amine (12) gives a ketoseamine (16).

An alternative mechanism for the Amadori rearrangement was proposed by Micheel and Dijong (1962), but Palm and Simon (1965) have shown that it is not valid for aldosylamines with free hydroxyl groups.

The mechanism of the Heyns rearrangement of a ketosylamine to an aldoseamine involves intermediates similar to those shown in Fig. 73. In order to account for the formation of a ketoseamine from a ketosylamine, Heyns et al. (1961) suggested that the aldoseamine (17) was converted to an aldosylamine (18); an Amadori rearrangement of (18)

CHOH	CHNRR'	CH_2NRR'
CHNRR'	CHNRR'	C = NRR'
CHOH	CHOH	CHOH
(17)	(18)	(19)

would give (19), from which a ketoseamine would be formed by hydrolysis (see Reynolds 1965).

Properties of Rearrangement Products

Ketoseamines and aminohexulosonic acids are stable in aqueous solution and in cold acid. They decompose rapidly in cold alkali, and consequently readily reduce cold alkaline ferricyanide and oxidation-reduction indicators, such as triphenyltetrazolium chloride and o-dinitrobenzene. These reactions have been used to determine ketoseamines, and to detect both types of compound on paper chromatograms. Aldoseamines are less reactive; their reducing properties resemble those of glucose, and they reduce hot alkaline ferricyanide.

When heated with dilute acid, ketoseamines gave the parent amine and degradation products. The aldoseamines, with $2N$ acetic acid, regenerated fructose and the amine, usually at $100°–115°C$.

All three types of rearrangement product reacted on paper chromatograms with alkaline silver nitrate and ninhydrin (ninhydrin-acetic acid in the case of aldoseamines). On cation exchange resins, the rearrangement products were readily separated from the parent base.

Factors Affecting Browning and Formation of Ketoseamines

Various factors have the same effect on nonenzymic browning and on the formation of ketoseamines. The activation energy of the formation of fructoseglycine was 26 Cal (Reynolds 1959), and the same value was calculated for the browning of dried apricots (Stadtman et al. 1946A) and of potatoes with 13–17% water content (Hendel et al. 1955).

An increase in pH increased the rate of formation of fructoseglycine and of browning, and both processes were assisted by acid-base catalysis.

The rate of browning and of formation of ketoseamines increased as the water content of the system decreased to about 18%, although the maximum rate of formation of ketoseamines occurred at lower water contents.

The formation of ketoseamines has been demonstrated in browning reactions, both in foods (e.g., Anet and Reynolds 1957) and in model systems. However, the stability of these compounds is such that it is necessary to postulate a more reactive intermediate; for example, some less stable form of the ketoseamine (cf. Hodge 1955), or an unstable derivative such as the diketoseamines discussed below.

Diketoseamines

A ketoseamine (20) derived from a primary amine can react with another molecule of aldose and the product can undergo an Amadori

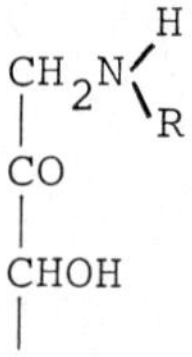

$$
\begin{array}{cc}
\begin{array}{l}
\text{CH}_2\text{N}\!\!\begin{array}{l}\text{H}\\ \text{R}\end{array} \\
\mid \\
\text{CO} \\
\mid \\
\text{CHOH} \\
\mid
\end{array}
&
\begin{array}{l}
\text{CH}_2\text{-N-CH}_2 \\
\mid \quad\ \mid \quad\ \mid \\
\text{CO} \quad \text{R} \quad \text{CO} \\
\mid \qquad\quad \mid \\
\text{CHOH} \quad \text{CHOH} \\
\mid \qquad\quad \mid
\end{array} \\
(20) & (21)
\end{array}
$$

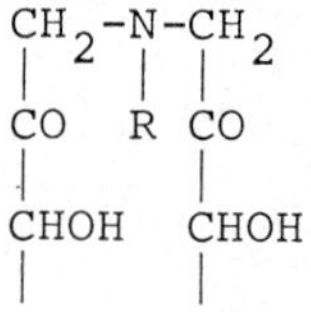

rearrangement to give a diketoseamine (21). The only crystalline compound of this type is difructoseglycine (22) (1,1'-(carboxymethyl-amino)bis-[1-deoxy-D-fructose]) (Anet 1959A). Difructoseglycine (22) was prepared by heating fructoseglycine (4), or glycine, with glu-

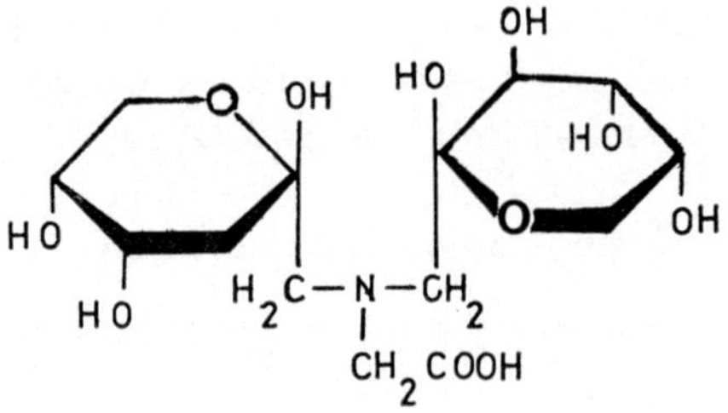

(22)

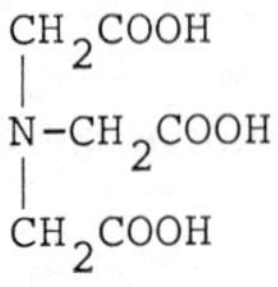

$$
\begin{array}{l}
\text{CH}_2\text{COOH} \\
\mid \\
\text{N-CH}_2\text{COOH} \\
\mid \\
\text{CH}_2\text{COOH}
\end{array}
$$

(23)

cose or mannose in a syrup containing 10–20% water. The products were separated on columns of a cation-exchange resin, difructoseglycine (22) emerging before fructoseglycine (4); the best yields, up to 20%, of diketoseamine were obtained when sodium hydroxide was added so that the pH of the reaction mixture did not fall below 8 (difructoseglycine being more stable at pH 8). Other primary amines, and other aldoses, gave diketoseamines, but they did not crystallize.

The structure assigned to difructoseglycine (22) was based on the following evidence:—(1) on periodate oxidation 8 molecules of reagent were consumed, 2 molecules of formaldehyde were formed, and tri(carboxymethyl)-amine (23) was isolated at the end of the reaction; (2) glucose and mannose gave the same compound; (3) hydrolysis gave no glucose or mannose, only degradation products; (4) there was no carbonyl band in the infrared.

The reactions of diketoseamines were the same as those described for ketoseamines, with the notable exception that diketoseamines decomposed readily in aqueous solution, as described below.

DEGRADATION OF KETOSEAMINES VIA 1,2-ENOLIZATION

In the mechanism shown for the Amadori rearrangement (Fig. 73), the ketoseamine (16) was formed, by a tautomeric shift, from the enolic amine (12). It will be shown that the reactions discussed in this section are initiated by the reverse shift, which is the 1,2-enolization of a ketoseamine.

Decomposition of Difructoseglycine in Water or Buffer

Although difructoseglycine (22) was sufficiently stable to be isolated by ion exchange chromatography and recrystallized from water, it was completely decomposed when an aqueous solution was heated at 100°C for 5 min; the products were fructoseglycine (4), "3-deoxyglucosulose," and unsaturated osuloses. The rate of decomposition of difructoseglycine (22) was higher at pH 5.5 than at pH 3.5 or pH 8; at pH 5.5 the reaction had an activation energy of 24 Cal (Anet 1959A).

Browning Reaction of Difructoseglycine with Glycine

Anet (1959B) found that the brown color (absorbance at 440 nm) produced in 9 days at 50°C by difructoseglycine and glycine (1 :2 moles, pH 5.5) was 15 times that given by fructoseglycine under the same conditions. Since difructoseglycine was completely decomposed in 17 hr at 50°C and pH 5.5, the browning was produced by the reaction between osuloses and glycine. Fructoseglycine decomposes in the same way as difructoseglycine, but the reaction is much slower, which explains the different rates of browning.

Degradation Products from Difructoseglycine

The nonionic degradation products from difructoseglycine (22) were separated on cellulose columns, after the removal of fructoseglycine (4) on a cation exchanger. The main product (80–90%) was a 3-deoxyosulose (24) (Anet 1960). Two unsaturated osuloses were formed; the *cis* form (25) in 5% yield, the *trans* form (26) in 1–10% yield depending on the pH (Anet 1962). There was a small yield (1–2%) of hydroxymethylfurfural (27), possibly largely formed by decomposition of the *cis* compound (25) on the cation exchange resin.

```
  CHO          CHO          CHO          CHO
   |            |            |            |
  CO           CO           CO           C ┐
   |            |            |           ‖  │
  CH2          CH           CH           CH │
   |           ‖            ‖            |  │ O
  HCOH         CH           HC           CH │
   |           |            |            ‖  │
  HCOH         HCOH         HCOH         C ┘
   |            |            |            |
  CH2OH        CH2OH        CH2OH        CH2OH
  (24)         (25)         (26)         (27)
```

The 3-deoxyosulose (24) was independently isolated by Kato (1960), from the action of acetic acid on glucosylbutylamine, and by Machell and Richards (1960), from the action of alkali on 3-O-benzylglucose, the 3-deoxyosulose (24) being an intermediate in the formation of metasaccharinic acids. The yields of 3-deoxyosulose from these reactions were low.

The *cis* compound (25) was also prepared through the action of limewater on a partially substituted glucose. When the sugar was methylated in the 2-position the formation of a 3-deoxyosulose (24) was blocked. The action of limewater at 50°C on 2,3-di-O-methyl-D-glucose (28) gave the 2-methyl ether (29) of the enolic form (30) of the 3-deoxyosulose (24). The action of 0.1N acid on the ether (29) gave the *cis* unsaturated osulose (25). Better yields of (25) were obtained from 2,3,4-tri-O-methyl-D-glucose (Anet 1963).

```
      CHO              CHO              CHO
       |                |                |
      HCOCH3           COCH3            COH
       |               ‖                ‖
   CH3OCH              CH               CH
       |                |                |
      HCOH             HCOH             HCOH
       |                |                |
      HCOH             HCOH             HCOH
       |                |                |
      CH2OH            CH2OH            CH2OH
      (28)             (29)             (30)
```

Properties of 3-Deoxy- and Unsaturated Osuloses

The 3-deoxyosulose (24) and the *trans* compound (26) were amorphous solids, the *cis* compound (25) was a liquid. All three compounds gave crystalline osazones without the aid of acids, and the osazones gave crystalline acetates. On paper chromatograms, the osuloses reacted with naphthoresorcinol and alkaline silver nitrate. The unsaturated osuloses gave single spots, but the 3-deoxyosulose gave a characteristic pattern of three spots, the main one greatly elongated, showing the existence of several forms.

The osuloses were converted readily to hydroxymethylfurfural (HMF) (27) by heating with dilute acids. The times for half-conversion to HMF at 100°C were as follows:—for the *cis* compound (25), 12 min in 0.03N acetic acid; for the *trans* compound (26) and the 3-deoxyosulose (24), 8 hr in 0.03N acetic acid or 2 hr in 2N acetic acid. The 3-deoxyosulose (24) was not converted to the *trans* compound (26), but it was converted to the *cis* compound (25) and thence to HMF (27) (Anet 1962).

It was shown by NMR spectroscopy (Anet 1962) that the *cis* compound (25) existed in the α,β-pyranose form (31) (3,4-dideoxy-D-glycero-hex-*cis*-3-enepyranosulose), and the *trans* compound (26) as an

(31)   (32)   (33)

open-chain hydrate (32) (3,4-dideoxy-D-*glycero*-hex-*trans*-3-enosulose hydrate). The 3-deoxyosulose (24) probably exists mainly in the α,β-pyranose form (33) (3-deoxy-D-*erythro*-hexopyranosulose).

Mechanism of Formation of Osuloses

A mechanism for the formation of osuloses from ketoseamines is shown in Fig. 74; it is based on mechanisms proposed by Isbell (1944), Anet (1960), Kato (1960), and Anet (1964).

Under acidic conditions the ketoseamine is in the salt form (34). The withdrawal of electrons from C-1 by the positively charged nitrogen atom assists the formation of the 1,2-enolamine (35); recent work with labeled compounds has shown that the enolization is reversible

(Simon and Heubach 1965; Anet 1968). The β-elimination of the hydroxyl from C-3 of (35), to give a Schiff's base (36), is assisted when the enolamine (35) is in the free base form. Hydrolysis of the Schiff's base (36) gives the enolic form (39) of a 3-deoxyosulose (38). Elimination of the hydroxyl from C-4 of (36) gives a Schiff's base (37); hydrolysis of (37) gives an unsaturated osulose (40), which will also be formed by elimination of the hydroxyl from C-4 of (39). Similarly, elimination of the hydroxyl from C-5 of the base (37) gives a Schiff's base (not shown) of a furfural (41).

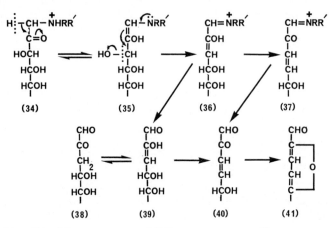

FIG. 74. MECHANISM OF 1,2-ENOLIZATION OF KETOSEAMINES AND OF FORMATION OF OSULOSES AND FURFURALS

Anet (1964) pointed out that the enolization (34) to (35), and the β-elimination (35) to (36), would proceed most easily with weak bases, and under conditions which gave approximately equal proportions of the salt and free base forms. The decomposition of difructose-glycine (22) gave the 3-deoxyosulose (24) in 80–90% yield, so that, in that instance, the main reaction was (36) → (39) → (38). The Schiff's bases (36) and (37), and the hydrolysis products, (39) and (40), could be *cis* or *trans* forms. However, the 3-deoxyosulose (24), when heated with dilute acids, gave the *cis* unsaturated osulose (25), but not the *trans* compound (26) (Anet 1962) (see section above on Properties of 3-Deoxy- and Unsaturated Osuloses). Therefore, the *trans* compound (26), isolated from the decomposition of difructose-glycine (22), was formed from the base (37), or from the 2-enose (39) before de-enolization of (39) to (38) occurred (see Anet 1964 for further discussion of these relationships).

The mechanism shown in Fig. 74 for the formation of 3-deoxyosulose (38) was confirmed by comparisons of the NMR spectra obtained when difructoseglycine was decomposed in water and deuterium oxide (Anet 1968). When a solution of difructoseglycine (42) in deuterium oxide was heated for 5 min at 95°C, the signals for the hydrogens of the C-1

$$
\begin{array}{cccc}
\text{CH}_2\text{COOH} & & \text{CH}_2\text{COOH} & \\
| & & | & \\
\text{CH}_2 - \text{N} - \text{CH}_2 & & \text{CD}_2 - \text{NH} & \text{DC} = \text{N.}\ddot{\text{N}}\text{HR} \\
| \qquad\qquad | & & | & | \\
\text{CO} \qquad\quad \text{CO} & & \text{CO} & \text{C} = \text{N.NHR} \\
| \qquad\qquad | & & | & | \\
\text{HOCH} \qquad \text{HOCH} & & \text{HOCH} & \text{DCH} \\
| \qquad\qquad | & & | & | \\
\text{HCOH} \qquad \text{HCOH} & & \text{HCOH} & \text{HCOH} \\
| \qquad\qquad | & & | & | \\
\text{HCOH} \qquad \text{HCOH} & & \text{HCOH} & \text{HCOH} \\
| \qquad\qquad | & & | & | \\
\text{CH}_2\text{OH} \quad\;\; \text{CH}_2\text{OH} & & \text{CH}_2\text{OH} & \text{CH}_2\text{OH}
\end{array}
$$

$$R = C_6H_3(NO_2)_2$$

(42)   (43)   (44)

methylene group of fructoseglycine, readily detected in water, disappeared. This showed that the deuterated compound (43) had been formed, and that the 1,2-enolization, (34) to (35), was reversible. The β-elimination of the hydroxyl on C-3, as in (35) to (36), was confirmed by the NMR spectra of an osazone of the 3-deoxyosulose (24) formed in the reaction. The NMR spectrum of the deuterated osazone (44) showed no signal for a hydrogen on C-1, and the signal for the hydrogens on C-3 was reduced by 50%.

When fructoseglycine was heated for 5 min at 95°C in deuterium oxide, only 50% of the deuterated form (43) was present, showing that the enolization was much slower than in difructoseglycine, and that both fructose residues in difructoseglycine (42) were highly deuterated before decomposition occurred. The NMR spectrum of fructoseglycine, after 45 min at 95°C, showed complete deuteration at C-1, but no other detectable change.

REACTIONS BETWEEN GLUCOSE AND BISULFITE

Sulfur dioxide and bisulfites are used to inhibit browning in dried fruits and vegetables. The reactions of sulfites in foods and model systems were reviewed by Reynolds (1965). Some reactions of bisulfites are discussed below.

Reactions at pH 4.5

At room temperature, sodium and potassium bisulfites form addition compounds with aldoses. These compounds crystallize when saturated aqueous solutions of an aldose and a bisulfite (pH about 4.5) are mixed. Braverman (1953) used sodium bisulfite; Ingles (1959A) found that the potassium bisulfite compounds crystallized more readily. Glucose, in the open-chain form, gives glucose potassium sulfonate (45). The

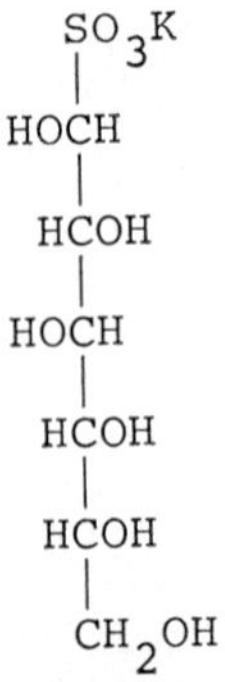

$$
\begin{array}{c}
SO_3K \\
| \\
HOCH \\
| \\
HCOH \\
| \\
HOCH \\
| \\
HCOH \\
| \\
HCOH \\
| \\
CH_2OH \\
(45)
\end{array}
$$

configuration at the new asymmetric center at C-1 was elucidated (Ingles 1959B); (45) is D-*glycero*-D-*ido*-1,2,3,4,5,6-hexahydroxyhexyl potassium sulfonate. The whole of the sulfite is released from these compounds as sulfur dioxide in the Monier-Williams determination.

The products of the reaction between glucose and bisulfites at 100°C were gluconic acid, glucose-6-sulfate, inorganic sulfate, and sulfur. These reactions occur because, at this pH and at this higher temperature, the bisulfite disproportionates.

Reactions at pH 6.5

The reaction at 100°C between glucose and an equimolar mixture of sodium bisulfite and sodium sulfite (pH 6.5) gave a mixture of acids. One acid, which was isolated as a crystalline brucine salt, was shown to be a 4-sulfohexosulose (46) (3,4-dideoxy-4-sulfo-D-*glycero*-hexosulose) (Ingles 1962). The sulfohexosulose (46) gave a crystalline osazone, and the osazone gave a crystalline diacetate. Ingles (1962) suggested that the 4-sulfohexosulose (46) was formed from the enolic form (30) of the 3-deoxyosulose (24) by a replacement reaction, or by the addition of bisulfite ion to the *cis* unsaturated osulose (25).

$$
\begin{array}{ccc}
\text{CHO} & \text{COOH} & \text{COOH} \\
| & | & | \\
\text{CO} & \text{CHOH} & \text{CO} \\
| & | & | \\
\text{CH}_2 & \text{CH}_2 & \text{CH}_2 \\
| & | & | \\
\text{CHSO}_3\text{H} & \text{CHSO}_3\text{H} & \text{CHSO}_3\text{H} \\
| & | & | \\
\text{HCOH} & \text{HCOH} & \text{HCOH} \\
| & | & | \\
\text{CH}_2\text{OH} & \text{CH}_2\text{OH} & \text{CH}_2\text{OH} \\
(46) & (47) & (48)
\end{array}
$$

Two carboxylic acids with a 4-sulfo group (47, 48) were also isolated from the reaction, and analogous products were obtained from xylose (Reynolds 1965).

Heating with hydrochloric acid, as in the Monier-Williams determination, does not release sulfur dioxide from these sulfonic acids.

Synthesis of a 4-Sulfohexosulose

The 4-sulfohexosulose (46), described above, was synthesized by Anet and Ingles (1964). A 10% solution of a 3-deoxyhex-2-enose (49), in water saturated with sulfur dioxide, was allowed to stand for 24 hr at 20°C. A bisulfite addition compound was formed, presumed to be

$$
\begin{array}{cc}
\text{CHO} & \text{SO}_3\text{H} \\
| & | \\
\text{COCH}_3 & \text{HOCH} \\
\| & | \\
\text{CH} & \text{CO} \\
| & | \\
\text{HCOCH}_3 & \text{CH}_2 \\
| & | \\
\text{HCOH} & \text{CHSO}_3\text{H} \\
| & | \\
\text{CH}_2\text{OH} & \text{HCOH} \\
 & | \\
 & \text{CH}_2\text{OH} \\
(49) & (50)
\end{array}
$$

the addition compound (50); steam distillation gave the 4-sulfohexosulose (46), which was identified by means of crystalline derivatives. These operations comprise a synthesis of the 4-sulfohexosulose (46) from the *cis* unsaturated osulose (25) that is the first product of the action of acid (in this case sulfurous acid) on the hex-2-enose (49) (see section above on the Degradation Products of Difructoseglycine). The 4-sulfohexosulose (46) was also synthesized by heating an aqueous solution of the 3-deoxyosulose (24) and sodium sulfite (pH 6.5) at 100°C for 4 hr (Lindberg et al. 1964).

Anet (1959B) showed that the addition of bisulfite (1 mole) to a mixture of difructoseglycine and glycine (1 : 2 moles) (see section above on Properties of 3-Deoxy- and Unsaturated Osuloses) prevented the formation of brown pigments. This was evidently due to the combination of bisulfite with osuloses to give the addition compound (50). This mechanism for the inhibition of browning bisulfite was proposed by Reynolds et al. (1962) and Anet and Ingles (1964), and discussed by Reynolds (1965).

The quantity of bisulfite that can be added to foods is limited by considerations of palatability and toxicity; this limits the period of inhibition, but the effect can still be due to the formation of an addition compound of a sulfohexosulose, such as (50), from the bisulfite available.

The 4-sulfohexosulose (46), being a dicarbonyl compound, could take part in browning reactions, but would be less reactive than the unsubstituted osuloses.

1,2-ENOLIZATION OF KETOSEAMINES IN FOODS AND MODEL SYSTEMS

It has been shown in the preceding sections, (1) that 1,2-enolization of ketoseamines leads to the formation of 3-deoxyosuloses, unsaturated osuloses, and furfurals (Fig. 74), (2) that these reactions proceed most readily with weak bases under conditions where both the salt and free base forms of the ketoseamine can exist, (3) that diketoseamines undergo these reactions with extreme ease, and (4) that bisulfite reacts with 3-deoxy- and unsaturated osuloses to give stable sulfonic acids. The occurrence of these reactions in foods and model systems is discussed below.

The formation of hydroxymethylfurfural (HMF) in dried and concentrated foods after heating or storage is well-known, and examples are given in many of the reviews listed in the introduction. The formation of 3-deoxyglycosuloses in soy sauce and miso was demonstrated by Kato (see Reynolds 1963).

A loss of glucose exceeding the loss of amino nitrogen, which would occur if diketoseamines were formed, was reported by Lea and Hannan (1950); a mixture of glucose and casein containing 4 moles of glucose per free amino nitrogen was stored at 37°C and 70% RH, and after 30 days the loss of glucose to amino nitrogen was 2.2 to 1. Hannan and Lea (1952) used a mixture of glucose and α-N-acetyl-ʟ-lysine (1.5 : 1 moles) to study the effect of the water content of the system on the loss of glucose and amino nitrogen (Table 19). The maximum loss of amino nitrogen occurred at 8% water content, but the maximum browning and loss of glucose occurred at 18% water content. This suggests that the maximum rate of formation of a diketoseamine occurs at a higher water content than the maximum rate of formation of a ketoseamine.

TABLE 19

EFFECT OF WATER CONTENT ON BROWNING AND LOSS OF GLUCOSE[1]

	20	40	60	80
Relative Humidity %	20	40	60	80
Water Content %	4.8	8.3	18.0	28.6
Amino nitrogen (% of initial)	3.7	6	12	37
Optical density[2] at 4500 Å	0.1	2.3	16.4	7.9
Ratio moles glucose lost/NH₂ groups lost	N.D.	1.08	1.37	1.27

Source: Hannan and Lea (1952).
[1] Reaction between α-N-acetyl-ʟ-lysine and glucose (1.5 moles/NH₂ group); 24 days at 37°C.
[2] For solution containing 1 mg original N/ml.

Reactions in Freeze-dried Fruit

Freeze-dried apricots and peaches were equilibrated at 70% RH (21% water content) and stored at 25°C. All the amino acids reacted with glucose to give fructoseamino acids, and difructoseamino acids formed from aspartic acid and asparagine were also detected (Anet and Reynolds 1957; Ingles and Reynolds 1958). Glucose, fructose, and sucrose formed monoesters with malic and citric acids (Anet and Reynolds 1957; Ingles and Reynolds 1959). The addition of phosphate and malate at pH 3.5 to 3.6 to syrups containing glucose and glycine, increased the rate of formation of fructoseglycine (Reynolds 1959). It was concluded that organic acids acted as catalysts of the glucose-amino acid reactions, but otherwise took little part in browning reactions when amino acids were present.

Reactions in Model Systems

The composition of model systems used in the work discussed below was related to that of some freeze-dried apricots (see preceding section). In the fruit, the proportions of glucose, amino acids, and organic acids were approximately 8 : 1 : 6 moles; the pH was 3.6 and the water content 21%. The model systems contained glucose, glycine, and citrate (or malate), $8:1:1$ moles; the water content was 23%, and the pH was adjusted to either pH 3.6 (citrate or malate), or pH 6.1 (citrate). The effects of bisulfite were studied by adding sodium bisulfite (1 mole/8 moles glucose) to mixtures buffered at pH 3.6. The reaction mixtures were held at 50°C, at which temperature they were mobile syrups. The components of the mixtures were determined by specific chemical methods; the results are being confirmed and extended by experiments with ^{14}C-labeled compounds.

Formation of Fructoseglycine and Loss of Glycine and Glucose.—At pH 3.6, the glycine content of the model system dropped to less than 0.2 moles in 10 days at 50°C, and was virtually zero after 20 days. The curve for fructoseglycine rose rapidly, flattened, and then fell slowly; the value after 10 days, 0.6 moles, was close to the maximum. The loss of glucose exceeded the loss of glycine, being 2 moles after 10 days, and nearly 4 moles after 30 days. These results suggest the formation of difructoseglycine. This diketoseamine was detected as small spots on paper chromatograms, but it did not accumulate. This showed that the rate of decomposition of difructoseglycine in this system was close to its rate of formation. (Reynolds *et al.*, unpublished; cf. Reynolds *et al.* 1962, Reynolds 1963).

Glucose Lost in Reactions with Glycine and Fructoseglycine.—The model systems containing glycine were those described above; exactly equivalent mixtures were prepared containing 1 mole of fructoseglycine in place of 1 mole each of glycine and glucose (glucose : fructoseglycine : citrate, $7:1:1$). At pH 3.6 the rate of loss of glucose was slightly higher with fructoseglycine than with glycine; at pH 6.1 the rates were the same. The loss of glucose after 15 days at 50°C was as follows:— at pH 3.6, 2.6 moles with glycine, 2.8 moles with fructoseglycine; at pH 6.1, 4.5 moles with glycine or fructoseglycine.

The formation of difructoseglycine in these systems is thus confirmed. It would be expected that the results obtained by Hannan and Lea (1952) (Table 19) would apply, at least qualitatively, to these systems. In that case, the rate of reaction of fructoseglycine with glucose, compared with that of glycine with glucose, would decrease as the water content decreased (Reynolds and Fenwick, unpublished).

Formation of Osuloses, Hydroxymethylfurfural and Brown Pigment.— The "osulose" was mainly 3-deoxyosulose (24), but the values obtained included any unsaturated osuloses present. The proportion of glucose converted to brown pigment was determined with ^{14}C-labeled glucose.

As would be expected from the rapid formation and decomposition of difructoseglycine, there was no lag in the formation of osulose. At pH 3.6, the osulose content reached a maximum of 0.9 moles after 18 to 21 days at 50°C. There was a lag in the formation of hydroxy-methylfurfural (HMF), as expected for a secondary product; the HMF content, at pH 3.6, was 0.6 moles after 20 days, and the curve was flattening at 30 days, when the experiment ended.

The curve for the brown pigment had a lag period similar to that shown by the HMF curve. However, the values fell on a straight line from 10 to 30 days; the pigment formed after 30 days at 50°C was derived from 1.3 moles of glucose (Reynolds and Fenwick, unpublished results).

The results outlined above are consistent with the view that the glucose residues in the pigment are derived mainly from unsaturated osuloses, which are, at first, formed directly from the decomposition of difructoseglycine, and then from the 3-deoxyosulose. Some HMF participates later, but it was estimated that less than 20% of the pigment was derived from HMF (Reynolds 1965).

Effect of Addition of Bisulfite.—The addition of bisulfite (1 mole/8 moles glucose) to the model system buffered at pH 3.6, reduced the rate of formation of fructoseglycine by less than 20%, and had little effect on the rate of formation of osulose, or the rate of loss of glucose. The effect of bisulfite on the formation of pigment is shown in Table 20, together with the loss of bisulfite (Reynolds and Fenwick, unpublished; see Reynolds 1965).

It is suggested above (see Synthesis of a 4-Sulfohexosulose) that the inhibition of browning by bisulfite could be due to the formation of the bisulfite addition compound (50) of the 4-sulfohexosulose (46). Since the formation of (50) from (46) is reversible, the presence of an excess

TABLE 20

EFFECT OF BISULFITE ON YIELD OF PIGMENT[1]

	Time Days at 50°C	Without Bisulfite[2]	With Bisulfite[3]
Absorbance at 440 nm[4]	14	350	0
	30	760	350
Pigment as moles glucose[5]	14	0.5	0
	30	1.3	0.5 (calc.)
Bisulfite – moles lost[6]	14	—	0.5
	30	—	0.8

[1] Reynolds and Fenwick unpublished.
[2] Glucose, glycine, citrate, 8 : 1 : 1 moles, pH 3.6, 23% water.
[3] As above, plus one mole bisulfite.
[4] For mixture diluted to one mole glycine (initial) per liter.
[5] Determined with ^{14}C-labeled glucose.
[6] Determined by Monier-Williams method

of osulose would lead to a decrease in the addition compound (50), and an increase in the sulfohexosulose (46).

The results in Table 20 support this hypothesis. The agreement between the deficit in pigment in the presence of bisulfite (0.5 moles after 14 days, 0.8 moles after 30 days) and the bisulfite lost suggests the formation of sulfohexosulose (46) in place of pigment. In the earlier stages, there was sufficient bisulfite present to form the addition compound (50).

Brown Pigments.—The brown pigments isolated from sugar-amine, or from aldehyde-amine, interactions contained nitrogen. The ratio of glucose to glycine residues was 1 : 1 in a pigment prepared by Wolfrom *et al.* (1953) from glucose and glycine (1 : 10 moles). Other pigments contained an excess of sugar, or aldehyde, residues. The ratio of glucose, or aldehyde, residues to amine residues in various pigments was as follows:—2-2.5 : 1 from glucose and glycine (1.7 : 1 moles) (see Reynolds 1963); 4 : 1 from acetaldehyde and ethylamine (1-4 : 1 moles) (Carson and Olcott 1954); 3 : 1 from furfural and butylamine (4 : 1 moles) (Reynolds unpublished). This characteristic was also found in pigments isolated by dialysis from glucose, glycine, and citrate (8 : 1 : 1 moles, pH 3.6 and 6.1). The ratio of glucose to glycine residues increased from 3 : 1 to 6 : 1 as the time of reaction was increased; the ratio of nondialysable to dialysable pigment also increased.

Infrared spectra of pigments isolated from the model systems containing glucose and glycine showed the presence of the following groups:— OH, CH_2, > C=O, —CH=CH—C=O and/or —CH=CH—CH=CH—, —COO⁻, —SO_3H (in pigments from sulfited systems). The pigments showed strong absorption in the ultraviolet, with indications of several overlapping peaks; there were no peaks in the visible region, and the absorption was much weaker than in the ultraviolet (Reynolds unpublished; see Reynolds 1963).

The pigments prepared from glucose, glycine, and citrate (8 : 1 : 1 moles, pH 3.6 and 6.1) contained 0.06 C-methyl groups per C_6 unit. Thus 3–6% of the glucose molecules involved in the formation of the pigment had been converted to compounds containing a C-methyl group; the actual percentage would depend upon the proportion of C-methyl compounds formed by rupture of the carbon chain of glucose.

None of the intermediates or products formed from the 1,2-enolization of ketoseamines (Fig. 74) contains a C-methyl group; nor is it easy to see how they might split to give compounds with C-methyl groups. The results discussed in the preceding sections showed the formation, in the model systems, of large quantities of products from the 1,2-enolization of ketoseamines. Therefore, it can be concluded that, in these systems, and by extension, in many types of foods, non-enzymic browning is essentially the result of the decomposition of ke-toseamines by 1,2-enolization.

DECOMPOSITION OF KETOSEAMINES *VIA* 2,3-ENOLIZATION

Isolation and Structure of Reductones

Ketoseamines derived from strongly basic secondary amines were degraded in basic media to crystalline "amino-hexose-reductones" (Hodge and Rist 1953; Hodge *et al.* 1963). The same reductones were formed from ketohexoses (Hodge *et al.* 1963), but there was a long lag period and the yield was lower (Simon and Heubach 1965). Various secondary amines were used, for example, piperidine, morpholine, dimethylamine, and N-benzylmethylamine (Hodge *et al.* 1963).

The best yields of reductone (about 35%) were obtained when an aldohexose was heated, in alcoholic solution, with 1–2 moles of the acetate salt of a secondary amine, with the addition of a tertiary amine to maintain basic conditions; the ketoseamine could be isolated in good yield after 1–2 hr, and the reductone was isolated after 24–48 hr.

The structure of the "amino-hexose-reductones," shown in Fig. 75 in one of the betaine forms (58), was established by Weygand and Hodge (Weygand *et al.* 1959) (discussed by Reynolds 1963).

A second reductone was isolated from the reaction between N-benzyl-methylamine and glucose. After crystallization of the "amino-hexose-reductone" (58; R = benzyl, R′ = methyl) (35% yield), the mother liquor was concentrated; the amino reductone (63; R = benzyl, R′ = methyl; keto form) crystallized in 5% yield (Hodge *et al.* 1963).

Preparation of Reductones from Labeled Glucose

Simon (1962) showed, by the preparation and degradation of ^{14}C-labeled reductone (58), that 25% of the C-methyl of (58) was derived from C-1 of glucose, and 72% from C-6. By the same methods, Simon and Heubach (1965) showed that the reductone (63) was derived from C-3 to C-6 of glucose.

Simon and Heubach (1965) also studied the enolization of fructose-piperidine. Glucose-3-T and piperidine acetate were heated in etha-nolic solution, and fructosepiperidine was isolated at intervals; the T-activity of the fructosepiperidine increased with time, showing that 2,3-enolization was not reversible. On the other hand, when fructose-

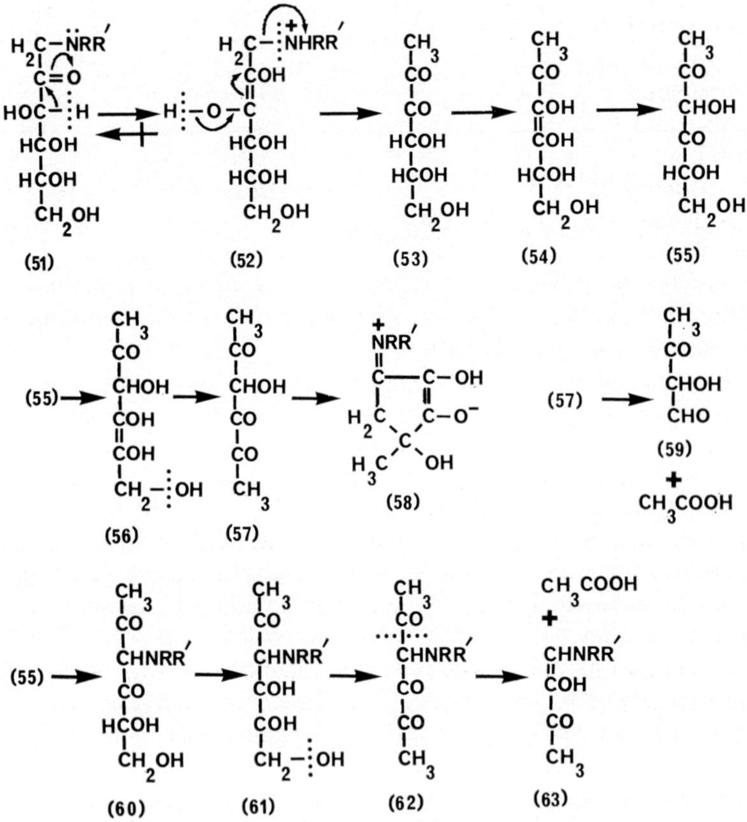

$$
\begin{array}{ccccc}
\text{(51)} & \text{(52)} & \text{(53)} & \text{(54)} & \text{(55)} \\
\begin{array}{l}
H_2C-NRR' \\
C=O \\
HOC-H \\
HCOH \\
HCOH \\
CH_2OH
\end{array}
&
\begin{array}{l}
H_2C-NHRR'^{+} \\
COH \\
C \\
HCOH \\
HCOH \\
CH_2OH
\end{array}
&
\begin{array}{l}
CH_3 \\
CO \\
CO \\
HCOH \\
HCOH \\
CH_2OH
\end{array}
&
\begin{array}{l}
CH_3 \\
CO \\
COH \\
COH \\
HCOH \\
CH_2OH
\end{array}
&
\begin{array}{l}
CH_3 \\
CO \\
CHOH \\
CO \\
HCOH \\
CH_2OH
\end{array}
\end{array}
$$

$$
\text{(55)} \rightarrow
\begin{array}{l}
CH_3 \\
CO \\
CHOH \\
COH \\
COH \\
CH_2-OH
\end{array}
\;(56) \rightarrow
\begin{array}{l}
CH_3 \\
CO \\
CHOH \\
CO \\
CO \\
CH_3
\end{array}
\;(57) \rightarrow
\begin{array}{l}
NRR'^{+} \\
C-C-OH \\
H_2C \quad C-O^{-} \\
\quad C \\
H_3C \quad OH
\end{array}
\;(58)
\qquad (57) \rightarrow
\begin{array}{l}
CH_3 \\
CO \\
CHOH \\
CHO \\
(59) \\
+ \\
CH_3COOH
\end{array}
$$

$$
\text{(55)} \rightarrow
\begin{array}{l}
CH_3 \\
CO \\
CHNRR' \\
CO \\
HCOH \\
CH_2OH
\end{array}
\;(60) \rightarrow
\begin{array}{l}
CH_3 \\
CO \\
CHNRR' \\
COH \\
COH \\
CH_2-OH
\end{array}
\;(61) \rightarrow
\begin{array}{l}
CH_3 \\
CO \\
CHNRR' \\
CO \\
CO \\
CH_3
\end{array}
\;(62) \rightarrow
\begin{array}{l}
CH_3COOH \\
+ \\
CHNRR' \\
COH \\
CO \\
CH_3
\end{array}
\;(63)
$$

Fig. 75. Mechanism of 2,3-Enolization of Ketoseamines and of Formation of Reductones and Other Degradation Products

piperidine and acetic acid were heated in ethanolic solution in the presence of C_2H_5OT, the T-activity of the fructosepiperidine did not increase with time; this showed that 1,2-enolization was reversible.

Mechanism of Formation of Reductones

The mechanism shown in Fig. 75 is based on that proposed by Simon (1962), and developed and extended by Simon and Heubach (1965). In the basic media used in the preparation of reductones, the ketose-amine (51) is in the free base form, and a flow of electrons from the nitrogen atom assists the formation of the 2,3-enol (52); this process is not reversible. Since the reaction mixture contained an excess of amine salt, the salt form of the enol (52) can be present; this assists the elimination of the amino group from C-1 (cf. Anet 1964) to give a 1-deoxy-2,3-diulose (53). Simon and Heubach (1965) found, for a model compound, that the β-elimination of the amino group was almost complete, and that little elimination of the β-hydroxyl group, on C-4, occurred.

The 1-deoxy-2,3-diulose (53) was synthesized from 1-deoxy-D-fructose by Ishizu et al. (1967); they found that the diulose (53) was unstable, except in acid solution, and decomposed when an aqueous solution was lyophilized. This behavior contrasts strongly with the considerable stability of the 3-deoxyosulose (24) formed by the 1,2-enolization of ketoseamines (Fig. 74).

It is presumed that the 2,3-diulose (53) enolizes rapidly to (54), and thus gives a 2,4-diulose (55). Further enolization makes easy the elimination of the hydroxyl from C-6 of (56) to give the triketone (57) (these steps can be catalyzed by the amine, or assisted by condensation with the amine, followed by its elimination).

The reductone (58) is formed from the triketone (57) by condensation between C-2 and C-6, or between C-1 and C-5; Simon (1962) found that the ratio between these reactions was approximately 3 : 1 (see preceding section).

The reaction mixture contained acetic acid derived from C-6 (Simon and Heubach 1965); this would be obtained if the triketone (57) split between C-4 and C-5 to give acetic acid, and a four-carbon aldehyde (59) whose further reactions are unknown.

Simon and Heubach (1965) showed several ways in which the amino reductone (63) could be formed, but considered that the following would be the principal route:—the 2,4-diulose (55) condenses with the amine to give the amino ketone (60); elimination of the hydroxyl from C-6 of (61) gives the amino triketone (62); splitting of (62) between C-2 and C-3 gives acetic acid derived from C-1 (found in the mixture) and an amino reductone (63) derived from C-3 to C-6 (as shown by reactions with labeled glucose).

Relationship Between 2,3-Enolization and Browning

All the reactions shown in Fig. 75 produce compounds containing a C-methyl group. Hodge (personal communication) found that the brown pigment, precipitated at the end of the reaction between glucose and piperidine, contained 0.6 C-methyl groups per C_6 unit. The pigment presumably accounts for some of the unidentified C-methyl derivatives in the reaction mixture (cf. Simon and Heubach 1965).

The value of only 0.6 C-methyl groups per C_6 unit suggests that the pigment also contained residues formed by 1,2-enolization, which does not produce compounds with C-methyl groups.

FLAVORS FROM SUGAR-AMINE INTERACTIONS

Some reactions that produce flavoring compounds of known structure are discussed briefly below. The production of flavors from nonenzymic browning reactions was discussed fully by Hodge (1967).

Flavors from 2,3-Enolization of Ketoseamines

When L-rhamnose was heated with piperidine acetate, under the conditions used to prepare "amino-hexose-reductones," the product was a furanone (65) (4-hydroxy-2,5-dimethyl-3(2H)-furanone) (Hodge and Fisher 1963B; Hodge *et al.* 1963; Hodge 1967). The structure (65) was confirmed by synthesis (Henry and Silverstein 1966). Hodge

(64)   (65)   (66)

et al. (1963) suggested that the furanone (65) was formed from a dicarbonyl intermediate (64) by furanose ring closure between C-2 and C-5 and dehydration; the intermediate (64) is the analogue of (55) in Fig. 75.

The furanone (65) was a pseudoreductone, which hydrolyzed, in water or dilute acid, to a reductone. The furanone (65) was a crystalline solid (m.p. about 80°C), but it was quite volatile at room temperature, and had an intense, fragrant odor. The most frequent description of the odor was "burnt sugar," but other terms included "fruity-caramel" and "burnt pineapple." The furanone (65) was unstable in air; within 1–2 days at 25°C it decomposed to a viscous liquid of altered odor (Henry and Silverstein 1966). The furanone (65) was isolated by preparative gas chromatography from pineapple flavor concentrates, and was described as the major "character impact" component (Rodin *et al.* 1965).

A compound considered to be the furanone (66) (4-hydroxy-5-methyl-3(2*H*)-furanone) was isolated in 0.25–0.4% yield when pentoses (3 moles) were heated in aqueous solution (14%) with isopropylamine acetate (1 mole) (Severin and Seilmeier 1967). The same compound was obtained with other strongly basic amines; the compound melted at 122–127°C, and gave a single band on thin layer chromatograms run in 2 solvent mixtures.

When lactose was heated with secondary amine salts, under the conditions used to prepare "amino-hexose-reductones," the product was O-galactosyl isomaltol (67, R = β-galactosyl); on dry distillation O-galactosyl isomaltol gave isomaltol (68), in 25% yield from lactose (Hodge and Nelson 1961). Fisher and Hodge (1964) proved that

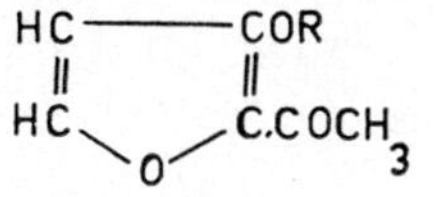

(67)

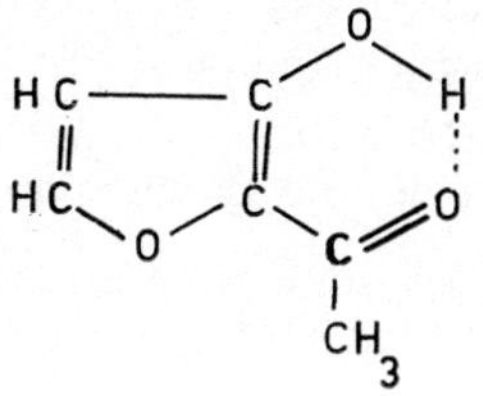

(68)

isomaltol was 3-hydroxy-2-furyl methyl ketone (68, one of the resonating forms). Hodge *et al.* (1963) suggested that O-galactosyl isomaltol (67) was formed from the 1-deoxy-2,3-diulose (69, R = β-galactosyl) (cf. Anet 1964). The diulose (69) is a derivative of (53) (Fig. 75); substitution of the hydroxyl on C-4 prevents the formation of the enol (54). Maltose gave isomaltol (68) in low yield (< 1%); O-glucosyl isomaltol (67, R = α-glucosyl) could not be isolated (Hodge and Nelson 1961). Pyrolysis of maltose hydrate with piperidine phosphate gave maltol (70) in 2% yield; the yield from lactose was 0.2%. Maltol was also considered to be derived from the diulose (69) (Hodge *et al.*

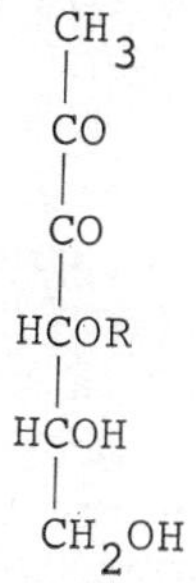

(69)

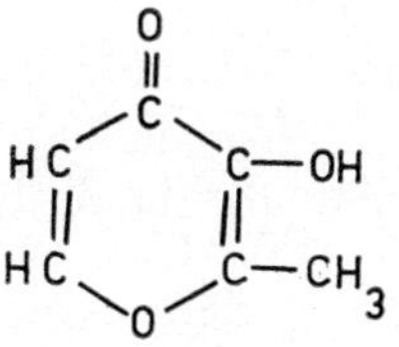

(70)

1963). Isomaltol had a fragrant, weakly pungent, burnt sugar odor. Maltol had a fragrant, caramel-like aroma; at low concentrations it enhanced the flavor and sweetness of carbohydrate-rich foods (Hodge 1967).

The conditions under which these flavoring compounds were prepared did not resemble those obtaining in foods; the amino compounds were also of a different type. However, these are optimum conditions. Under what were probably the least favorable conditions, namely, heating with dilute acid in the absence of amine, fructose gave traces of several compounds derived from 2,3-enolization (Shaw *et al.* 1967).

It can be expected that the range of flavoring compounds derived from the 2,3-enolization of ketoseamines will be extended, and there can be no doubt that such compounds make a major contribution to the flavors of foods.

Flavors from 1,2-Enolization of Ketoseamines

Furfurals and pyrroles are considered to contribute to flavors in bread (Linko and Johnson 1963) and coffee (Gianturco *et al.* 1964). No other types of compound derived from the 1,2-enolization of ketoseamines have been implicated in desirable flavors.

Furfural is derived from pentoses, and methylfurfural from rhamnose, by the mechanism shown in Fig. 74. From the reaction between D-xylose, butylamine, and acetic acid, Kato (1967) isolated 1-butylpyrrole-2-aldehyde (71), as the 2,4-dinitro-phenylhydrazone; similarly, the hydrazone of 1-butyl-5-methylpyrrole-2-aldehyde (72) was obtained from L-rhamnose and butylamine.

(71)   (72)

Nonenzymic browning in foods is accompanied by the development of off-flavors. The nature, and origin, of the compounds involved is unknown; they may be derived from the 1,2-enolization of ketoseamines.

Flavors from Strecker Degradation of Amino Acids

Any α-dicarbonyl compound can react with an α-amino acid to bring about the Strecker degradation. Since ketoseamines are degraded to

α-dicarbonyl compounds (Figs. 74 and 75), a Strecker degradation will normally follow. The reaction may be represented by the following equation:

$$\begin{matrix} -CO \\ | \\ -CO \end{matrix} \quad + \quad \begin{matrix} NH_2 \\ | \\ RCHCOOH \end{matrix} \quad \longrightarrow \quad \begin{matrix} -CNH_2 \\ \| \\ -COH \end{matrix} \quad + \; RCHO \; + \; CO_2$$

In the case of only a few amino acids, is the aldehyde thus produced an important contributor to flavor (Hodge 1967; Streuli 1967). However, further reactions readily occur. Since the Strecker degradation, and its consequences, have been reviewed by Hodge (1967) and Streuli (1967), only one example is given.

When glucose was heated with glycine, or phenylalanine, 2,5-dimethylpyrazine (73) was separated from about 30 other volatile products, and identified (Dawes and Edwards 1966). This dimethylpyrazine (73) was identified among the volatile flavor compounds of

(73)   (74)

potato chips (Deck and Chang 1965). Dawes and Edwards (1966) suggested that pyruvaldehyde participated in a Strecker degradation, forming an amino carbonyl compound that condensed to 2,5-3,6-dihydropyrazine (74); oxidation of (74) leads to the pyrazine (73).

EFFECT OF OXYGEN ON SUGAR-AMINE INTERACTIONS

It is possible that some flavors formed in nonenzymic browning are promoted by oxidation. Dawes and Edwards (1966) suggested that 2,5-dimethylpyrazine (73) was formed by oxidation of the dihydropyrazine (74).

The effect of oxygen on nonenzymic browning in foods is linked with both the water content of the system, and the temperature of storage. When the water content of potatoes exceeded 25%, the rate of browning in air was greater than the rate in vacuum; the difference between these rates increased as the water content increased from 25 to 79%, although the rate of browning decreased (Hendel et al. 1955). The maximum rate of browning occurred at 13–17% water content, and oxygen had little effect. These experiments were carried out at

65°C. Stadtman *et al.* (1946B) found that the rate of browning of sulfited apricots was increased by oxygen when the moisture content exceeded 20%, but only when the temperature exceeded 43°C (cf. Reynolds 1965). Effects of this kind should be considered when setting up model systems.

A ketoseamine derivative that is readily oxidized was prepared by Kitaoka and Onodera (1962). Whereas ketoseamines are oxidized in alkaline solution, the N-glycoside of a ketoseamine (75) was oxidized in methanolic solution, or in the solid state, with the formation of a formamide and an arabonamide (77). It is likely that the intermediate

$$
\begin{array}{ccc}
\mathrm{CH_2NHR} & \mathrm{CHNHR} & \mathrm{O{=}CHNHR} \\
\mid & \parallel & + \\
\mathrm{C{\diagdown}^{NHR'}} & \mathrm{CNHR'} & \mathrm{O{=}CNHR'} \\
\mid & \mid & \mid \\
\mathrm{HOCH \quad O} & \mathrm{HOCH} & \mathrm{HOCH} \\
\mid \quad \mid & \mid & \mid \\
(75) & (76) & (77)
\end{array}
$$

in the oxidation is an enediamine (76). In the presence of an abundance of oxygen, the oxidation products (77) were formed in good yield, with little browning. Severe browning occurred in air, implying some oxidative degradation followed by a nonoxidative browning reaction (Kitaoka and Onodera 1962).

CONCLUSION

The effects of oxygen on sugar-amine interactions are examples of unsolved problems in this field. There are many other unsolved problems, but these were selected for two reasons. First, they have received little attention in the past, as was inevitable when the core reactions were still relatively unknown. Second, they illustrate some effects of water content and temperature, not only on the rate of browning, but also on the nature of the reactions. In devising model systems for the study of these, and other, sugar-amine interactions, it is necessary to select the water content and the reaction temperature, as well as the reactants, the catalysts, and the hydrogen ion concentration.

DISCUSSION

M. L. Wolfrom.—I think that if you make a search of the literature you will find that the rearrangement (of a ketose and amine) to an aminodeoxyaldose was first reported by Carson of the Western Regional Laboratory.

T. M. Reynolds.—Carson and Heyns independently reported the reaction between ketoses and alkylamines in 1955, but Heyns reported the formation of glucosamine from fructose and ammonia in 1952 and 1953. In calling this reaction the Heyns rearrangement, I have followed the terminology used by other workers (Reynolds 1965).

J. Spinelli.—You did not mention anything about phosphorylated sugars entering into the browning reaction. Can you comment on the rates of phosphorylated sugars vs. free sugars in the browning reaction?

T. M. Reynolds.—There are no recent results. Glucose-6-phosphate is thought to contribute to browning in meat and fish. Schwimmer and Olcott showed that glucose-6-phosphate gave more brown color with glycine than did glucose, but this work was not followed up (Reynolds 1963).

W. A. Gortner.—Could you say a little more about the role of the amine? Generally you are using glycine as your model. Have you used peptides or other secondary amines?

T. M. Reynolds.—We used glycine because the system was a model for foods containing primary amino acids. The results with secondary amino acids would depend on the pH of the system. What pH had you in mind?

W. A. Gortner.—One reason for asking the question is that you mentioned the change in nutritive value associated with these reactions. It has been demonstrated that many of the amino acids of proteins, as well as lysine, may become relatively unavailable to a microbiological assay through reactions with carbohydrate in foods. How do these amino acids react?

T. M. Reynolds.—Those reactions have not been studied. I know of no experimental work since that done by Lea on dried milk and casein (Reynolds 1965). It has been suggested (Reynolds 1965) that the guanidyl groups of arginine could react with a 3-deoxyosulose. I think that model systems containing appropriate peptides should be studied; further information is certainly needed in this field.

A. Neuberger.—One would expect that the primary reaction between a carbonyl group and an amino group would involve the unprotonated form of the amine. At the pH described (pH 3.5), the relative concentration of the unprotonated form would lie between one in 100,000 and one in a million. That does not negate or contradict the mechanism, but one would expect a much greater difference than you have found between the rates of reaction at pH 3.5 and pH 6.5. Would you like to comment on this?

T. M. Reynolds.—I certainly agree with your comments. I think that the reason for the results obtained is that there is throughout the process a balance between the two stages of the reaction. The free base form of the amine is involved in the reaction with the carbonyl group, and the salt form of the amine is required for the rearrangement. The yield of rearrangement product is not quantitative and, particularly with amino acids, both the rate of reaction and the yield are dependent on the balance between the two forms.

J. E. Hodge.—By what mechanism do you propose the formation of 3- and 4-carbon fragments, such as pyruvaldehyde and diacetyl, from difructoseglycine?

T. M. Reynolds.—There is no evidence that these compounds are formed from difructoseglycine under the conditions considered in this paper. I think that they can be formed in small quantities by reactions involving 2,3-enolization, even when the predominant reactions involve 1,2-enolization.

J. E. Hodge.—We have some work in progress on the pyrolysis at low temperatures (subliming conditions) of browning intermediates. Have you done this type of work with difructoseglycine?

T. M. Reynolds.—No, we have not attempted to study any volatile breakdown products that might be formed.

BIBLIOGRAPHY

Anet, E. F. L. J. 1959A. Chemistry of non-enzymic browning. VII. Crystalline di-D-fructose-glycine and some related compounds. Australian J. Chem. *12*, 280–287.

Anet, E. F. L. J. 1959B. Chemistry of non-enzymic browning. X. Di-fructose-amino acids as intermediates in browning reactions. Australian J. Chem. *12*, 491–496.

Anet, E. F. L. J. 1960. Degradation of carbohydrates. I. Isolation of 3-deoxyhexosones. Australian J. Chem. *13*, 396–403.

Anet, E. F. L. J. 1962. Degradation of carbohydrates. III. Unsaturated hexosones. Australian J. Chem. *15*, 503–509.

Anet, E. F. L. J. 1963. Degradation of carbohydrates. IV. Formation of *cis*-unsaturated hexosones. Australian J. Chem. *16*, 270–277.

Anet, E. F. L. J. 1964. 3-Deoxyglycosuloses (3-deoxyglycosones) and the degradation of carbohydrates. Advan Carbohydrate Chem. *19*, 181–218.

Anet, E. F. L. J. 1968. Mechanism of formation of 3-deoxyglycosuloses. Tetrahedron Letters. No. 31, 3525–3528.

Anet, E. F. L. J., and Ingles, D. L. 1964. Mechanism of inhibition of non-enzymic browning by sulphite. Chem. Ind. (London) *1964*, 1319.

Anet, E. F. L. J., and Reynolds, T. M. 1957. Chemistry of non-enzymic browning. I. Reactions between amino acids, organic acids, and sugars in freeze-dried apricots and peaches. Australian J. Chem. *10*, 182–192.

Braverman, J. B. S. 1953. The mechanism of the interaction of sulphur dioxide with certain sugars. J. Sci. Food Agr. *4*, 540–547.

CARSON, J. F., and OLCOTT, H. S. 1954. Brown condensation products from acetaldehyde and aliphatic amines. J. Am. Chem. Soc. 76, 2257–2259.

DANEHY, J. P., and PIGMAN, W. W. 1951. Reactions between sugars and nitrogenous compounds and their relationship to certain food problems. Advan. Food Res. 3, 241–290.

DAWES, I. W., and EDWARDS, R. A. 1966. Methyl substituted pyrazines as volatile reaction products of heated aqueous aldose, amino-acid mixtures. Chem. Ind. (London) 1966, 2203.

DECK, R. E., and CHANG, S. S. 1965. Identification of 2,5-dimethyl-pyrazine in the volatile flavour compounds of potato chips. Chem. Ind. (London) 1965, 1343–1344.

ELLIS, G. P. 1959. The Maillard reaction. Advan. Carbohydrate Chem. 14, 63–134.

ELLIS, G. P., and HONEYMAN, J. 1955. Glycosylamines. Advan. Carbohydrate Chem. 10, 95–168.

FISHER, B. E., and HODGE, J. E. 1964. The structure of isomaltol. J. Org. Chem. 29, 776–781.

GIANTURCO, M. A., GIAMMARINO, A. S., FRIEDEL, P., and FLANAGAN, V. 1964. The volatile constituents of coffee. IV. Furanic and pyrrolic compounds. Tetrahedron 20, 2951–2961.

HANNAN, R. S., and LEA, C. H. 1952. Studies of the reaction between proteins and reducing sugars in the "dry" state. VI. The reactivity of the terminal amino groups of lysine in model systems. Biochim. Biophys. Acta 9, 293–305.

HENDEL, C. E., SILVEIRA, V. G., and HARRINGTON, W. O. 1955. Rates of non-enzymic browning of white potato during dehydration. Food Technol. 9, 433–438.

HENRY, D. W., and SILVERSTEIN, R. M. 1966. Rational synthesis of 4-hydroxy-2, 5-dimethyl-3(2H)-furanone, a flavor component of pineapple. J. Org. Chem. 31, 2391–2394.

HEYNS, K., and PAULSEN, H. 1960. On the chemical basis of the Maillard reaction. Wiss. Veroeffentl. Deut. Ges. Ernaehrung 5, 15–42. (German)

HEYNS, K., PAULSEN, H., and SCHROEDER, H. 1961. Reaction of ketohexoses with secondary amino acids and secondary amines. Tetrahedron 13, 247–257. (German)

HODGE, J. E. 1953. Dehydrated foods. Chemistry of browning reactions in model systems. J. Agr. Food Chem. 1, 928–943.

HODGE, J. E. 1955. The Amadori rearrangement. Advan. Carbohydrate Chem. 10, 169–205.

HODGE, J. E. 1967. Origin of flavor in foods. Nonenzymatic browning reactions. In Symp. Foods: The Chemistry and Physiology of Flavors, H. W. Schultz (Editor). AVI Publishing Co., Westport, Conn.

HODGE, J. E., and FISHER, B. E. 1963A. Amadori rearrangement products. In Methods in Carbohydrate Chemistry, Vol. II, R. L. Whistler, and M. L. Wolfrom (Editors). Academic Press, New York.

HODGE, J. E., and FISHER, B. E. 1963B. The structure of a pseudoreductone from L-rhamnose. Abstr. Papers Am. Chem. Soc. 145th Meeting 3D.

HODGE, J. E., and NELSON, E. C. 1961. Preparation and properties of D-galactosylisomaltol and isomaltol. Cereal Chem. *38*, 207–221.

HODGE, J. E., and RIST, C. E. 1953. The Amadori rearrangement under new conditions and its significance for non-enzymatic browning reactions. J. Am. Chem. Soc. *75*, 316–322.

HODGE, J. E., FISHER, B. E., and NELSON, E. C. 1963. Dicarbonyls, reductones, and heterocyclics produced by reactions of reducing sugars with secondary amine salts. Am. Soc. Brewing Chemists *1963*, 84–92.

INGLES, D. L. 1959A. Chemistry of non-enzymic browning. V. The preparation of aldose-potassium bisulphite addition compounds and some amine derivatives. Australian J. Chem. *12*, 97–101.

INGLES, D. L. 1959B. Chemistry of non-enzymic browning. VIII. The hydrolytic reactions of aldose bisulphite addition compounds. Australian J. Chem. *12*, 288–295.

INGLES, D. L. 1962. The formation of sulphonic acids from the reactions of reducing sugars with sulphite. Australian J. Chem. *15*, 342–349.

INGLES, D. L., and REYNOLDS, T. M. 1958. Chemistry of non-enzymic browning. IV. Determination of amino acids and amino acid-deoxyfructoses in browned freeze-dried apricots. Australian J. Chem. *11*, 575–580.

INGLES, D. L., and REYNOLDS, T. M. 1959. Chemistry of non-enzymic browning. IX. Studies of sugar mono-esters of malic acid found in browned freeze-dried apricots. Australian J. Chem. *12*, 483–490.

ISBELL, H. S. 1944. Interpretation of some reactions in the carbohydrate field in terms of consecutive electron displacement. J. Res. Natl. Bur. Std. *32*, 45–59.

ISBELL, H. S., and FRUSH, H. L. 1958. Mutarotation, hydrolysis, and rearrangement reactions of glycosylamines. J. Org. Chem. *23*, 1309–1319.

ISHIZU, A., LINDBERG, B., and THEANDER, O. 1967. 1-Deoxy-D-*erythro*-2,3-hexodiulose, an intermediate in the formation of D-glucosaccharinic acid. Carbohydrate Res. *5*, 329–334.

KATO, H. 1960. Studies on browning reactions between sugars and amino compounds. V. Isolation and characterization of new carbonyl compounds, 3-deoxy-osones formed from *N*-glycosides and their significance in the browning reaction. Bull. Agr. Chem. Soc. (Japan) *24*, 1–12.

KATO, H. 1967. Chemical studies on amino-carbonyl reaction. III. Formation of substituted pyrrole-2-aldehydes by reaction of aldoses with alkylamines. Agr. Biol. Chem. (Tokyo) *31*, 1086–1090.

KITAOKA, S., and ONODERA, K. 1962. Oxidative cleavages of 1,2-diamino sugars and their significance in the mechanism of the amino-carbonyl reaction. Agr. Biol. Chem. (Tokyo) *26*, 572–580.

LEA, C. H., and HANNAN, R. S. 1950. Studies of the reaction between proteins and reducing sugars in the 'dry' state. II. Further observations on the formation of the casein-glucose complex. Biochim. Biophys. Acta *4*, 518–531.

LINDBERG, B., TANAKA, J., and THEANDER, O. 1964. Reaction between D-glucose and sulphite. Acta Chem. Scand. *18*, 1164–1170.

LINKO, Y.-Y., and JOHNSON, J. A. 1963. Changes in amino acids and formation of carbonyl compounds during baking. J. Agr. Food Chem. *11*, 150–152.

MACHELL, G., and RICHARDS, G. N. 1960. Mechanism of saccharinic acid formation. III. The α-keto-aldehyde intermediate in formation of D-glucometasaccharinic acid. J. Chem. Soc. *1960*, 1938–1944.

MICHEEL, F., and DIJONG, I. 1962. The mechanism of the Amadori rearrangement. Ann. Chem. *658*, 120–127. (German)

PALM, D., and SIMON, H. 1965. Mechanism of the Amadori rearrangement. Z. Naturforsch. *20b*, 32–35. (German)

PAULSEN, H. 1966. Carbohydrates containing nitrogen or sulfur in the hemiacetal ring. Angew. Chem., Intern. Ed. Engl. *5*, 495–511.

REYNOLDS, T. M. 1959. Chemistry of nonenzymic browning. III. Effect of bisulphite, phosphate, and malate on the reaction of glycine and glucose. Australian J. Chem. *12*, 265–274.

REYNOLDS, T. M. 1963. Chemistry of nonenzymic browning. I. The reaction between aldoses and amines. Advan. Food Res. *12*, 1–52.

REYNOLDS, T. M. 1965. Chemistry of nonenzymic browning. II. Advan. Food Res. *14*, 167–283.

REYNOLDS, T. M., ANET, E. F. L. J., and INGLES, D. L. 1962. Mechanism of nonenzymic browning reactions and of their inhibition by sulphur dioxide. Proc. 1st Int. Congr., Food Sci. Technol., London, (1968), Vol. 1., p. 209–216.

RODIN, J. O. *et al.* 1965. Volatile flavor and aroma constituents of pineapple. I. Isolation and tentative identification of 2,5-dimethyl-4-hydroxy-3(2*H*)-furanone. J. Food Sci. *30*, 280–285.

SEVERIN, T., and SEILMEIER, W. 1967. Studies on the Maillard Reaction. II. Rearrangement of pentoses under the influence of amine acetates. Z. Lebensm. Untersuch.–Forsch. *134*, 230–232. (German)

SHAW, P. E., TATUM, J. H., and BERRY, R. E. 1967. Acid-catalyzed degradation of D-fructose. Carbohyd. Res. *5*, 266–273.

SIMON, H. 1962. Mechanism of formation of piperidino-hexose reductone. Chem. Ber. *95*, 1003–1008. (German)

SIMON, H., and HEUBACH, G. 1965. Alicyclic, open-chain, N-containing reductones by the reaction between secondary amino salts and monosaccharides. Chem. Ber. *98*, 3703–3711. (German)

STADTMAN, E. R., BARKER, H. A., HAAS, V., and MRAK, E. M. 1946A. Storage of dried fruit. Influence of temperature on deterioration of apricots. Ind. Eng. Chem. *38*, 541–543.

STADTMAN, E. R., BARKER, H. A., HAAS, V., MRAK, E. M., and MACKINNEY, G. 1946B. Storage of dried fruit. Gas changes during storage of dried apricots and influence of oxygen on rate of deterioration. Ind. Eng. Chem. *38*, 324–329.

STREULI, H. 1967. Aromas from roasting. (Röstaromen). *In* Aroma-und Geschmacksstoffe in Lebensmitteln, J. Solms, and H. Neukon (Editors). Forster-Verlag A. G., Zürich. (German)

WEYGAND, F. 1940. N-Glycosides. II. Amadori rearrangements. Chem. Ber. *73B*, 1259–1278. (German)

WEYGAND, F. *et al.* 1959. Structure of piperidino-hexose-reductone. Tetrahedron *6*, 123–138.

WOLFROM, M. L., SCHLICHT, R. C., LANGER, A. W. JR., and ROONEY, C. S. 1953. Chemical interactions of amino compounds and sugars. VI. The repeating unit in browning polymers. J. Am. Chem. Soc. *75*, 1013.

CHAPTER 13

Robert M. Horowitz
Bruno Gentili

Glycosidic Pigments
and Their Reactions[1]

INTRODUCTION

Glycosidic pigments occur throughout the plant kingdom. Without doubt the anthocyanins are the most typical, widespread and important of these coloring matters in plants. Indeed, the deep red, blue, or violet coloration of flowers, fruits and berries is almost always due to the presence of anthocyanins. A notable exception occurs in the order Centrospermae, which includes, among other plants, beets, bougainvillea and cactus. Here the color is due to the presence of the betacyanins, a group of nitrogenous glycosides whose structure has been worked out only within the past few years. Some typical examples of these two types of pigment are shown in Fig. 76. A third but very limited group of glycosidic pigments is made up of the carotenoid

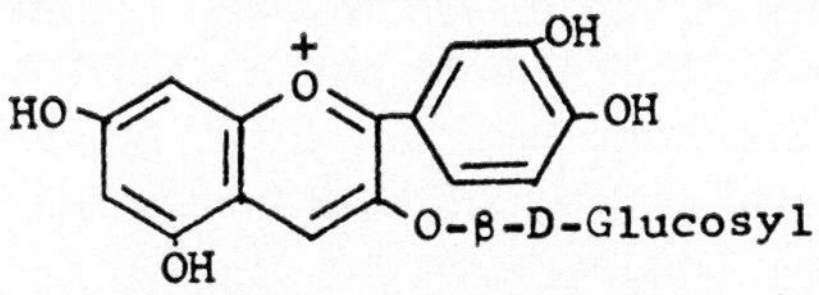

CYANIDIN 3-GLUCOSIDE

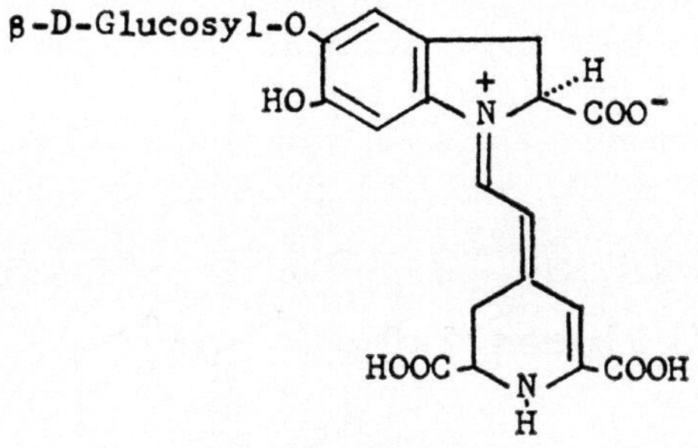

BETANIN

FIG. 76. STRUCTURE OF AN ANTHOCYANIN AND A BETA-
CYANIN

[1] Acknowledgement—We should like to thank Dr. Roger Albach, Weslaco, Texas, Dr. A. N. Booth, Albany, California, and Dr. W. Gaffield, Albany, California for permission to cite their work in advance of publication.

glycosides which were discovered recently and which thus far have been found only in microbes (Hertzberg and Jensen 1967).

Though these groups of glycosides are obviously of prime importance as pigments, it is not our intent to discuss their chemistry or reactions. This is because the theme of the Symposium is the properties and chemistry of carbohydrates, and the chemistry of these compounds is to a large extent the chemistry of the aglycone pigments. Historically the sugar part of anthocyanin molecules was often regarded as a nuisance to be discarded or forgotten without proper identification, and it is only within recent years that much has been learned regarding the structure, distribution, and inheritance of the glycosyl portions of anthocyanins (cf. Pridham 1965; Harborne 1967). We shall, therefore, discuss a group of plant glycosides in which the sugar component, no less than the aglycone portion, is of importance in determining the properties of the compound. The compounds in question are flavanone glycosides, particularly those of citrus fruits. A number of unusual taste phenomena are associated with these glycosides and they are also useful in studies of the chemotaxonomy of citrus. In the narrowest sense of the word they should perhaps not even be called pigments, since their contribution to fruit color is minimal. Nevertheless, they are closely related to and are easily converted into other compounds that are strongly colored.

FLAVANONE GLYCOSIDES OF CITRUS

Let us consider the two key citrus flavonoids, hesperidin and naringin, both of which have been known for more than a century. Naringin is the major flavonoid in grapefruit and hesperidin the major flavonoid in oranges and lemons. The main structural features of these compounds are shown in Fig. 77. The glycosides are made up of three

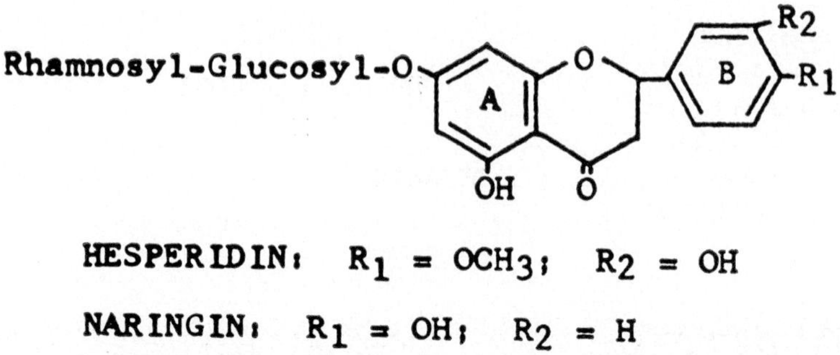

HESPERIDIN: R_1 = OCH_3; R_2 = OH

NARINGIN: R_1 = OH; R_2 = H

FIG. 77. GENERAL STRUCTURE OF HESPERIDIN AND NARINGIN

fragments: a rhamnose unit joined to a glucose unit which in turn is joined to a flavanone aglycone. At first glance it would appear that the only difference between the 2 compounds lies in the substituents at the 3'- and 4'-positions. However, when we scrutinize their structures more critically we find, contrary to the opinion of earlier workers, that there is also a difference in the way rhamnose is linked to glucose. In hesperidin the rhamnose is linked α to the C-6 hydroxy group of glucose, as shown in the studies of Zemplén and Gerecs (1934, 1935) and Gorin and Perlin (1959). This disaccharide is named rutinose. In contrast to this we discovered that in naringin rhamnose is linked α to the C-2 hydroxy group of glucose, as determined by methylation-hydrolysis experiments that yielded 3,4,6-tri-O-methyl-D-glucopyranose and 2,3,4-

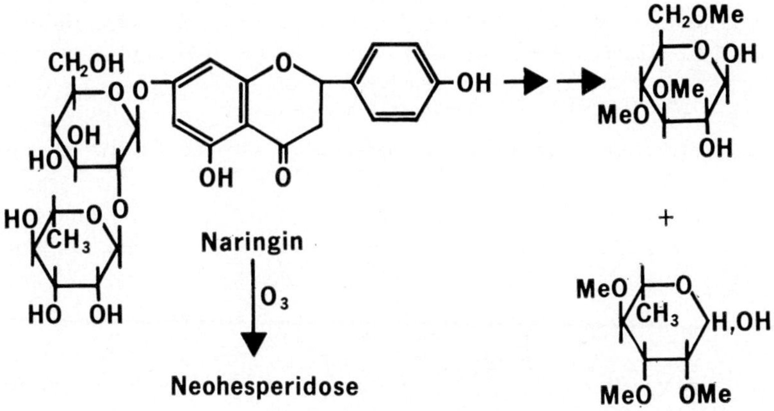

FIG. 78.   METHYLATION-HYDROLYSIS AND OZONOLYSIS OF NARINGIN

tri-O-methyl-L-rhamnopyranose (Fig. 78) (Horowitz and Gentili 1963). Later we synthesized 2-O-α-L-rhamnopyranosyl-D-glucopyranose by the reactions shown in Fig. 79 and found it identical with the disaccharide obtained from the cleavage of naringin with ozone (Fig. 78) (Horowitz et al. 1964). This disaccharide is named neohesperidose, since it was first recognized in a third citrus flavanone, neohesperidin, which is a major component of Seville oranges (Kolle and Gloppe 1936). We can now write the structures of hesperidin, naringin, and neohesperidin as shown in Fig. 80.

RELATION OF STRUCTURE TO TASTE

The results outlined thus far would probably not have been viewed with anything more than routine interest had they not suddenly brought

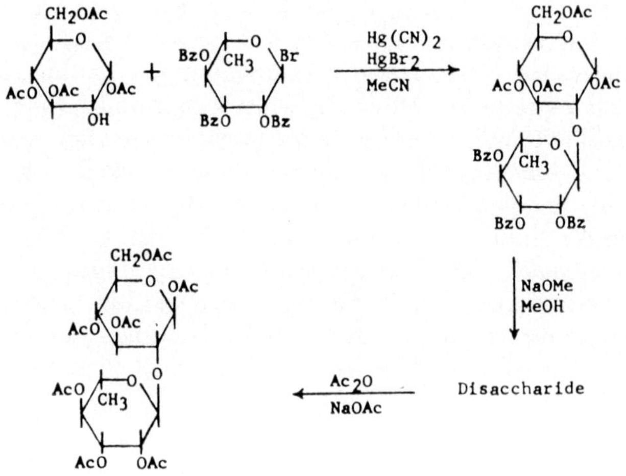

M.p. 151°; $[\alpha]_D^{25}$ +0.32° (CHCl$_3$).
Identical with neohesperidose β-heptaacetate.

FIG. 79. SYNTHESIS OF NEOHESPERIDOSE AND ITS β-HEPTA-
ACETATE

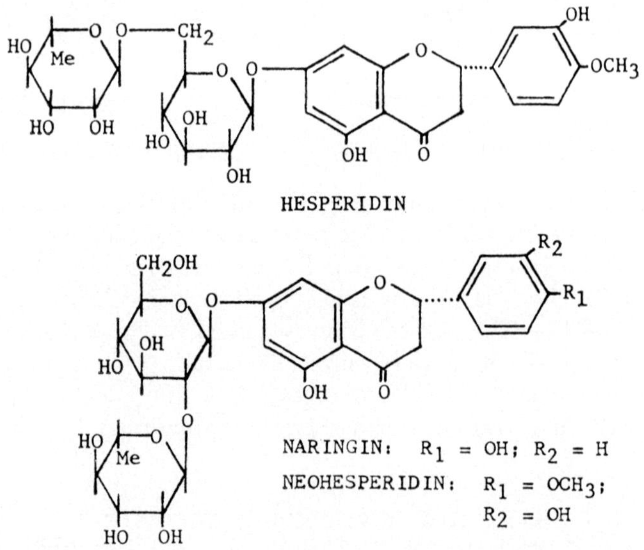

HESPERIDIN

NARINGIN: R$_1$ = OH; R$_2$ = H
NEOHESPERIDIN: R$_1$ = OCH$_3$;
R$_2$ = OH

FIG. 80. DETAILED STRUCTURES OF HESPERIDIN, NARINGIN AND
NEOHESPERIDIN

into focus the long known observation that hesperidin is a tasteless compound and naringin an extremely bitter one. Moreover, it turned out that neohesperidin was also bitter. It appeared, therefore, that the point of attachment of rhamnose to glucose was of crucial importance in determining whether compounds in this series are tasteless or bitter (Horowitz and Gentili 1961; Horowitz 1964). At the time this work was done the three compounds we have been discussing, together with a fourth, poncirin, were the only well characterized flavanone glycosides known in citrus. Subsequently some rather painstaking isolation studies enabled us to extend the list and to provide structures for four new glycosides. The presently known compounds are shown in Table 21.

TABLE 21

FLAVANONE GLYCOSIDES OF CITRUS FRUITS

X = Rutinosyl	X = Neohesperidosyl	R_1	R_2
Hesperidin	Neohesperidin	OMe	OH
Naringenin Rutinoside	Naringin	OH	H
Isosakuranetin Rutinoside	Poncirin	OMe	H
Eriocitrin	Neoeriocitrin	OH	OH

The most interesting feature in this table is that of the four different flavanone aglycones shown, each is represented by both a β-rutinosyl and a β-neohesperidosyl derivative. Furthermore, the taste-structure relations that we had arrived at earlier by examining only a few compounds stand up remarkably well throughout the series. Thus, all the flavanone rutinosides in Table 21 are tasteless and all the flavanone neohesperidosides are bitter, though there is considerable individual variation in degree of bitterness between members of the series. The relative bitterness of these compounds is shown in Table 22.

After the relation between structure and bitterness had been established, we could not resist the temptation to explore a little further along this path. Initially our intent was simply to determine whether the presence of neohesperidose is necessary or sufficient for bitterness and to pinpoint, if possible, some of the major structural requirements for taste. The work in our own and other laboratories has now branched out considerably and we will attempt here only to summarize what we think are the more salient features of this research.

The general approach has been to modify in various ways the naturally occurring flavanone glycosides. These modifications have taken the following forms: (1) the removal of large fragments of the molecule; (2) the conversion of the flavanone aglycone into other flavonoid types; and (3) the alteration of substituent groups. In general, modifications of types 1 or 3 give rise only to quantitative changes in the existing taste, while modifications of type 2 may, in certain cases, give rise to qualitative changes. Some examples of each of these kinds of modification will be discussed.

(1) Removal of large fragments.—By this we mean loss of one or both sugars, or loss of the B-ring together with one or more carbon atoms of the heterocyclic ring. Typical examples of the effect of losing rhamnose are seen in naringenin 7-β-D-glucoside (prunin) and hesperetin 7-β-D-glucoside (Fig. 81), each of which retains at least some of the

TABLE 22

MOLAR CONCENTRATIONS OF ISOBITTER SOLUTIONS OF FLAVANONE NEOHESPERIDOSIDES AND QUININE

Compound	Molarity	Relative Bitterness
Neoeriocitrin	$>5 \times 10^{-4}$	<2
Neohesperidin	5×10^{-4}	2
Naringin	5×10^{-5}	20
Poncirin	5×10^{-5}	20
Quinine dihydrochloride	1×10^{-5}	100

bitterness of the parent compound. The fact that these compounds are bitter at all shows that neohesperidose is not a prerequisite for bitterness. From the available quantitative data we judge that the effect of adding an α-L-rhamnosyl residue to the C-2 hydroxyl of D-glucose (to give neohesperidosyl) is to enhance or, at least, maintain bitterness, while the effect of adding it to the C-6 hydroxyl (to give rutinosyl) is to abolish it entirely. When both rhamnose *and* glucose are lost, the solubility of the aglycone is greatly diminished and the taste is usually nil.

The loss of the B-ring under alkaline conditions affords phloracetophenone 4'-β-neohesperidoside or, with certain compounds, phloroglucinol β-neohesperidoside (Fig. 81). The former compound is intensely bitter; the latter is tasteless. We infer from these and other data that the carbonyl group is needed for bitterness.

(2) Conversion to other flavonoid types.—A number of interesting compounds have resulted from modifications of this type. The conversion of naringin and neohesperidin to the corresponding flavones,

rhoifolin, and neodiosmin (Fig. 82), gives products which exhibit none of the bitterness of the parent substances. A solution of naringin containing a large amount of rhoifolin is less bitter than naringin alone.

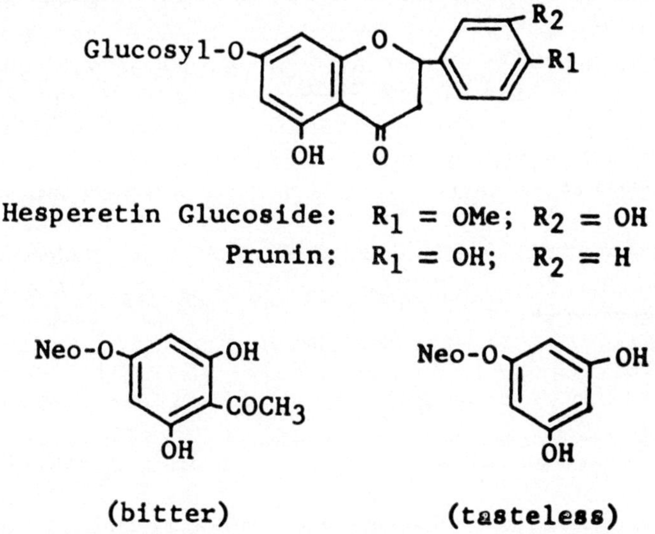

Hesperetin Glucoside: R_1 = OMe; R_2 = OH

Prunin: R_1 = OH; R_2 = H

(bitter)   (tasteless)

FIG. 81. STRUCTURES OF SOME FLAVANONE GLUCOSIDES AND DEGRADATION PRODUCTS OF FLAVANONE NEOHESPERIDOSIDES

This suggests that rhoifolin is able to compete with naringin for sites on the taste receptors, though it does not produce a taste response of its own.

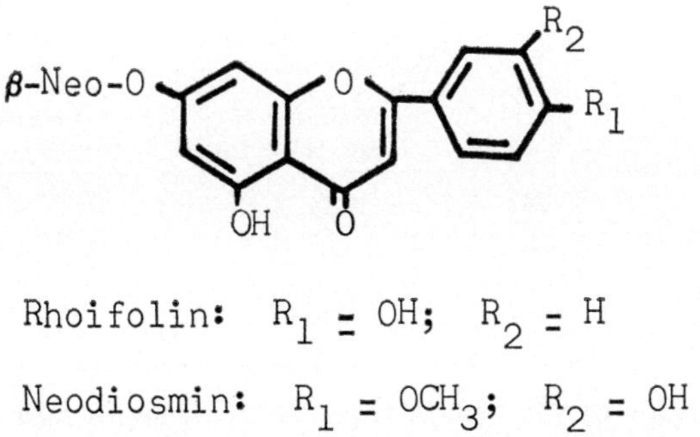

Rhoifolin: R_1 = OH; R_2 = H

Neodiosmin: R_1 = OCH$_3$; R_2 = OH

FIG. 82. STRUCTURE OF RHOIFOLIN AND NEODIOSMIN

We were thus led to the hypothesis that highly conjugated, planar aglycones tend to abolish taste responses, while less conjugated, nonplanar aglycones favor them. To check this point we prepared the chalcone and dihydrochalcone corresponding to naringin, our expectation being that the planar chalcone would be tasteless and the nonplanar dihydrochalcone would be bitter. The reactions involved are shown in Fig. 83. In fact, both the chalcone and dihydrochalcone turned out

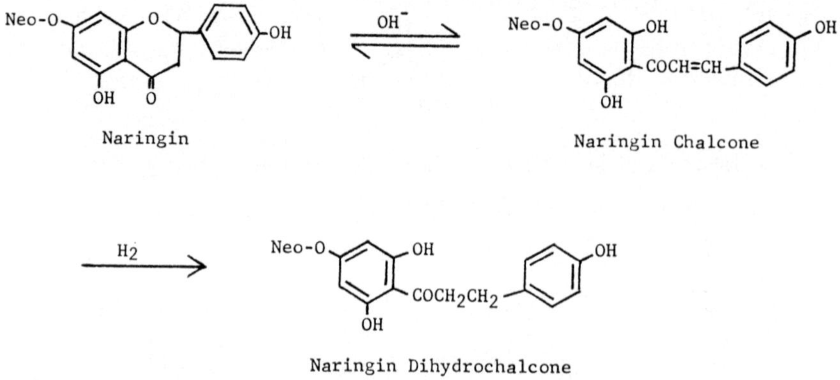

Naringin   Naringin Chalcone

Naringin Dihydrochalcone

FIG. 83. CONVERSION OF NARINGIN TO ITS CHALCONE AND DIHYDROCHALCONE

to be intensely sweet. Obviously, more subtle explanations than mere planarity or nonplanarity are required here.

Because of the unexpectedness of this finding, the remaining bitter flavanone neohesperidosides listed in Table 21 were also converted to their dihydrochalcones. The results for the four compounds are shown in Table 23.

TABLE 23

TASTE AND RELATIVE SWEETNESS OF DIHYDROCHALCONE NEOHESPERIDOSIDES
AND SACCHARIN

Compound	Taste	Molarity of Isosweet Soln.	Relative Sweetness (Molar)	(Weight)
Naringin DHC	Sweet	2×10^{-4}	1	0.4
Neohesperidin DHC	Sweet	1×10^{-5}	20	7
Neoeriocitrin DHC	Sl. sweet	—	—	—
Poncirin DHC	Sl. bitter	—	—	—
Saccharin (Na)	Sweet	2×10^{-4}	1	1

Two of the dihydrochalcones, those from naringin and neohesperidin, have exceedingly high levels of sweetness, inasmuch as saccharin, the substance used for comparison, is said to be about 300 times sweeter than sucrose. Poncirin dihydrochalcone is slightly bitter and is, in fact, the only nonsweet compound in the group. We have found in subsequent work that at least one hydroxyl group in the B-ring is required for sweetness in the dihydrochalcones.

In contrast to the results with flavanone neohesperidosides, flavanone rutinosides such as hesperidin or naringenin 7-β-rutinoside (Table 21) yield only tasteless dihydrochalcones. Thus, the influence of the rutinosyl radical in abolishing the taste response is very strong and is manifested in the dihydrochalcones as well as in the flavanones. We can, however, produce a sapid substance from hesperidin dihydrochalcone by carrying out a modification of type 1, i.e., hydrolyzing rhamnose to give hesperetin dihydrochalcone glucoside (HDG) (Fig. 84). This

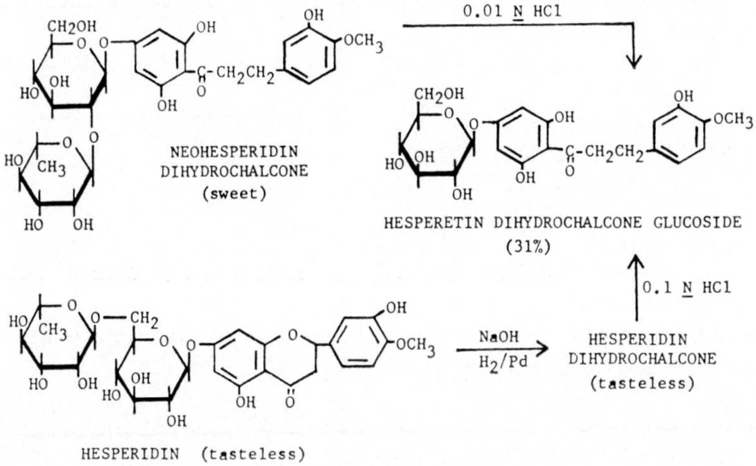

FIG. 84. PREPARATION OF HESPERETIN DIHYDROCHALCONE GLUCOSIDE (HDG)

compound can also be made by partial hydrolysis of neohesperidin dihydrochalcone. HDG is about 1/20 as sweet as the latter compound and is much less soluble.

To recapitulate, most of the taste phenomena we have been discussing are exhibited to some extent by the simple β-D-glucosides of the phenolic aglycones. The attachment of α-L-rhamnose to the C-2 hydroxyl of glucose usually enhances the taste and increases the water solubility, while its attachment to the C-6 hydroxyl destroys the taste. We note in passing that the free disaccharide, neohesperidose, is itself only very slightly sweet, and free rutinose is essentially tasteless.

(3) Alteration of substituent groups.—Most of these alterations involve the alkylation of free A- or B-ring phenolic hydroxyl groups. As a rule, methylation or ethylation of these groups causes a decrease in bitterness in the case of flavanones and a decrease in sweetness in the case of dihydrochalcones. An interesting exception has been un-

covered by Krbechek *et al.* (1968), who studied the effect of length-ening the chain of the 4-alkoxy group in neohesperidin dihydrochalcone. Replacement of the 4-methoxy with a 4-ethoxy gave little change, but replacement with a 4-*n*-propoxy gave a twofold increase in sweetness. We have prepared the 4-isopropoxy derivative and find it less sweet than any of the others in the series. The relevant structures and order of sweetness are given in Fig. 85. It is worthwhile comparing this series with the corresponding 4-nitro-2-aminophenyl alkyl ethers (Fig. 85), which are reported to be intensely sweet (cf. Moncrieff 1967). One might surmise that the compounds in the two series act on the same set of taste receptors.

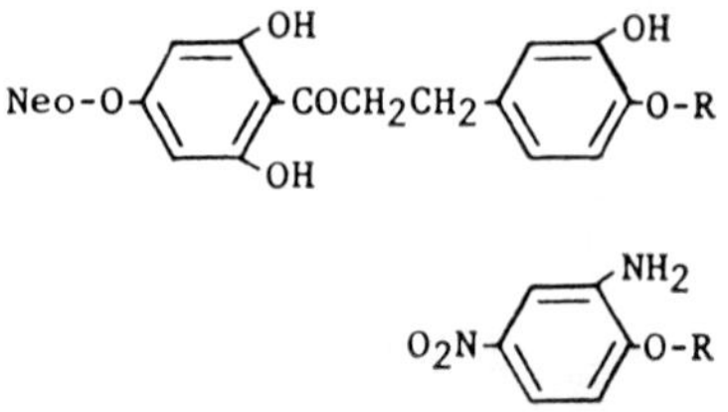

FIG. 85. RELATIVE SWEETNESS OF A SERIES OF DIHYDROCHALCONES AND 4-NITRO-2-AMINOPHENYL ALKYL ETHERS

Instead of altering the phenolic hydroxyl groups one can alter the sugar hydroxyls, though this requires more tedious synthetic procedures. It is conceivable that experiments of this sort will throw light on the challenging question of the difference in the taste properties of the rutinose and neohesperidose substituted compounds. As mentioned earlier, the taste responses produced by glucosides and neohesperido-sides are qualitatively similar. A feature shared by these 2 groups of glycosides is the presence of the free C-6 hydroxyl group in glucose. This is the only primary hydroxyl in the molecules and it is, of course, absent in the rutinosides where it is blocked by rhamnose. It seemed reasonable to suppose that this primary hydroxyl group must be more than casually involved in taste stimulation, possibly because it becomes bound to a taste site by hydrogen bonding. We therefore set about to methylate the glucose C-6 hydroxyl in neohesperidin dihydrochalcone to see whether blocking the group would abolish the sweet taste (Fig. 86). To our surprise the product appeared both qualitatively and quantitatively to be the same as the parent compound. From these results we have reached some tentative conclusions about structural specificity in the sugar part of the molecule:

(1) Neither the C-2 nor C-6 hydroxyl of glucose is required for taste, since taste is not abolished by blocking them with a 2-O-α-L-rhamnosyl substituent or a 6-O-methyl substituent. Furthermore, the C-2 and C-6 hydroxyls can be absent simultaneously without affecting the taste (see Fig. 86).

(2) From this we infer that the structural features most directly involved in taste are the C-3 and C-4 hydroxyl groups of glucose. A similar conclusion was reached by Evans (1963) in studies on taste

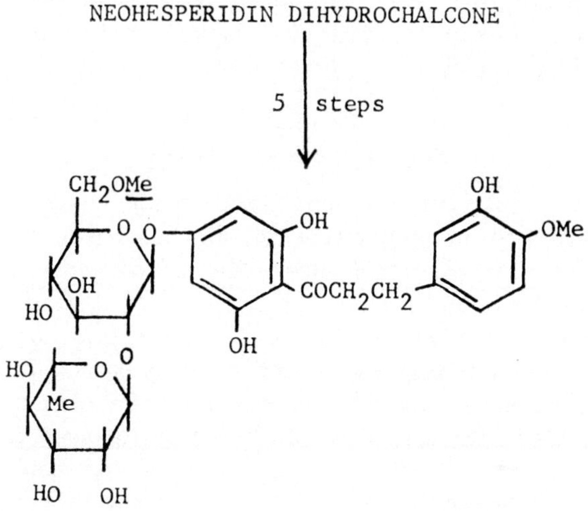

NEOHESPERIDIN DIHYDROCHALCONE

5 | steps

Fig. 86. 6″-O-Methylneohesperidin Dihydrochalcone

receptor stimulation in the blowfly. When a series of D-glucose derivatives was tested in this insect, the results were consistent with the idea that the C-3 and C-4 hydroxyl groups alone are responsible for taste stimulation. There is also an analogy in the antibody-antigen like reactions of the plant agglutinins with the blood group substances. These reactions may be inhibited by mono- or oligosaccharides and the extent of inhibition is determined mainly by the stereochemistry of the C-3 and C-4 hydroxyl groups of the terminal sugar (cf. Boyd 1962). If there is validity in this view of the importance of the C-3 and C-4 hydroxyl groups one might expect that epimerizing or alkylating at one or both of these positions would have a marked effect on taste responses.

(3) A bulky substituent such as L-rhamnosyl can have opposing effects depending on where it is attached to glucose. When linked at

the C-2 position the overall shape of the molecule is favorable for attachment to the taste receptor site. In addition, the hydroxyl groups of the rhamnose probably enhance binding to the site and consequently the taste. At the C-6 position the overall shape is unfavorable to the extent that attachment to the receptors is strongly inhibited and the taste abolished. On the other hand, a C-6 methyl group is small enough so that it does not interfere with binding and consequently the taste of 6″-O-methylneohesperidin dihydrochalcone (Fig. 86) is similar to that of the parent compound. If these views are correct we would expect that substituting other pentoses or hexoses for rhamnose would neither seriously impair the taste if attached at C-2 nor produce a taste if attached at C-6.

POSSIBLE APPLICATIONS

The discovery that the intensely bitter flavanone neohesperidosides can be converted to intensely sweet dihydrochalcone neohesperidosides conceivably has practical significance. The presently accepted artificial sweeteners, saccharin and cyclamate, have enjoyed a steadily increasing demand during the last decade (Fig. 87). Neither they nor the dihydrochalcones are without their own peculiar set of flaws. In the case of the dihydrochalcones, although the sweetness is intense and is not marred by a bitter aftertaste, it is rather slow in its onset, is felt mainly

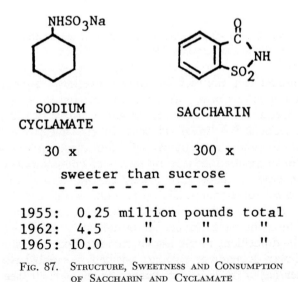

SODIUM CYCLAMATE	SACCHARIN
30 x	300 x

sweeter than sucrose

- - - - - - - - - - -

1955: 0.25 million pounds total
1962: 4.5   "   "   "
1965: 10.0   "   "   "

FIG. 87. STRUCTURE, SWEETNESS AND CONSUMPTION
OF SACCHARIN AND CYCLAMATE

in the back part of the mouth, and is of very long duration. It is described by some as having a slight licorice or menthol-like quality. On the other hand, the dihydrochalcone sweeteners seem to be remarkably free of toxicity in laboratory animals. In experiments carried out by A. N. Booth (unpublished) when neohesperidin dihydrochalcone or naringin dihydrochalcone was fed to rats for 170 days at the very high level of 5% of the diet, no abnormal pathology, changes in growth rate, or toxic effect of any kind were observed. This is in line with the fact that flavonoids as a group are, if nothing else, innocuous. It is also in line with the fact that flavonoids are metabolized by the intestinal flora to carbon dioxide (from the A-ring) and various aromatic acids (from the B-ring), none of which can be considered harmful. Another favorable aspect of the dihydrochalcones is the fact that they contain neither nitrogen nor sulfur, elements that are commonly present in toxic compounds.

It is conceivable, then, that the dihydrochalcones will eventually be used in combination with other sweeteners or in special applications requiring long-lasting sweetness. Whether it will be possible to make derivatives of these compounds having improved taste characteristics while retaining the nontoxic characteristic remains to be seen. Much effort is being directed to this end in various laboratories.

OCCURRENCE AND METABOLISM OF FLAVANONES IN THE PLANT

Early in our studies on citrus flavanones it became apparent that those citrus fruits about which anything was known seemed to contain either all rutinose or all neohesperidose derivatives. In only a few cases, one of which was the grapefruit, was it clear that both rutinosides and neohesperidosides are present. These results have recently been extended and put on a firmer theoretical footing by R. Albach (unpublished) who made a detailed chromatographic survey of a large number of citrus varieties and hybrids. In brief, he found that bonafide citrus species contain either all rutinosyl or all neohesperidosyl flavanones, while hybrids originating from a rutinosyl and a neohesperidosyl parent contain *both* types of flavanone. These findings have important implications in plant breeding and in the very complex field of citrus taxonomy. They also throw some light on possible mechanisms of debittering in citrus fruit. Thus, the bitterness of grapefruit, which is thought to be due chiefly to the presence of naringin, begins to diminish as the fruit reaches maturity. Earlier we had speculated that this was due to either (a) a transfer of rhamnose from the 2- to the 6-position

of glucose to give the tasteless naringenin 7-β-rutinoside or (b) oxidation of naringin to the corresponding flavone, rhoifolin (Fig. 82), which
is also tasteless. The operation of path (a) now seems unlikely. The
fact that grapefruit is a hybrid containing both rutinosides and neohesperidosides merely reflects its origin from one rutinosyl and one
neohesperidosyl parent [presumably *C. sinensis* (oranges) and *C.
grandis* (pummelos)]. It seems improbable that a mechanism for
transferring rhamnose would develop in the progeny when it is absent
in the parents. Whether path (b) is the main route for debittering
remains to be demonstrated.

Quite unexpected was a recent finding by W. Gaffield (unpublished)
on the apparent change in configuration of the aglycone portion of
naringin from 2(S) in small, green grapefruit to a mixture of 2(S) and
2(R) in ripe fruit (Fig. 88). Ordinarily the natural flavanones are

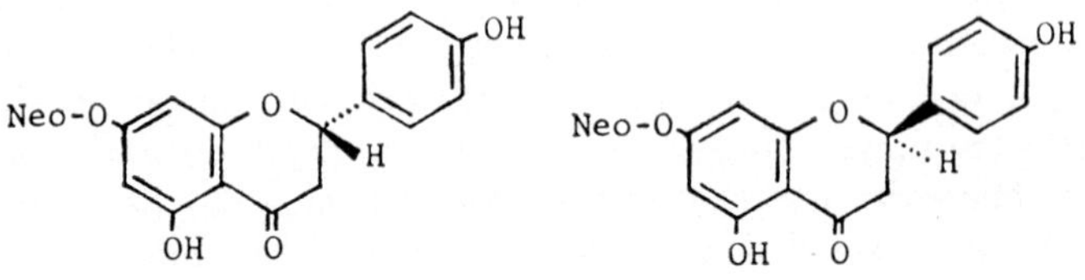

(−)-2(S)-Naringin   (−)-2(R)-Naringin
(green fruit)   (ripe fruit)

FIG. 88. 2(S)- AND 2(R)-CONFIGURATIONS OF NARINGIN

thought to occur entirely in the 2(S)-configuration (Gaffield and Waiss
1968). The change in configuration doubtless has biosynthetic implications but it does not seem to be a factor in debittering, since 2(S)-
and 2(R)-naringin appear to be equally bitter.

SUMMARY

We have tried to show how the two types of flavanone glycosides
occurring in citrus have a well defined relation to bitterness; how certain
of these glycosides can be transformed into intensely sweet substances;
and how modifications of the compounds enable one to formulate tentative conclusions with regard to structure-activity relations. We have
pointed out possible applications of these compounds as sweetening
agents and have reiterated the potential importance of flavanone glycosides in the chemotaxonomy of citrus. It can be expected that the

taste-structure theme will lead to further synthetic variations, but whether this simple approach will bring clarity or confusion remains to be seen.

DISCUSSION

D. French.—I think a sweet taste receptor protein has been isolated from the taste buds of a cow. The response of this protein to various substances depends on whether they have a sweet taste or not. That suggests that the mechanism of tasting involves a combination of the material with a certain receptor protein, perhaps in the taste bud. If this be the case it might not be unreasonable that various substances could compete with one another for sites on the taste protein. Is it possible that in the development of the grapefruit one has the concomitant development of another substance, perhaps another glycoside or maybe a very closely related glycoside which could so effectively compete on the surface of the protein that it excludes the bitter substance? For a concrete test do you have any information as regards the masking of these bitter or sweet substances by other substances? What happens if you have a mixture of a bitter and a sweet glycoside and you taste this mixture?

R. M. Horowitz.—We have only one result along this line. If you take a solution of naringin at threshold concentration so that you can just detect the bitterness and mix it with the corresponding flavone, rhoifolin, which also occurs in grapefruit and which is tasteless, the bitterness disappears. However, it takes a very large amount of the tasteless compound to mask the bitterness. The tasteless compound does not mask the sweetness of the corresponding dihydrochalcone. Although it is conceivable that the formation of taste inhibitors could account for the debittering of grapefruit, we have never isolated any really effective ones from that source.

J. L. Hickson.—When you get this tasting figured out we will know a lot more than we now do. There are several observations which add to this so that, Dexter, your hypothesis is interesting but not too good. There was a paper in SCIENCE several years ago from Monsanto where a wide variety of saccharin derivatives were prepared pure and the sweetness and the bitterness appeared to be concomitant in the same molecule. So it is not necessarily a case of one interfering with the other. This structure of concomitant taste is very interesting and of course one of the things that the sugar industry leans back on.

BIBLIOGRAPHY

BOYD, W. C. 1962. Introduction to Immunochemical Specificity. Interscience Publishers, New York.

EVANS, D. R. 1963. Chemical structure and stimulation by carbohydrates. *In* Olfaction and Taste, Y. Zotterman (Editor). The Macmillan Co., New York.

GAFFIELD, W., and WAISS, A. C. 1968. Optical rotatory dispersion and absolute configuration of flavanones, 3-hydroxyflavanones, and their glycosides. Chem. Comm. *1968*, 29–31.

GORIN, P. A. J., and PERLIN, A. S. 1959. Configuration of glycosidic linkages in oligosaccharides. VIII. Synthesis of α-D-mannopyranosyl- and α-L-rhamnopyranosyl-disaccharides by the Königs-Knorr reaction. Can. J. Chem. 37, 1930–1933.

HARBORNE, J. B. 1967. Comparative Biochemistry of the Flavonoids. Academic Press, New York.

HERTZBERG, S., and JENSEN, S. L. 1967. Bacterial carotenoids. XX. The carotenoids of *Mycobacterium phlei* strain Vera 2. The structures of the *phlei*-xanthophylls—two novel tertiary glucosides. Acta Chem. Scand. *21*, 15–41.

HOROWITZ, R. M. 1964. Relations between the taste and structure of some phenolic glycosides. *In* Biochemistry of Phenolic Compounds, J. B. Harborne (Editor). Academic Press, New York.

HOROWITZ, R. M., and GENTILI, B. 1961. Phenolic glycosides of grapefruit: a relation between bitterness and structure. Arch. Biochem. Biophys. 92, 191–192.

HOROWITZ, R. M., and GENTILI, B. 1963. Flavonoids of citrus. VI. The structure of neohesperidose. Tetrahedron, *19*, 773–782.

HOROWITZ, R. M., GENTILI, B., and HAND, E. S. 1964. Synthesis and characterization of neohesperidose. Abstr. of Papers, IUPAC Symp., Kyoto, Japan, 158–159.

KOLLE, F., and GLOPPE, K. 1936. A new hesperidin. Pharm. Zbl. 77, 421–425.

KRBECHEK, L. *et al.* 1968. Dihydrochalcones. Synthesis of potential sweetening agents. J. Agr. Food Chem. *16*, 108–112.

MONCRIEFF, R. W. 1967. The Chemical Senses. CRC Press, Cleveland, Ohio.

PRIDHAM, J. B. 1965. Phenol-carbohydrate derivatives in higher plants. *In* Advances in Carbohydrate Chemistry, Vol. 20. M. L. Wolfrom, and R. S. Tipson (Editors). Academic Press, New York.

ZEMPLÉN, G., and GERECS, A. 1934. Effect of mercury salts on aceto-halogen sugars. IX. Synthesis of derivatives of β-1-L-rhamnosido-6-D-glucose. Ber. Dtsch. Chem. Ges. 67, 2049–2051.

ZEMPLÉN, G., and GERECS, A. 1935. Constitution and synthesis of rutinose, the biose of rutin. Ber. Dtsch. Chem. Ges. 68, 1318–1321.

CHAPTER 14

Louis M. Massey, Jr.
Miklos Faust[1]

Irradiation—Effects
on Polysaccharides

INTRODUCTION

Radiation research at this laboratory has been concerned largely with the influence of relatively large doses of ionizing radiation on fruits and vegetables, and fruit and vegetable products for possible beneficial effects for food purposes. Although the potential of irradiation treatment for protecting these products from deterioration by microorganisms and autolytic enzymes is great, and there is every indication that such control of product deterioration is possible, adverse effects of the treatment upon most plant materials has been experienced at doses large enough to bring about indefinite stabilization. The principal deleterious effect of radiation upon these foodstuffs is that of loss in texture. As the texture of most fruits and vegetables is intimately related to the polysaccharide chemistry of the tissues, much of our interest has therefore centered around the radiochemistry of these compounds.

This review will concern itself only with the effects of ionizing radiation upon the polysaccharides of plants, with particular emphasis upon the major polysaccharide constituents relating to cell wall structure and texture. Industrial applications of radiation used for graft-polymerization of cellulose and starch with artificial polymers, or crosslinking of unnatural substituted celluloses, etc., will not be covered but have been reviewed elsewhere (Arthur and Blovin 1962; and Adler 1963). In terms of our own interests, the most important experiments are those which have been conducted *in vivo*.

The environment within the living tissues is extremely complex and even the most ingeniously designed model experiments are far from *in situ* conditions. Nevertheless, to elucidate certain effects, polysaccharides have been irradiated either in pure dry form or in relatively simple well defined solutions in which they contribute to our understanding of the mechanism of radiation-induced changes. Some of this work will be included.

Although 30 yr have passed since the degrading effect of ionizing radiation on polysaccharides was first observed (Schoepfle and Connell 1929), most of the knowledge of this field has been accumulated in the past decade. Recent progress is considerable. However, a precise

[1] Present address: U. S. Dept. of Agr., Agr. Res. Serv., Crops Res. Div., Plant Ind. Sta., Beltsville, Md.

model for the mechanism of action of radiation on polysaccharides still does not exist. From a chemical standpoint, the radiation-induced changes in polysaccharides result from physical absorption of radiation leading to ionization or excitation of individual atoms or molecules which undergo alteration. Transmission of energy to the molecule may occur either directly by the release of energy within the structure of the molecule itself, or indirectly by the formation of highly reactive agents (radicals) from water or other cell constituents which then diffuse to critical sites on the molecule and there react. In addition to these chemical effects which occur during or soon after irradiation, there are also indirect biochemical or physiological effects which can be caused by radiation through a disturbance of the normal synthetic or degradative metabolic processes. The latter effects may manifest themselves long after radiation treatment. Hence they may become important in plant organs stored or utilized for extended times after irradiation. Thus the complete picture of radiation-induced changes of *in vivo* polysaccharides is complex.

IRRADIATION EFFECTS

In evaluating the effects of irradiation upon polysaccharides it is difficult to draw qualitative and quantitative comparisons from published data. As will be brought out in the subsequent discussion, there are a number of factors which are known to contribute to this lack of precision.

Nowhere is this matter of comparative precision more evident than in relating the radiation susceptibility of various polysaccharides. In an effort to develop such information we conducted a study with a number of polysaccharides from various sources (Kertesz 1968). Samples representing a number of types and molecular sizes were gamma irradiated in air with a ^{60}Co source in the air-dry state and assayed for loss in specific viscosity as soon as possible following treatment. The threshold dose for loss in viscosity (the lowest dose following which the effect of irradiation is just measurable) was used as a measure of radiation susceptibility. A general positive correlation was found between the size of the molecule as measured by the intrinsic viscosity of the unirradiated material regardless of type and its susceptibility to radiation damage (Fig. 89). Although considerable variation within some of the classes of polysaccharides used for this study were obtained, comparison between the NF pectin series which had been pretreated by dry 100°C heat for various periods of time (a known

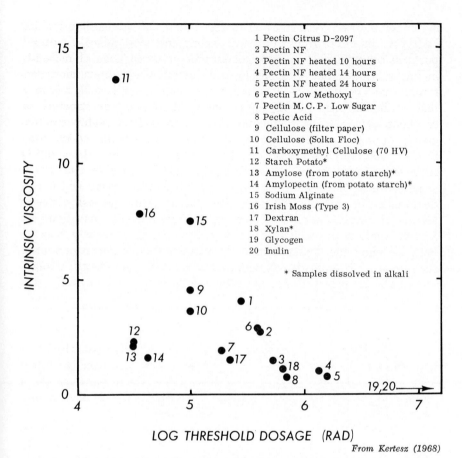

1 Pectin Citrus D-2097
2 Pectin NF
3 Pectin NF heated 10 hours
4 Pectin NF heated 14 hours
5 Pectin NF heated 24 hours
6 Pectin Low Methoxyl
7 Pectin M.C.P. Low Sugar
8 Pectic Acid
9 Cellulose (filter paper)
10 Cellulose (Solka Floc)
11 Carboxymethyl Cellulose (70 HV)
12 Starch Potato*
13 Amylose (from potato starch)*
14 Amylopectin (from potato starch)*
15 Sodium Alginate
16 Irish Moss (Type 3)
17 Dextran
18 Xylan*
19 Glycogen
20 Inulin

* Samples dissolved in alkali

LOG THRESHOLD DOSAGE (RAD)

From Kertesz (1968)

FIG. 89. COMPARATIVE GAMMA RADIATION SUSCEPTIBILITY OF POLYSACCHARIDES

manner of reducing average molecular weight) bears out this relation. Obviously there are discrepancies in this relationship. Let us consider some of the various polysaccharides in detail to illuminate the cause of some of these variations.

Pectin

This structural polysaccharide is believed to contribute significantly to the texture of fruit and vegetable tissues, and its degradation by radiation thus held to be a principal cause of the undesirable softening. For this reason, this polysaccharide has been a prime object of investigation. Unfortunately the *in situ* structure of pectin is considerably more complicated than that of many other polysaccharides making interpretation of the results of many experiments most difficult. Briefly, pectin is composed of α-D(1,4) galacturonic acid residues linked mainly in linear fashion, but with considerable crosslinking with other galacturonic acid chains as well as other compounds. It has been shown that during the course of normal softening of fruits, a depolymerization of pectin occurs. The highly methylated, long-chain protopectin decreases in methyl content, becomes shorter in chain length, and increases in water solubility. More details of the nature of pectin are covered in reviews by Kertesz (1951) and Henglein (1961).

Although we have no direct evidence for the radiation-induced *in situ* degradation of pectin being responsible for tissue texture loss, we have plenty of circumstantial evidence of its involvement. As shown in Fig. 90, there is an excellent correlation between the threshold dose for both softening and degradation of the extracted tissue pectin as well as the rate at which both change with increasing dose (Kertesz *et al.* 1964). It will also be observed that the radiation-induced breakdown of cellu-

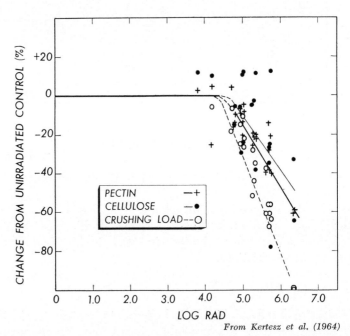

From Kertesz *et al.* (1964)

FIG. 90. OVERALL RELATION BETWEEN SOFTENING OF IRRADIATED APPLE TISSUES AS MEASURED BY THE CRUSHING LOAD AND DEGRADATION OF TISSUE PECTINS AND CELLULOSE AS MEASURED BY CHANGES IN VISCOSITY

lose relates directly to texture in both factors.   Most investigators place emphasis upon the former as the single most important cause of radiation softening (Kertesz *et al.* 1959).

The natural depolymerization of pectin is facilitated by three enzymes: polygalacturonase, capable of breaking the 1,4 linkage and resulting in a shortening of the chain; pectin esterase, capable of reducing the methylesterification (Kertesz 1951; Joslyn 1962); and pectin transeliminase, capable of both (Albersheim *et al.* 1960).   Irradiation of pectin *in situ* results in an effect somewhat similar to that of enzymatic depolymerization.   The results shown in Fig. 91 indicate a radiation-dependent decrease in the water-insoluble fraction with an increase in

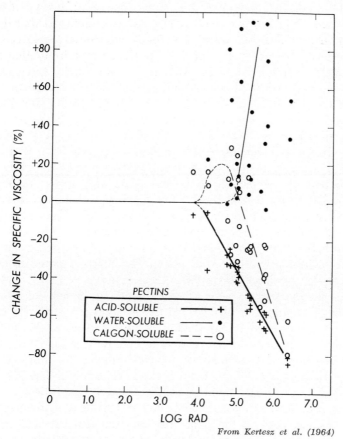

From Kertesz et al. (1964)

FIG. 91. CHANGES IN THE VISCOSITIES OF THREE DIFFERENT
PECTIN EXTRACTS FROM IRRADIATED APPLE TISSUES

the water-soluble fraction isolated from the irradiated tissues (McArdle and Nehemias 1956; Kertesz *et al.* 1964; Massey *et al.* 1964; Massey *et al.* 1965; Rouse *et al.* 1966). Part of the *in situ* irradiated pectin is degraded to such small units that it is lost during subsequent analytical extraction. The loss in total pectin however may be dependent on the tissue. A 41% loss in total pectins was observed when apples were exposed to 1,500,000 rad (1.5 Mrad) of radiation, but only a 17% loss was observed in carrots irradiated at the same dose (McArdle and Nehemias 1956). A similar difference is expressed in the threshold of pectin degradation which is about 80,000 rad (80 Krad) for apples, and 170 Krad in carrots (Kertesz *et al.* 1964). Model experiments conducted with purified pectin show that the radiation-induced degradation proceeds by random fissure of the glycosidic linkages (Dwight and Kersten 1938; Kertesz *et al.* 1956; Skinner and Kertesz 1960), presumably in a manner similar to that found with other polysaccharides.

In contrast, there seems to be little evidence that irradiation results in significant demethylation of pectin. Using model experiments involving purified pectin, Skinner and Kertesz (1960) observed the movement of the irradiated product on electrophoretic plates and found no radiation-induced changes in mobility of the pectin boundary. As a change in esterfication would have changed mobility, they concluded that irradiation did not affect the degree of methylation. Contrary to this observation, Somogyi and Romani (1964) found a slightly lower ester content of pectin extracted from pears and apples immediately after irradiation up to 600 Krad. The maximum change observed was only 2% in pears and less in apples. A similar observation was made in papayas (Young and Lin 1966) and in isolated pectin (Wu 1963). It is probable that a random fissure of the methylester groups may result from irradiation, but it is apparently minor compared to changes in the degree of depolymerization.

Pectin methylesterase activity was found to increase immediately following irradiation in cherries where it appeared that the enzyme activity was increased by irradiation in unripe fruit where activity is generally low, but not in ripe ones where activity is high (Al-Delaimy 1964). An increased enzyme activity was reported to be accompanied by a small decrease in the degree of esterification a few days after irradiation in pears and apples (Somogyi and Romani 1964). In this work enzyme activity measurements and methyl content determinations were not conducted on the same fruit and the real significance of the temporary increase in activity immediately after radiation is difficult to assess. Dennison *et al.* (1967) reported an increase in the activity of pectin methylesterase in juice of oranges expressed immediately after irradiation followed by a decrease in 4 to 7 days. It is not clear whether the increased enzyme activity resulted from an activation of the enzyme or a radiation-induced increase in the extractability of the enzyme. Either could explain the observed results. Pectin trans-eliminase activity has been found to decrease immediately following irradiation (Romani and Sommer 1963). The possibility that increased enzymatic demethylation in the irradiated fruit may contribute largely to pectin degradation and softening of fruits, however, has been ruled out by Maxie and Sommer (1966). They found that the rate of change and the magnitude of the effect in the tissue softening are not compatible with enzymatic capabilities. The radiation effect is temperature-insensitive, and there is no measurable after-effect immediately following treatment in living plant parts.

Despite the possible change in polygalacturonase activity following irradiation, there is evidence that pectin is not metabolized subsequent to irradiation. Massey et al. (1964) found that there was essentially no change in the ratio of water soluble to insoluble pectin in irradiated apples during 5 months of storage, but found a 271% increase in the control fruit from an early harvest date and 164% for fruit from a later harvest date. Similar results were obtained by Clarke (1961) with apples, and by Somogyi and Romani (1964) with both pears and peaches during much shorter postirradiation storage. The latter found that in unirradiated fruit total pectin and protopectin decreased and water soluble pectin increased during storage, but that these fractions remained unchanged in pears at every radiation level and in peaches at higher radiation levels.

In living tissues, in addition to the direct effect of radiation where energy is released within the target molecule itself, pectin may be subject to an indirect effect in which degradation is caused by the production of free radicals through the radiolysis of cell water. Several attempts have been made to elucidate the importance of water in radiation-induced degradation. Because of difficulties in the conduct of in vivo experiments with the relatively complex pectin molecule, reliance must be made on model in vitro experiments including both pectin and the structurally simpler cellulose and starch.

The first indication that water may be protective came from the observation of Glegg and Kertesz (1956) who found that the addition of small amounts of water to dry purified pectin and cellulose completely prevented the "after-effect" due to long lived radiation-induced free radicals. A similar observation was made by Ehrenberg et al.

(1957) who found that the number of breaks produced in the amylopectin polysaccharide chains per unit of energy absorbed decreased with increasing water content up to about 20%. On the other hand, polysaccharides in very dilute solutions were observed to degrade faster than in more concentrated solutions (Bourne *et al.* 1956). Wahba and Massey (1966) investigated the whole range using *in vitro* experiments with purified pectin. Changes in intrinsic viscosity of dilute pectin solutions were used as a measure of the degree of degradation.

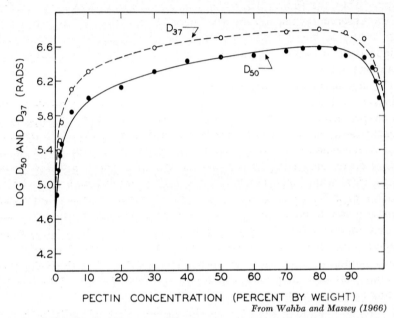

From Wahba and Massey (1966)

Fig. 92. Effectiveness of Radiation in Pectin Degradation at Various Moisture Levels as Determined by D_{50} and D_{37} Values for a Decrease in μ

Pectin was found to be least susceptible to degradation in the moisture range of approximately 10 to 95% moisture. Below 10% moisture, degradation decreased sharply with increasing moisture content, while above 95% moisture, the opposite was true (Fig. 92). Thus the indirect effect of radiation in dilute aqueous solutions decreases progressively with increasing pectin concentration, while the direct effect of radiation at the other end of the scale was decreased by the addition of small amounts of water. Oreshko (1960) observed a similar protective effect in low moisture content starch. Equations to predict the radiolysis at a given moisture level were developed by both investigations.

The significance of these observations to *in vivo* conditions remains uncertain due to lack of information concerning the moisture level in the immediate vicinity of the polysaccharide. It is probable that neither moisture extreme is applicable. In most mature cells the water content of the whole tissue only rarely approaches 95%, and much of it is located in the cell vacuole. Consequently the surroundings of cellulose and pectin in the cell wall and the starch granules in the cytoplasm probably contain too little water to permit the operation of the indirect effect. That the indirect effect is not operative may also be deduced by the observation that the *in situ* pectin degradation is independent of dose rate. It has been shown that the direct effect of radiation is independent of and the indirect effect is dependent on dose rate (Gunckel 1965).

The significance of the protective effect of water on the radiolysis of polysaccharides is more difficult to assess. The exact mechanism by which protection is afforded by water is in itself difficult to explain. A rapid decay of free radicals formed upon irradiation was observed on corn starch when moisture was present (O'Meara and Shaw 1957), the rate of decay increasing by addition of water to 15%. A similar decay up to 15% water was reported in cellulose (Florin and Wall 1963), and up to 20% water in the starch-rich seeds of *Agrostis stolonifera* (Dwight and Kersten 1938). These critical moisture levels are close to that found to be the upper limit of protective effect of water and suggest some connection between the two effects. Florin and Wall (1963) proposed that the principal role of water is to open the crystalline structure of cellulose to diffusion and the long life radicals trapped in this structure may be free to drift away. This theory is supported by observations reported by Mishina (1962A) that the crystalline part of amylose served as a radical trap and that heat and deuterium damaged the crystalline structure causing a decay of radicals. Lee and Chen (1965) reported that the decay of free radicals in irradiated wheat flour starch was brought about by water via a second order reaction presumably through the same type of mechanism. Reuschl and Guilbot (1964, 1966) studied the degradation products resulting from the exposure of potato starch to 30 Mrad of gamma radiation. They concluded from the presence of glucose and hydroxymethyl furfural among the breakdown products in dry starch that a radiation-induced hydrolytic splitting takes place in starch through intramolecular elimination reactions. They postulated that water interfered with this radiolytic reaction by (1) protecting against oxidation reactions which are most evident when hydration is low, (2) protecting against chain rupture when water content is above 21%, and (3) interfering with glucose formation in proportion to hydration. The precise mechanism by which water protects against radiation degradation at these highly concentrated conditions is still largely a matter of conjecture.

A few observations indicate that calcium may play a role in the radiation-induced degradation of pectin. The presence of inorganic ions, especially calcium, protected pectin from degradation when exposed to radiation (Deshpande 1965). However, when calcium was combined with pectic acid in the form of calcium pectate, some of the calcium was split by radiation (Al Jasin and Markakis 1965). Shah (1966) also reported that calcium could be released from plant tissues as well as from model calcium pectate systems by high doses of radiation. These observations have not been completely substantiated by experiments in our own laboratory. The exact role of calcium in the native plant cell wall is still open to question (Ito and Fujiwara 1967), including possible involvement with cell wall proteins (Ginsberg 1961), and the known influence of divalent cations on cell membrane permeability and resulting tissue turgor (Massey 1968). Although calcium ion appears to be related to radiation-induced softening of plant tissues, the involvement with *in situ* pectin under these circumstances is not nearly as clear.

In general, radiation-induced crosslinking occurs less frequently in polysaccharides than in many synthetic polymers. Low doses of radiation, however, can form a thermo-reversible gel of pectin solutions in a narrow pH range (Table 24). Oxygen, nitric oxide, and other free-radical scavengers prevent this crosslinking (Wahba *et al.* 1963). Pectin gels, induced by radiation are similar to gels formed by hydrogen bonding through the addition of multivalent ions such as calcium. The

TABLE 24

VISCOSITY OF GELS FORMED BY IRRADIATION UNDER VACUUM OF 2% AQUEOUS PECTIN SOLUTIONS AT 20°C AFTER ADJUSTMENT OF PH WITH HCL.[1]

pH	Gel Viscosity[2] (10^{-3} Cp)	pH	Gel Viscosity[2] (10^{-3} Cp)
1.0	No gel	1.9	144
1.2	No gel	2.0	132
1.4	70	2.2	120
1.6	124	2.4	82
1.7	136	2.6	48
1.8	146	2.8	No gel
		3.0	No gel

Source: Wahba *et al.* (1966).
[1] The samples were exposed to 630 Krad of gamma radiation for a period of 90 min.
[2] Viscosity measurements were made at 25°C immediately after irradiation.

thermoreversibility of the radiation-induced gels suggests that the linkages in the gel are weak, hydrogen, or other secondary bonding. Pectin gels formed by principal valence-bonds are not thermo-reversible (Deuel *et al.* 1953). Furthermore, the narrow pH range of formation, the action of free-radical scavengers and the low dose required for formation indicate that the crosslinking of pectin is related to the production of free radical intermediates perhaps as the first step in radiation-induced breakdown of pectin. A similar two-step breakdown was postulated for cellulose by Emamura (1963) who found that crosslinks appeared in cellulose before breakdown occurred.

The possibility that crosslinking of polysaccharides occurs *in vivo* as a result of radiation has to be considered. Loos (1962) reported that the toughness of sapwood of *Liriodendron tulipifera* was slightly increased by a low level of radiation (0.1–1 Mrad) but was considerably decreased by higher doses. A slight firming of fruits also has been observed in our laboratory at doses immediately below the degradation threshold (Kertesz *et al.* 1964). Although hardening is not a good indication of crosslinking, it is the only sign so far observed that crosslinking of polysaccharides may go on at low doses *in situ*.

Cellulose

This structural polysaccharide is a linear polysaccharide consisting of 1,4 linked α-D-glucopyranose units. The chains of molecules are generally juxtaposed to form linear crystals, or amorphous microfibrils (Heritage *et al.* 1963). This structure is disturbed by ionizing radiation proportional to the dose received. Schoepfle and Connell (1929) observed that partially purified dry cellulose lost its fibrous structure when irradiated. The loss of structure was found to be the result of extensive depolymerization of the structure (Little, 1952; Saeman *et al.* 1952; Lawton *et al.* 1953; McBurney and Siu 1954; Arthur 1966). The depolymerization was just as rapid for cotton linters as for pulp wood cellulose indicating that the depolymerization by radiation occurs as readily in the crystalline as in the amorphous fractions of cellulose. Polcin (1967) reported that, although gamma irradiation-induced degradation of cellulose occurred above 10 Mrad, as evidenced by significant change in the macromolecular structure with concomitant increase in aldehydes, carbonyls, and peroxidized groups, the physical bonds are sufficient to retain the hypomolecular system intact, although obviously in a weakened state.

We have investigated radiation-induced *in vivo* cellulose degradation in reference to the observed softening of apples, carrots, and beets (Fig.

90). The viscosity of the cellulose extracted from these tissues decreased with increasing radiation dosages indicating the degradation of the cellulose *in situ*. The threshold of degradation occurred in all these tissues at about 80 Krad. Although there was no difference in the threshold for cellulose degradation, the molecules were depolymerized much faster with increasing doses of radiation in apples than in carrots or beets. The threshold of the latter was found to be the same with both tissues (Kertesz *et al.* 1964).

Irradiated cellulose has been shown to contain high energy long-lived radicals within the molecule itself which can be detected readily by electron spin resonance (ESR). Two types of long-lived radicals were reported in irradiated cellulose by Bernard and Gagnaire (1964). One type involved only carbon atom number 6, and the other involved carbon atoms numbers 2 and 3. These radicals appear to be relatively unreactive as evidenced by their relative stability to SO_2 gas (Kuri and Ueda 1961) and long life (Florin and Wall 1963; Neal and Kraessig 1963). Some of the radicals may react after irradiation resulting in either scission or crosslinking. Glegg (1957) and Glegg and Kertesz (1956, 1957) showed that both cellulose and pectin exhibited an "after-effect" following exposure to gamma radiation. Recent studies of the radiation graft-polymerization of cellulose and starch with other polymers also indicate that these irradiated polysaccharides are in an "activated" state and may crosslink (Arthur and Blovin 1962; Mishina 1962B). Differences in crystallinity of cellulose have no obvious effect on yield, or on the nature of radicals determined by ESR spectra (Florin and Wall 1963). The G value for radical formation was found to be 2.8, but the concentration of radicals levels off at 10 Mrad and remains essentially constant at higher dosages (Dalton *et al.* 1963; Florin and Wall 1963).

Saeman *et al.* (1952) calculated from viscosity data that a dose of 100 Mrep might be expected to break about 14 to 15% of the bonds in dry cellulose. As they also found that 100 Mrad caused a 14% loss in glucose resulting from postirradiation hydrolysis, it appears that for every chain fracture one glucosyl unit is destroyed at high dosages (Bovey 1958). Identification of breakdown products indicated several that were smaller than 6 carbon units. Among the gaseous breakdown products, H_2, CO, and CO_2 were identified (Flynn *et al.* 1958; Arthur and Blovin 1962; Dilli and Garnett 1963; Lueck and Dell 1963). In the nongaseous fraction, glucose, cellobiose, cellodextrin, 2-ketogluconic acid and 2-ketocellobionic acid were found, along with several unidentified carbonyl- and carboxy- compounds (McBurney and Siu 1954; Emamura 1963; Lueck and Dell 1963), as well as water (Dilli and Garnett 1963).

It was suspected that the 1,4 glycosidic bond of cellulose was the least stable to breakage by ionizing radiation (Bopp and Sisman 1953). Subsequent work has shown that this is indeed the linkage which is broken most often by irradiation of the cellulose chain (Charlesby 1955; Bovey 1958; Lueck and Dell 1963). Charlesby calculated that 9 ev energy is required per scission in dry cellulose. This value is very close to that required to break a similar bond in dry starch (Ehrenberg and Ehrenberg 1958). The energy requirements for gas, acid, and aldehyde production from cellulose also have been measured. G values for H_2 production were found by Dalton and Houlton (1966) to be 6 at the dose of 0.1 Mrad and decreased as doses increased to a minimum at about 10 Mrad. On the other hand, G values for CO and CO_2 production increased with increasing doses indicating that these gases are probably secondary products (Dalton et al. 1963). G values for aldehyde production suggested that one aldehyde group was produced initially per chain fracture. Acid production, however, was only ⅕ that for chain fracture at 20 Mrad. By interpreting the energy requirement for the various breakdown products, Dalton and Houlton (1966) proposed two distinct but simultaneous reactions for scission of dry cellulose chains. One reaction causes a chain fracture, an aldehyde end group, a hydrogen molecule, and an acidic fragment. The second, which is not yet understood, produces one aldehyde end group for every chain fracture without the formation of hydrogen gas or acid group. They postulated that the second mechanism may be hydrolytic, utilizing the elements of water produced by radiolysis.

Irradiation also makes the cellulose more susceptible to enzymatic degradation. Cellulose was made more fermentable by rumen bacteria (Lawton et al. 1951; Mater 1957) and was found to be more sensitive to both cellulase and cellulose-splitting microorganisms (Kenaga and Cowling 1959; Lueck and Dell 1963). The rate of acid hydrolysis of irradiated cellulose was found to be greater than those of the unirradiated cellulose (Saeman et al. 1952). Florin and Wall (1963) proposed that the disorganization of structure of cellulose caused by irradiation makes access of water to the crystalline or micellar structure much easier. The observed increase in rate of water uptake by irradiated cellulose (Lawton et al. 1951; Kenaga and Cowling 1959; Susler et al. 1964) seems to support this theory.

Starch

The starch content of fruit and vegetable tissues is not directly related to tissue texture and hence of secondary interest to this laboratory in terms of our study of softening as the principal detrimental effect of radiation treatment. However, as starch-containing tissues are being considered seriously for radiation preservation treatment, the effect of radiation upon this reserve carbohydrate is of obvious importance. Flour may ultimately be milled from cereal grains exposed to radiation disinfestation; potatoes, onions, or carrots may be irradiated to prevent sprouting during storage; and starch-containing fruits such as bananas are considered worthy candidates for radiation treatment to prevent untimely ripening during shipment.

Chemically starch is a mixture of amylose and amylopectin polymers of α-D-glycopyranose units, the former being essentially linear chains possessing only 1,4 glucosidic linkages, but the latter being branched by virtue of possessing a limited number of 1,6 linkages. Although some investigators have utilized models of the unresolved starch for these studies, others have used amylose or amylopectin. Kertesz *et al.* (1959A) found that the radiation response of starch and both amylose and amylopectin fraction from the same sample of starch were identical. General conclusions may therefore be drawn from investigations using any of the three systems.

The degradation of starch resulting from radiation may be measured by several methods. The viscosity of starch, amylose, and amylopectin has been found to decrease upon irradiation over a certain threshold (Fig. 93). The threshold value for all 3 materials irradiated in the dry state is about 50 Krad, above which the breakdown is very rapid amounting to about 80% loss at 1 Mrad (Kertesz *et al.* 1959A). Similar observations have been made by others (Khenokh 1950; Brasch *et al.* 1952; Wolfrom 1955, 1958; Long and Lirot 1957; Samec 1958; Mishina and Nikuni 1959A,B; Oreshko and Korotchenko 1960; Renner *et al.* 1963; Sosedov and Shabolenko 1964).

Another method of following degradation is to measure change in the number of end-groups in the molecule as determined by reducing power. The reducing power of starch has been shown to increase upon irradiation (Khenokh 1947; Brasch *et al.* 1952; Wolfrom 1955; Bourne *et al.* 1956; Samec 1958, 1960; Kertesz *et al.* 1959A; Mishina and Nikuni 1959A; Oreshko and Korotchenko 1960), but more slowly than does viscosity values, especially at low levels of radiation. This increase, however, is not as uniform and easy to interpret as the decrease in viscosity.

Changes in iodine-staining power is another measure of the radiolysis of starch. The point of starch degradation at which no iodine-complex formation occurs is called the "achroic point" and the reducing power where this is observed, the "achroic-R-value." This value was reached when the radiation-induced reducing power approached 10% of that obtainable on complete hydrolysis (Bourne *et al.* 1956). A progressive shift of the absorption peak of the iodine-complex to shorter wave-

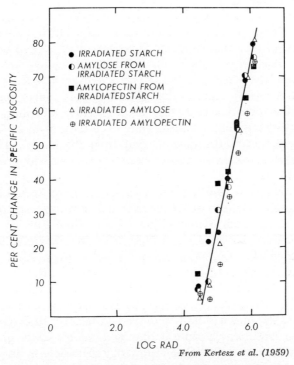

From Kertesz et al. (1959)

FIG. 93. EFFECT OF GAMMA IRRADIATION OF DRY
STARCH AND STARCH FRACTIONS ON SOLUTION
VISCOSITIES IN $1N$ KOH

lengths was also observed when amylose was irradiated in a dilute solution. A nitrogen or vacuum atmosphere increased this shift as compared to an oxygen atmosphere. On the contrary, Kertesz *et al.* (1959A) did not observe a shift in iodine-complex absorption when they irradiated dry amylose in air. The wavelength of the absorption maximum of iodine complex largely depends on the degree of poly-merization of the amylose and it is possible that the shift in the absorption maximum is only a reflection of faster depolymerization of amylose occurring in dilute solution.

Additional measures of starch degradation also have been utilized. Visible changes occurred in starch grains at radiation doses above 5 Mrad (Roberts and Proctor 1955; Roberts 1956). Gelation of starch was prevented by a dose of 1 Mrad (Milner and Finney 1959), and paste-viscosity determination with the amylograph indicated a decrease in molecular size of starch (Milner and Yen 1956; Milner 1957; Milner and Finney 1959).

Development of carboxyl groups has been found when starch was exposed to radiation either in dry form or in solution (Khenokh 1947, 1950; Bourne et al. 1956, Samec 1960; Kertesz et al. 1959A; Sosedov and Shabolenko 1964). G values for carboxyl production in amylose in 0.1% solution have been reported to be 1.5 in oxygen, and 1.4 in vacuum (Bourne et al. 1956). This value is considerably higher than that required for the oxidation of hexose to uronic acid (Phillips 1954) and probably represents mainly the oxidation of the polymer chain rather than the oxidation of the glucose split from the chain. Theories for oxidative decomposition of starch have been advanced (Reuschl and Guilbot 1964, 1966).

Products of starch resulting from radiolysis have been identified as glucose, maltose, maltotriose, formaldehyde, dihydroxyacetone, glyoxal, gluconic acid, and hydroxymethyl furfural (Wolfrom 1955; Bourne et al. 1956; Mishina and Nikuni 1959, 1959A, 1960; Deschreider 1960; Reuschl and Guilbot 1964, 1966; Putilova and Traubenberg 1965). Concomitantly, H_2, CO, and CO_2 gases are also produced by radiation (Mishina and Nikuni 1960; Oreshko and Korotchenko 1960; Oreshko et al. 1962; Dilli and Garnett 1963). An increased absorption in the ultraviolet range of 220–265 mμ has been observed (Khenokh 1955; Oreshko and Korotchenko 1960) which could be due to a number of the above breakdown products. An interesting aspect of the radiolysis of polysaccharides illustrative of the complexity of the problem is brought out by Lofroth and Ehrenberg (1960) who reported finding starch was decomposed when mixed with unidentified decomposition products obtained by irradiating solutions of glucose, itself a degradation product of starch.

Radiation-induced changes of *in vivo* starch have also been investigated. Results of irradiating air-dry cereal grain indicate that the radiation effects on tissue starch are very similar to that described in dry isolated form (Milner and Yen 1956; Milner and Finney 1959; Lai *et al.* 1959; Takaoka 1960; Doguchi *et al.* 1961). From a biochemical standpoint, this may result in a number of interacting effects. Radiation has been reported to increase the susceptibility of starch to the hydrolytic enzymes, diastase (Sosedov and Vakar 1961; Deschreider 1966) and β-amylase (Milner 1957; Milner and Finney 1959; Lee 1959; Mishina and Nikuni 1959A,B), but not to α-amylase (Okada *et al.* 1960; Linko and Milner 1960; Faust and Massey 1966). The enzymes themselves may also be inactivated. Both α- and β-amylase have been found partially inactivated in irradiated wheat (Bure 1959), and phosphorylase was severely affected when irradiated in solution (Phillips and Griffiths 1965). Radiation has also been shown to cause a severe biochemical lesion in the synthesis of α-amylase in barley endosperm amounting to complete inhibition of enzyme synthesis. Consequently, less starch was hydrolyzed in irradiated endosperms when they were exposed to conditions in which starch hydrolysis is known to occur (Faust and Massey 1966).

It is even more difficult to evaluate the effect of radiation on starch in large storage tissues of plants. While starch in apples is degraded as a result of irradiation, no change in starch of onion, garlic (Salkova 1963), or chestnut (Iwata and Ogota 1959, 1961) was found immediately after irradiation. Metabolism of starch in irradiated potatoes has also been investigated since the accumulation of sugars in potatoes during storage is undesirable. Investigators have tried to correlate the decrease in starch resulting from irradiation with the increase in sugar content. Several studies have found an increase in sugar content soon after irradiation (Brownell *et al.* 1957; Schreiber and Highlands 1958; Cloutier *et al.* 1959; Truelsen 1964), others only after long storage (Rubin and Metlitsky 1961). There was little or no difference in total starch content found as a result of irradiation (Sereno *et al.* 1957; Schreiber and Highlands 1958; Perdomo *et al.* 1965). Parks *et al.* (1958) reported that temperature and storage time contributed to reversing the rise in sugar content induced by radiation. This finding emphasizes the importance of plant metabolism in evaluating the *in vivo* effects of radiation. Difficulties in the interpretation of data obtained from research with potatoes are twofold. Most dosages given to desprout potatoes are very low (below 20 Krad) and the effect on starch was investigated only as a quality characteristic. No experiments have been reported with potatoes which study the effect of radiation on starch *per se*. Further, the comparison of starch and sugar content during storage of any tissue is meaningless from a physiological standpoint without study of other factors which may utilize sugars, such as respiration, etc.

Dextran

This polysaccharide is synthesized by a number of species of lactic-
acid bacteria of which *Leuconostoc mesenteroides* is the best known.
It is composed of D-glucopyranose residues joined principally by 1,6-
linkages, but possessing various degrees of branching through 1,3 linkages. Natural dextran has a molecular weight range from 20 to 500
million.

Dextran is degraded as a result of irradiation, and this process has
been used to produce a dextran of molecular weight between 50 to 75
thousand suitable for a blood plasma extender (General Electric Co.
1956). This decrease in molecular-weight is reflected by a decrease in
viscosity (Price *et al.* 1954; Rickets and Rowe 1954; General Electric Co.
1956), by light scattering (Price *et al.* 1954), and by an increase in
reducing groups at 100 Mrad dose, also accompanied by an increase
in branching. They calculated that out of 5 ion pairs, 1.0 produces a
new branch, 1.1 produces a break in the chain, and the rest produce
rupture of the glucose ring without either degradation or branching.
Branching also was calculated by Bovey (1958) who found the apparent
energy requirement of 130 ev per scission too high for a simple breakage of chain. Assuming that the value of 9 ev found for cellulose by
Charlesby (1955) and for starch by Ehrenberg and Ehrenberg (1958)
also held for dextran, he calculated that the β/α ratio (the probability
of crosslinking) was 2.14. Whenever this ratio is larger than 2, gel
will not set regardless of crosslinking. Flynn *et al.* (1967) found that
crosslinking predominated over chain scission in concentrated aqueous
solutions of dextran in the absence of oxygen. Postirradiation viscosity
degradation was due to hydrogen peroxide from the radiolysis of water.
Irradiation of solid dextran resulted in primarily degradative reactions.

Several breakdown products resulting from the radiolysis of dextran
have been identified, and found to be very similar to those found resulting from the radiolysis of starch. These include D-glucose, isomaltose, isomaltotriose, D-gluconic acid, D-glucoronic acid, glyoxal, and
erythrose (Phillips and Moody 1958). A 6% loss in carbon at the dose
of 10 Mrad, and 15% at 100 Mrad also was observed (Price *et al.*
1954). This loss might be accounted for by CO and CO_2 gas formation,
although these gases were not identified. In the case of dextran as
with other polysaccharides, 2 independent scission processes were postulated (Phillips 1961). One is hydrolytic, the other is oxidative. The
presence of glyoxal and erythrose was thought to result from secondary
breakdown (Phillips *et al.* 1958).

Other Polysaccharides

The radiation-induced breakdown of a number of other polysaccharides such as agar (Kersten and Dwight 1937; Khenokh 1950; Glegg and Kertesz 1957; Iwata and Ogota 1961), alginic acid (Feinstein and Nejelski 1955), various gums (Khenokh 1950), and inulin (Jansen and MacDonnell 1945; Khenokh 1950; Wolfrom 1958; Iwata and Ogota 1961) also have been reported. Doshi and Rao (1967) reported that seaweed irradiated with low doses produced agar with better gelling power which they attributed to the radiolytic splitting of sulfur from the molecule.

From these brief investigations it appears that the effect of radiation on these compounds does not differ greatly from those discussed previously, but at this time there is little evidence upon which to build a case.

CONCLUSION

Most of our knowledge about the radiation-induced breakdown of plant polysaccharides has come from investigations on isolated polysaccharides which were subjected to *in vitro* radiation either in solution or in dry form. Two mechanisms have been proposed for degradation, one of them oxidative and the other hydrolytic. Many of the breakdown products have been identified. Radical formation and many effects of water on the breakdown have been elucidated. It is also possible to detect radiation-induced crosslinking in most polysaccharides.

In contrast, very little is known about the radiation-induced breakdown of the same polysaccharides *in situ*. They are apparently degraded but the mechanism is unknown. It can only be speculated that the indirect effect of radiation may not be operative in the *in situ* condition. A few observations indicate that crosslinking may occur and that this may be responsible for the decreased metabolism of polysaccharides following irradiation. Radiation may also interrupt the synthesis of hydrolytic enzymes and thus affect polysaccharide breakdown *in vivo*, but each of these points must still be confirmed.

The extension of knowledge into this latter area of investigation may not be easy. Few if any techniques existing today are readily applicable for such studies. There is much to learn using existing techniques, such as the study of the enzymatic mechanism involved in the utilization of irradiated polysaccharides in irradiated tissues, or the influence of irradiation on cofactors or inhibitors involved in enzymatic activities. However, the elucidation of some of the more involved mechanisms of radiolysis within the specialized environment exisiting at the cellular level must await the development of specialized techniques. It is hoped that such progress may be forthcoming in the not-too-distant future, for it is then and only then that the real evaluation of the practicality of irradiation technology in the food industry may be made.

DISCUSSION

K. Ward.—Do you think that it is possible that the protective action of water may be due to increased mobility of the macromolecules with a resulting disappearance of radicals by combination? Could you get a recombination and a termination of your radicals by a combination of two of them?

L. M. Massey.—In the case of low-moisture protection, I don't think so. The concentration/protection curve is in the wrong direction for such an effect. Also, in our gelation study, the cross-linking of pectin was extremely pH specific. Although I would not say it could not be possible, I would look to another explanation, such as an unfolding or a distortion permitting migration of high energy particles from the molecule.

H. Cory.—Can you tell me whether in the pectin solutions, which I would expect to be nonNewtonian, radiation affected their nonNewtonian character? I am referring in particular to viscosity which is essentially meaningless if nonNewtonian. Pectins might show a yield region or a yield point at low rates of shear and then become rather fluid. As these materials are worked, they may become increasingly fluid. Is this property altered by radiation?

L. M. Massey.—Although there has been some observations with pectin gels which might suggest such an effect, our study was not carried this far. In terms of dilute pectin solutions, viscosity curves for irradiated pectins are similar in shape to those of nonirradiated pectins, and as far as I know, there is no evidence for that type of reaction.

P. Muneta.—With the very well known detrimental effects of the presence of oxygen, in some of your radiation experiments did you try the irradiation in the presence and absence of oxygen?

L. M. Massey.—Yes. From a cross-linking standpoint, oxygen prevents cross-linking. From a degradation standpoint, it has little or no effect.

A. S. Perlin.—You had carboxymethylcellulose on your first slide. Could you tell me the significance of that?

L. M. Massey.—We have done some work with carboxymethylcellulose which I did not discuss. Carboxymethylcellulose is interesting from the standpoint that it forms a very stable gel with low dosages of radiation within a rather narrow pH range. This is a very stable gel. Unfortunately we get a lot of interference with free radical scavengers including sugar.

A. S. Perlin.—In that connection, too, you mentioned the increase in cellulolytic response of irradiated cellulose. In this regard, how far does the cellulolytic breakdown go? Is this just a question of an observed change in viscosity, or does one really get down to a high yield of low molecular weight fragments? The point I am interested in is if you modify the sugar moieties to any large extent, you interfere with the specificity of the enzyme.

L. M. Massey.—Radiation does favor the rate of cellulolytic action. It seems to me that the explanation made by the authors of the works cited here, of the opening up of the molecule making water more available to hydration, is pretty fair. You have an interesting point, however, but I am afraid I cannot contribute to your thinking.

P. Nordin.—What type of linkage is postulated in the cross-linking?

L. M. Massey.—With pectin, cross-linking is probably a secondary hydrogen bond or another thermolabile link. With carboxymethyl-cellulose, it seems to be a more substantial linkage.

P. Nordin.—How is the cross-linking demonstrated?

L. M. Massey.—We have partially characterized the bond by studying the effect of various additives which are known to influence the formation of cross-linking, with various degrees of success. But the principal demonstration of this phenomenon is an increase in viscosity including gelation.

P. C. Markakis.—Would you care to comment on the alleged toxicity of irradiated sucrose?

L. M. Massey.—When I prepared this presentation, I considered the possibility of including this point. The fact that sugars are degraded by radiation has been known for a long time, and a lot of the degradation products have been known for a long time. A couple of years ago, some work was published which indicated that some of these degradation products might have some toxic effects when tested under certain specific conditions. This has lead to quite a flurry of research activity, but to date I have seen few of the results in press. When more data are published a review of this nature should include it. Some toxic effects have been demonstrated under very specialized conditions. Whether or not this could or should be extrapolated to the food industry at its present status, I don't know.

R. A. Gallop.—The use of irradiation to soften dried beans is being investigated. How can we possibly explain the differential resistance to radiation of soft cellulose walls vs. the firm starch granule which occur concurrently in the cotyledons of beans?

L. M. Massey.—We too have shown this with irradiating a number of dehydrated foods. I do not know why it is that radiation picks out the harder portion of the bean to soften before it does the softer.

BIBLIOGRAPHY

ADLER, G.   1963.   Cross-linking of polymers by radiation.   Science *141*, 321–329.

AL-DELAIMY, K. A.   1964.   Pectic substances and pectic enzymes of fresh and processed Montmorency cherries.   Thesis, Mich. State Univ. Dissertation Abstr. *25*, 4.

AL JASIM, H., and MARKAKIS, P.   1965.   Radiation induced calcium release from plant tissues.   Mich. State Univ. Quart. Bull. *47*, 505–507.

ALBERSHEIM, P., NEUKOM, H., and DEUEL, H.   1960.   Formation of unsaturated degradation products from pectin degrading enzymes.   Helv. Chem. Acta *43*, 1422–1426.   (German)

ARTHUR, J. C.   1966.   Intramolecular energy transfer in cellulose and related model compounds.   *In* Energy Transfer in Radiation Processes, G. O. Phillips, (Editor).   Elsevier Publishing Co. New York.

ARTHUR, J. C., and BLOVIN, F. A.   1962.   Radiation induced graft copolymers of cellulose.   U.S. At. Energy Comm. *TID-7643*, 319–334.

BERNARD, O., and GAGNAIRE, D.   1964.   Contribution to the study of irradiation effects on cellulose by EPR.   *In* Electronic Magnetic Resonance and Solid Dielectrics.   North-Holland Publishing Co. Amsterdam. Nucl. Sci. Abstr. *19*, 9237 (1965).

BOPP, C. D., and SISMAN, O.   1953.   Radiation stability of plastics and elastomers.   Rept. *ORNL-1373* (Suppl. to *ORNL-928*, 1951) Nucl. Sci. Abstr. *8*, 2792 (1954).

BOURNE, E. J., STACEY, M., and VAUGHAN, G.   1956.   The action of gamma radiation on dilute aqueous solutions of amylose.   Chem. Ind. (London), 573–574.

BOVEY, F. A.   1958.   The effect of ionizing radiation on natural and synthetic high polymers.   Polymer Rev., Vol. I.   Interscience–John Wiley and Sons, New York.

BRASCH, A., HUBER, W., and WALY, A.   1952.   Radiation effects as a function of dose rate.   Arch. Biochem. Biophys. *39*, 245–247.

BROWNELL, L. E. *et al.*   1957.   Storage properties of gamma irradiated potatoes.   Food Technol. *11*, 306–312.

BURE, J.   1959.   Milling tests for the treatment of flour with ionizing rays.   Getreide und Mehl *9*, 133–135.   (German)

CHARLESBY, A.   1955.   The degradation of cellulose by ionizing radiation.   J. Polymer Sci. *15*, 263–270.

CLARKE, I. D.   1961.   Some effects of gamma radiation on the chemical and physiological changes in fruits.   *In* Recent Advances in Botany, Sect. 11, University of Toronto Press, Toronto, Canada.

CLOUTIER, J. *et al.*   1959.   The effect of storage on the carbohydrate content of two varieties of potatoes grown in Canada and treated with gamma radiation.   Food Res. *24*, 659–664.

DALTON, F. L., and HOULTON, M. R.   1966.   Product yields from electron irradiated cotton cellulose.   Radiation Res. *28*, 576–584.

DALTON, F. L., HOULTON, M. R., and SYKES, J. A.   1963.   Gas yield from electron irradiated cotton cellulose.   Nature *200*, 862–864.

DENNISON, R. A., AHMED, E. M., and MARTIN, F. G.   1967.   Pectin-esterase activity in irradiated "Valencia" oranges. Proc. Am. Soc. Hort. Sci. *91*, 163–168.

DESCHREIDER, A. R. 1960. Changes in starch and its decomposition products after treatment of wheat flour with gamma rays. Starke *12*, 197–201. (French)

DESCHREIDER, A. R. 1966. Effects of gamma radiation on the constituents of wheat flour. *In* Food Irradiation, Proc. Series. Intern. At. Energy Agency, Vienna, Austria. (French)

DESHPANDE, S. N. 1965. Degradation of pectinic acid by ionizing radiations and by pectic enzymes. Thesis. Purdue Univ. Dissertation Abstr. *26*, 607.

DEUEL, H., SOLMS, J., and ALTERMATT, H. 1953. Pectins and their properties. Viertelahrschr. Naturforsch. Ges. Zurich. *98*, 49–86. Chem. Abstr. *47*, 9523i (1953).

DILLI, S., and GARNETT, J. L. 1963. Effect of ionizing radiation on crystalline carbohydrates. Chem. Ind. (London), 409–410.

DOGUCHI, M., YOKOYAMA, Y., and OKADA, O. 1961. The effect of gamma irradiation on wheat starch. Kogyo Kagaku Zasshi *64*, 2001–2005.

DOSHI, Y. A., and RAO, P. S. 1967. Stable agar by gamma irradiation. Nature *216*, 931–932.

DWIGHT, C. H., and KERSTEN, H. 1938. The viscosity of sols made from X-irradiated pectin. J. Phys. Chem. *42*, 1167–1169.

EHRENBERG, A., and EHRENBERG, L. 1958. The decay of x ray induced free radicals in plant seeds and starch. Arkiv Fysik *14*, 133–141.

EHRENBERG, L., JAARMA, M., and ZIMMER, E. C. 1957. The influence of water content on the action of ionizing radiation on starch. Acta Chem. Scand. *11*, 950–956.

EMAMURA, I. 1963. Effects of radioactivity on cellulose. Genshiryoku Kogyo *9*, 62–68. (Japanese) Nucl. Sci. Abstr. *19*, 13431 (1965).

FAUST, M., and MASSEY, L. M. JR. 1966. The effect of ionizing radiation on starch breakdown in barley endosperm. Rad. Res. *29*, 33–38.

FEINSTEIN, R. N., and NEJELSKI, L. L. 1955. Alginic acid as a model for testing the effects of radiomimetic agents. Rad. Res. *2*, 8–14.

FLORIN, R. E., and WALL, L. A. 1963. Electron spin resonance of gamma irradiated cellulose. J. Polymer Sci. (A-1) *1*, 1163–1173.

FLYNN, J. H., WALL, L. A., and MORROW, W. L. 1967. Irradiation of dextran and its aqueous solutions with Cobalt-60 gamma rays. J. Res. Natl. Bur. Std. *71A*, 25–31.

FLYNN, J. H., WILSON, W. K., and MORROW, W. L. 1958. Degradation of cellulose in a vacuum with ultraviolet light. J. Res. Natl. Bur. Std. *60*, 229–233.

GENERAL ELECTRIC CO. 1956. British Patent 764,547. Exposed Dec. 28.

GINSBURG, B. Z. 1961. Evidence for a protein gel structure cross-linked by metal cations in the intercellular cement of plant tissues. J. Exptl. Bot. *12*, 85–107.

GLEGG, R. E. 1957. The influence of oxygen and water on the after effect in cellulose degradation by gamma rays. Rad. Res. *6*, 469–473.

GLEGG, R. E., and KERTESZ, Z. I. 1956. After effect in the degradation of cellulose and pectin by gamma rays. Science *124*, 893–894.

GLEGG, R. E., and KERTESZ, Z. I. 1957. Effect of gamma radiation on cellulose. J. Polymer Sci. 26, 289–297.

GUNCKEL, J. E. 1965. Modification of plant growth and development induced by ionizing radiations. In Encyclopedia of Plant Physiology, Vol. 15/2. W. Ruhland (Editor). Springer-Verlag New York, Inc., New York.

HENGLEIN, F. A. 1961. The uronides and polyuronic acids. In Encyclopedia of Plant Physiology, Vol. 6. W. Ruhland (Editor). Springer-Verlag New York, Inc., New York.

HERITAGE, K. J., MANN, J., and ROLDAN-GONZALEZ, L. 1963. Crystallinity and the structure of celluloses. J. Polymer Sci. (A-1) 1, 671–685.

ITO, A., and FUJIWARA, A. 1967. Functions of calcium in the cell wall of rice leaves. Plant Cell Physiol. 8, 409–422.

IWATA, T., and OGOTA, K. 1959. Studies in the storage of chestnuts treated with gamma radiation. Bull. Univ. Osaka Prefect. Ser. B. 9, 59–65.

IWATA, T., and OGOTA, K. 1961. Studies in the storage of chestnuts treated with radiation II. Bull. Inst. Chem. Res. Kyoto Univ. 39, 112–119.

JANSEN, E. F., and MACDONNELL, L. R. 1945. Influence of methoxyl content of pectic substances on the action of polygalacturonase. Arch. Biochem. 8, 97–112.

JOSLYN, M. A. 1962. The chemistry of protopectin. A critical review of historical data and recent developments. Advan. Food Res. 11, 1–107.

KENAGA, D. L., and COWLING, E. B. 1959. Effect of gamma radiation on ponderosa pine. Forest Prod. J. 9, 112–116.

KERSTEN, H., and DWIGHT, C. H. 1937. The viscosity of sols made from X-irradiated agar. J. Phys. Chem. 41, 687–689.

KERTESZ, Z. I. 1951. The Pectic Substances. Interscience—John Wiley and Sons, New York.

KERTESZ, Z. I. 1968. Unpublished data. Cornell Univ., Geneva, N.Y.

KERTESZ, Z. I., EUCARE, M., and FOX, G. 1959. A study of apple cellulose. Food Res. 24, 14–19.

KERTESZ, Z. I., MORGAN, B. H., TUTTLE, L. W., and LAVIN, M. 1956. Effect of ionizing radiation on pectin. Rad. Res. 5, 372–381.

KERTESZ, Z. I., SCHULZ, E. R., FOX, G., and GIBSON, M. 1959A. Effect of ionizing radiation on plant tissues. IV. Some effects of gamma radiation on starch and starch fractions. Food Res. 24, 609–617.

KERTESZ, Z. I. et al. 1964. Effect of ionizing radiation on plant tissues. III. Softening and changes in pectins and cellulose of apples, carrots and beets. J. Food Sci. 29, 40–48.

KHENOKH, M. A. 1947. The influence of penetrating radium on the colloid chemical properties of starch sols. J. Gen. Chem. (USSR) 17, 1024–1029. (Russian) Chem. Abstr. 42, 3665h (1948).

KHENOKH, M. A. 1950. Cleavage of macromolecules of natural high molecular substances under the action of gamma rays. J. Gen. Chem. (USSR) 20, 1560–1567. (Russian) Chem. Abstr. 45, 963d (1951).

KHENOKH, M. A. 1955. Effects of γ-radiation of cobalt-60 on carbohydrates. Doklady Akad. Nauk (SSSR) 104, 746–749. (Russian) Chem. Abstr. 50, 7603a (1956).

KURI, Z., and UEDA, H. 1961. Electron spin resonance studies of the radicals produced in high polymers by γ irradiation and their reaction with sulfur dioxide. J. Polymer Sci. 50, 349–359.

LAI, SING PING, FINNEY, K. F., and MILNER, M. 1959. Treatment of wheat with ionizing radiation. IV. Oxidative physical, and biochemical changes. Cereal Chem. 36, 401–411.

LAWTON, E. J., BUECHE, A. M., and BALWIT, J. S. 1953. Irradiation of polymers by high energy electrons. Nature 172, 76–77.

LAWTON, E. J. et al. 1951. Some effects of high velocity electrons on wood. Science 113, 380–382.

LEE, C. C. 1959. The baking quality and maltose value of flour irradiated with ⁶⁰Co gamma rays. Cereal Chem. 36, 70–77.

LEE, C. C., and CHEN, C. 1965. A note on the disappearance of radicals trapped in gamma-irradiated starch and gluten. Cereal Chem. 42, 573–576.

LINKO, P., and MILNER, M. 1960. Treatment of wheat with ionizing radiations. V. Effect of gamma radiation on some enzyme systems. Cereal Chem. 37, 223–227.

LITTLE, K. 1952. Irradiation of linear high polymers. Nature 170, 1075–1076.

LOFROTH, G., and EHRENBERG, L. 1960. Degradation of starch produced by indirect effect of γ-irradiated glucose. In Proceedings 1st Nordic Meeting on Food Processing by Ionizing Radiation. Riso 16, 37–38. Nucl. Sci. Abstr. 15, (1960, 1961).

LONG, L. JR., and LIROT, S. J. 1957. Action of ionizing radiation on carbohydrates and polysaccharides. In Radiation Preservation of Food. U.S. Army Res. Develop. Ser. I, 126–132. U.S. Gov. Printing Office, Washington, D.C.

LOOS, W. E. 1962. Effect of gamma radiation on the toughness of wood. Forest. Prod. J. 12, 261–264.

LUECK, H., and DELL, F. 1963. Radiation effect on cellulose and plant odorous substances. Chimia (Switz.) 17, 1–8. Nucl. Sci. Abstr. 17, 14325 (1963).

MASSEY, L. M., JR. 1968. Tissue texture and intermediary metabolism of fresh fruits and vegetables. In Preservation of Fruit and Vegetables by Radiation. Panel Proc. Ser., 105–124. Intern. At. Energy Agency, Vienna, Austria.

MASSEY, L. M., JR., PARSONS, G. F., and SMOCK, R. M. 1964. Some effects of gamma radiation on the keeping quality of apples. J. Agr. Food Chem. 12, 268–274.

MASSEY, L. M., JR. et al. 1965. Effect of gamma radiation on cherries. J. Food Sci. 30, 759–765.

MATER, J. 1957. Chemical effects of high energy irradiation of wood. Forest Prod. J. 7, 208–209.

MAXIE, E. C., and SOMMER, N. F. 1966. Changes in some chemical constituents in irradiated fruits and vegetables. In Preservation of Fruit and Vegetables by Radiation. Panel Proc. Ser., 39–56. Intern. At. Energy Agency, Vienna, Austria.

McARDLE, F. J., and NEHEMIAS, J. V. 1956. Effects of gamma radiation on the pectic constituents of fruits and vegetables. Food Technol. 10, 599–601.

McBurney, L. F., and Siu, R. G. H. 1954. Degradation of cellulose. *In* High Polymers, Part I. Cellulose and Cellulose Derivatives, 2nd Edition, Vol. 4, E. Ott, and H. M. Spurlin (Editors). Interscience–John Wiley and Sons, New York.

Milner, M. 1957. Some effects of gamma irradiation on the biochemical, storage and breadmaking properties of wheat. Cereal Sci. Today 2, 130–133.

Milner, M., and Finney, K. F. 1959. Changes in wheat from gamma radiation. Getreide Mehl 9, 110–112. (German)

Milner, M., and Yen, Y. C. 1956. Treatment of wheat with ionizing radiation. III. The effect on breadmaking and related properties. Food Technol. 10, 528–531.

Mishina, A. 1962A. Graft copolymerization of amylose with acrylamide by preirradiation in vacuo. Nippon Nogeikagaku Kaishi 36, 617–620.

Mishina, A. 1962B. Electron spin resonance of free radicals formed in amylose by gamma irradiation. Nippon Nogeikagaku Kaishi 36, 620–623. Chem. Abstr. 62, 3555h, 1965.

Mishina, A., and Nikuni, Z. 1959A. Physical and chromatographical observations of γ-irradiated potato starch granules. Nature 184, 1867.

Mishina, A., and Nikuni, Z. 1959B. The change of properties of potato starch granules by gamma irradiation. Nippon Nogeikagaku Kaishi 33, 931–936.

Mishina, A., and Nikuni, Z. 1960. The effect of gamma rays on potato starch. Mem. Inst. Sci. Ind. Res., Osaka Univ. 17, 215–218.

Neal, J. L., and Kraessig, H. A. 1963. Degradation of cellulose with megavolt electrons. Tappi 46, 70–72.

Okada, S., Kraunz, R., and Gassner, E. 1960. Radiation induced changes in susceptibility of substrates to enzymatic degradation. Rad. Res. 12, 607–612.

O'Meara, J. P., and Shaw, T. M. 1957. Detection of free radicals in irradiated food constituents by electron paramagnetic resonance. Food Technol. 11, 132–136.

Oreshko, V. F. 1960. The protective action of water in the radiolysis of starch. Zhur. Fiz. Khim. (USSR) 34, 2396–2398. (Russian) Nucl. Sci. Abstr. 15, 8859 (1961).

Oreshko, V. F., and Korotchenko, K. A. 1960. Destruction of starch depending on the dose of ionizing γ-radiation. Doklady Akad. Nauk (USSR) 133, 1219–1222. (Russian) Chem. Abstr. 54, 23385g (1960).

Oreshko, V. F., Gorin, L. F., and Rudenko, N. V. 1962. Composition of gaseous products of starch radiolysis. Zhur. Fiz. Khim. (USSR) 36, 1084–1085. (Russian) Nucl. Sci. Abstr. 17, 208 (1963).

Parks, N., Cloutier, J. A., and MacQueen, K. F. 1958. Gamma irradiation of potatoes. Proc. Intern. Hort. Congr. Nice. 1, 223–241.

Perdomo, M. A., Hernandez, J. A., and Sanin, S. J. 1965. Gamma radiation preservation of potatoes. *In* 5th Inter-American Symp. Peaceful Application of Nuclear Energy, 277–283, Washington, D.C.

Phillips, G. O. 1954. Action of ionizing radiation on aqueous solutions of carbohydrates. Nature, 173, 1044–1045.

Phillips, G. O. 1961. Radiation chemistry of carbohydrates. *In* Advan.

Carbohydrate Chem., Vol. 16, M. L. Wolfrom (Editor). Academic Press, New York.

PHILLIPS, G. O., and GRIFFITHS, W. 1965. Radiation inactivation of potato phosphorylase. Rad. Res. 26, 363–377.

PHILLIPS, G. O., and MOODY, G. J. 1958. Radiation chemistry of carbohydrates. II. Irradiation of aqueous solutions of dextran with gamma radiation. J. Chem. Soc., 3534–3539.

PHILLIPS, G. O., MOODY, G. J., and MATTOK, G. I. 1958. Radiation chemistry of carbohydrates. I. The action of ionizing radiations on aqueous solutions of D-glucose. J. Chem. Soc., 3522–3534.

POLCIN, I. 1967. Radiation effects in chemical and super-molecular structure of cellulose. Jad. Energ. 13, 145–149. (Czech.) Nucl. Sci. Abstr. 21, 30466 (1967).

PRICE, F. P., BELLAMY, W. D., and LAWTON, E. J. 1954. Effect of high velocity electrons on dry dextran. J. Phys. Chem. 58, 821–824.

PUTILOVA, I. N., and TRAUBENBERG, S. E. 1965. Effects of γ-irradiation on starch acidity. Prikl. Biokhim. Mikrobiol. 1, 538–543. (Russian) Nucl. Sci. Abstr. 20, 33367 (1966).

RENNER, K., SEIFERT, J., and GERHARDS, K. P. 1963. Dose dependent changes to the amylose molecules of a 2.5% starch solution after ^{60}Co irradiation. Strahlentherapie 120, 81–86. Nucl. Sci. Abstr. 17, 16087 (1963).

REUSCHL, H., and GUILBOT, A. 1964. A contribution to study of the action of ionizing radiations on starch. Effects of gamma radiations on the water holding capacity of potato starch. Ann. Technol. Agr. 13, 399–448. (French)

REUSCHL, H., and GUILBOT, A. 1966. The influence of gamma rays on potato starch in relation to its water content. Starke 18, 73–77.

RICKETS, C. R., and ROWE, C. E. 1954. The effect of gamma rays upon dextran. Chem. Ind. (London), 189–190.

ROBERTS, E. A. 1956. The Microstructure of Starch Grain. Published by the author. Salem, Mass.

ROBERTS, E. A., and PROCTER, B. E. 1955. The comparative effect of ionizing radiations and heat upon starch containing cells of the potato tuber. Food Res. 20, 254–263.

ROMANI, R. J., and SOMMER, N. F. 1963. The biochemistry of radiation and fruit maturation. Summary of Progress for 1963, Contract AT(11-1)–34, Project 73. U.S. At. Energy Comm. Clearinghouse for Fed. Sci. and Tech. Inform., U.S. Dept. Comm., Springfield, Va.

ROUSE, A. H., DENNISON, R. A., and ATKINS, C. D. 1966. Irradiation effects on juices extracted from treated 'Valencia' oranges and 'Duncan' grapefruit. Proc. Florida State Hort. Soc. 79, 292–296.

RUBIN, B. A., and METLITSKY, L. V. 1961. Biochemical basis of the use of γ-rays to extend the storage life of potatoes. In Proc. Intern. Congr. Biochem. 5th, Moscow. Intern. Union Biochem. Symp. Ser., Vol. 28, Biochemical Principles of the Food Industry.

SAEMAN, J. F., MILLETT, M. A., and LAWTON, E. J. 1952. Effect of high energy cathode rays on cellulose. Ind. Eng. Chem. 44, 2848–2852.

SALKOVA, E. G. 1963. Change in carbohydrate composition in storage organs of plants under the influence of gamma irradiation. Doklady Akad. Nauk. (USSR) *149*, 1203–1205. Eng. Transl. *149*, 311–314 (1963).

SAMEC, M. 1958. Transformation of potato starch by ionizing radiation. Starke *10*, 76–79.

SAMEC, M. 1960. Some properties of gamma irradiated starches and their electrodialytic separation. J. Appl. Polymer Sci. *3*, 224–226.

SCHOEPFLE, C. S., and CONNELL, L. H. 1929. Effect of cathode rays on hydrocarbon oils and on paper. Ind. Eng. Chem. *21*, 529–537.

SCHREIBER, J. S., and HIGHLANDS, M. E. 1958. A study of the biochemistry of irradiated potatoes stored under commercial conditions. Food Res. *23*, 464–472.

SERENO, N. M., HIGHLANDS, M. E., CUNNINGHAM, C. E., and GETCHELL, J. S. 1957. The effect of irradiation and subsequent controlled storage upon the composition of Maine Katahdin potatoes. Maine Agr. Exp. Sta. Bull. *563*.

SHAH, J. 1966. Radiation-induced calcium release and its relation to post-irradiation textural changes in fruits and vegetables. Nature *211*, 776–777.

SKINNER, E. R., and KERTESZ, Z. I. 1960. The effect of gamma radiation on the structure of pectin. An electrophoretic study. J. Polymer Sci. *47*, 99–109.

SOMOGYI, L. R., and ROMANI, R. J. 1964. Irradiation induced textural change in fruits and its relation to pectin metabolism. J. Food Sci. *29*, 366–371.

SOSEDOV, N. I., and SHABOLENKO, V. P. 1964. Characteristic changes from ionizing irradiation of sols and gels of starch. Biochem. Zernai. Khlebopechnia *7*, 195–201. (Russian) Chem. Abstr. *62*, 4201f (1965).

SOSEDOV, N. I., and VAKAR, A. B. 1961. Effect of γ-rays on the biochemical properties of wheat. Proc. Intern. Congr. Biochem., 5th, Moscow. Intern. Union Biochem. Symp. Ser. Vol. 28. Biochemical Principles of the Food Industry.

SUSLER, H., MOHLER, H., and LUETHY, H. 1964. Physical chemical studies on cellulose altered by gamma radiation. Mitt. Geb. Lebensmitteluntersuch Hyg. *55*, 264–273. (German) Nucl. Sci. Abstr. *20*, 23131 (1966).

TAKAOKA, K. 1960. Gamma ray processing of rice. I. The effect of γ-irradiation on imported rice starch. Hakko Kogaku Zasshi *38*, 88–91. (Japanese). Nucl. Sci. Abstr. *16*, 27520 (1962).

TRUELSEN, T. A. 1964. Irradiation of potatoes. Effect on sugar content, blackening and growth substances. IVA Meddelande *138*, 55–62. Nucl. Sci. Abstr. *19*, 13834 (1965).

WAHBA, I. J., and MASSEY, L. M. JR. 1966. Degradation of pectin by γ-radiation under various moisture conditions. J. Polymer Sci. (A-1) *4*, 1759–1771.

WAHBA, I. J., TALLMAN, D. F., and MASSEY, L. M., JR. 1963. Radiation induced gelation of dilute aqueous pectin solutions. Science *139*, 1297–1298.

WOLFROM, M. L. 1955. Chemical and organoleptic changes in carbohydrates and proteins produced by radiation sterilization. Final Rept. Contract DA44–109–QM–1772. Dept. Comm. Publ., U.S. Gov. Printing Office, Washington, D.C.

WOLFROM, M. L. 1958. The characterization of products from irradiated carbohydrates. Final Rept. Contract DA19–129–QM–932. U.S. Dept. Comm. Publ., U.S. Gov. Printing Office, Washington, D.C.

WU, Y. 1963. Light scattering properties of pectin after x-ray irradiation. Bull. Inst. Chem. Acad. Sinica. 7, 61–72.

YOUNG, R., and LIN, M. 1966. Radiation effects on protein and pectin in papaya. Rept. *UH–235P5–1*, 7–13. Nucl. Sci. Abstr. *20*, 18246 (1966).

Carbohydrates in Nutrition and Disease

CHAPTER 15

A. E. Harper

Carbohydrates in Human Nutrition

INTRODUCTION

Various criteria can be used in assessing the significance of a substance in human nutrition. One might use a quantitative criterion—how much of it is ordinarily consumed by man? Or a qualitative criterion—is it essential? Other criteria might also be used. How critical is a deficiency of it? Can other substances be substituted for it? How important is its function in relation to the functions of other nutrients? The assessment of significance might well differ with the criterion used as the basis for the assessment.

QUANTITATIVE SIGNIFICANCE OF CARBOHYDRATE

Figure ninety-four, from information compiled by the Food and Agriculture Organization of the United Nations (Périssé 1968) shows that carbohydrate is the major component in the diet of man and comprises from 43 to 76% of the calories consumed by man. The figure also shows clearly that as the per capita income of a country decreases, dependence on carbohydrate increases. This is as true for individuals within a country as it is among countries (Cathcart and Murray 1931), for foods that are rich in carbohydrate, such as cereal grains and certain root crops, give the highest yields per unit of cultivated land and are the cheapest sources of food.

When the diets of most of the world's population contain more than 65% of carbohydrate there can be no question, from a quantitative viewpoint, about the great significance of carbohydrate in human nutrition. But this is more an economic and social assessment than a biological assessment, for there are many people in rich countries, hunters in primitive areas, and small groups in areas such as the Arctic where plants cannot be grown, who subsist on very low carbohydrate diets. Hence, it would appear that an assessment from a biological viewpoint might lead to conclusions concerning the nutritional significance of carbohydrates that are quite different from those based on a socio-economic analysis.

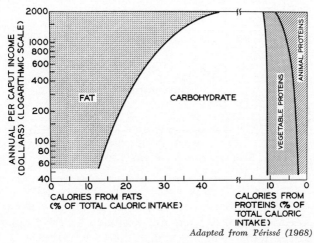

Adapted from Périssé (1968)

FIG. 94. CALORIES FROM FATS, CARBOHYDRATES AND PROTEINS AS PERCENT OF TOTAL CALORIE INTAKE IN RELATION TO PER CAPITA INCOME OF DIFFERENT COUNTRIES AND REGIONS IN 1962

CARBOHYDRATE IN RELATION TO OTHER REQUIREMENTS

In assessing nutritional significance from a biological viewpoint it is well to view the requirement for a specific nutrient in perspective by considering the need for it in relation to the other needs of the organism. As shown in Fig. 95, living systems have three major requirements: a relatively constant environment, a supply of building blocks, and a supply of energy. When we move from the general to the particular it immediately becomes evident that the specific requirements differ greatly with the organism.

The relatively constant environment needed for the development of life was presumably provided originally by the primeval ocean which had constant temperature, pH and osmotic pressure, a supply of nutrients, a huge bulk in which wastes were readily diluted and which, except along its perimeter, was relatively little affected by atmospheric

REQUIREMENTS OF LIVING SYSTEMS

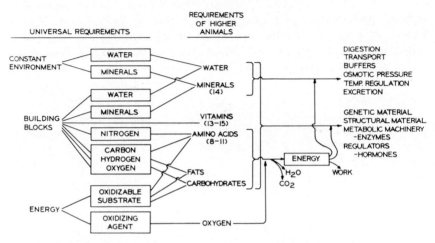

FIG. 95. SCHEMATIC REPRESENTATION OF THE REQUIREMENTS OF LIVING SYSTEMS

disturbances owing to its high viscosity and specific gravity. Development of complex multicellular organisms and their movement away from the ocean depended upon the evolution of mechanisms by the more primitive forms which enabled them to maintain a relatively constant internal environment despite fluctuations in the external environment.

In higher forms the blood and body fluids provide an environment of constant pH, osmotic pressure, temperature, and a means of transporting nutrients and removing wastes. To maintain this constant internal environment and the structures necessary for it to function efficiently requires an adequate supply of nutrients. The greatest freedom from the restrictions imposed by a fluctuating external environment, as seen in man, has been accomplished through the evolution of highly effective regulatory mechanisms, including the development of great mental capacity, which among other things has made possible the organized growing and storage of supplies of nutriment—the source of the other two major requirements, building blocks and energy.

The second general requirement, that for building blocks, is strictly a nutritional requirement. The specific substances that can be used to satisfy this requirement differ greatly with the organism and depend upon its synthetic capacity. Tissue components that cannot be synthesized from simpler substances must be obtained preformed. Water and mineral elements are universal requirements, but the required sources of carbon, hydrogen, oxygen and nitrogen differ from species to species. For man, essential fatty acids, some 15 vitamins, and 8 amino acids are required preformed in the diet; and for the synthesis of the other organic components of tissues, carbohydrate, fat, and protein—but especially carbohydrate—provide the basic elements in utilizable form.

The third general requirement is for energy. To carry out all of the activities needed for the maintenance of life; to maintain the body in what, chemically, is a highly unique state; to build it and repair it; to do the work necessary for its support; requires a continuous supply of energy. This requirement is, in essence, a requirement for an oxidizing agent and an oxidizable substrate. The oxidizing agent, oxygen, is provided abundantly in the air, and the internal transport system to carry it from the lungs to the sites where oxidation reactions occur is normally highly efficient. The oxidizable substrate for higher animals must be provided as relatively reduced organic matter, primarily in the form of carbohydrate and fat, and to a lesser extent, the amino acids of protein. For man the major source of oxidizable substrate, and therefore of energy, is carbohydrate.

CARBOHYDRATE AS A SUBJECT OF STUDY IN NUTRITION AND METABOLISM

Utilization of carbohydrate as an energy source was examined in great detail during the period when the law of conservation of energy was a relatively new concept and the question as to whether living systems obeyed this law had not been answered. However, when the science of nutrition received its greatest impetus at the beginning of this century from discoveries, one after another, of the essentiality of individual nutrients, little evidence accrued to indicate that a lack of carbohydrate resulted in any characteristic deficiency syndrome. In fact, it appeared that even as a source of energy, carbohydrate could be replaced, largely, if not entirely, by a mixture of protein and fat. As a result, despite its importance as a carbon and energy source, carbohydrate has had only limited appeal as a subject of study in nutrition.

Just the reverse is true, however, when one considers nutritional needs within the body rather than exogenous requirements. In relation to the nutrition of the cells, in relation to effects of metabolic diseases on the nutrition of tissues and organs, probably no substance has received as much attention as carbohydrate. It seems appropriate, therefore, to consider initially the essentiality of carbohydrate for various body processes.

METABOLIC ESSENTIALITY OF GLUCOSE

Starch is the major component of most human diets but only the free glucose released upon complete digestion of starch is absorbed into the blood. Lactose, which yields galactose as well as glucose, is the major dietary carbohydrate of infants. Sucrose, which yields fructose and glucose, is an important component of the diets of people in rich countries. But galactose and fructose, as well as other sugars that may be used by man, are converted to glucose mainly in the liver, although fructose may be converted, at least to some extent, in the intestine. Thus, despite the form in which carbohydrate is eaten, glucose is the sugar of the blood and body fluids. Nevertheless, there are some interesting metabolic effects from the ingestion of fructose and galactose, especially fructose (Ashida 1963).

The essentiality of an adequate supply of glucose for normal functioning of the body is evident from the effects of hypoglycemia—low blood sugar. A drastic fall in blood glucose results in convulsions and, if glucose is not supplied within a short time, either directly or through metabolic reactions within the body, leads to coma and death. The liver is the organ most intimately involved in regulation of blood glucose and in hepatectomized animals—animals from which the liver has been surgically removed—blood glucose concentration falls rapidly. If glucose is provided by intravenous infusions, hepatectomized animals will survive for some time and die only later from causes unrelated to hypoglycemia (Bollman et al. 1924).

The immediate effect of an inadequate supply of glucose is thus seen to be malfunction of the central nervous system. This is because, under normal conditions, glucose is its main source of energy. The rate of metabolism of brain is high—oxidation by brain may represent as much as 25% of the resting metabolism of man (Cahill et al. 1968). Since nervous tissue does not utilize fatty acids as a source of energy, this represents a substantial requirement specifically for carbohydrate. When the supply of glucose is inadequate the brain does oxidize some other substances probably mainly amino acids, but either the capacity for their oxidation is too low, or the amounts that enter the brain are inadequate for them to substitute efficiently for glucose. A continuous supply of glucose is therefore essential for the nervous system.

Although glucose is essential for the central nervous system, is it essential that it be provided in the diet? Osborne and Mendel (1921), almost 50 yr ago, demonstrated that the rat could be grown to maturity on a diet containing negligible amounts of carbohydrate. Stefansson as a result of his studies of the Canadian Eskimos and his own experience during his Arctic explorations concluded that man also could survive on a diet containing little or no carbohydrate (McClellan and Dubois 1930). Further, the human body does not contain enough carbohydrate to supply one day's energy expenditure, yet man can survive prolonged periods of starvation; and this despite the fact that the central nervous system requires about 140 gm of glucose a day, and the red blood cells and a few other tissues close to another 40 gm (Cahill et al. 1968).

Although all tissues have the ability to utilize glucose as an energy source, tissues other than the central nervous system, the red blood cell, and a few others have an equally great or greater capacity to utilize fatty acids (Cahill et al. 1968). But even assuming that these tissues depend entirely on fatty acid oxidation during periods of prolonged starvation, the body must obtain about 180 gm of glucose per day to maintain the tissues that do not utilize fatty acids. This again is a function primarily of the liver and to a lesser extent of the kidney. These two organs have the full complement of enzymes required for the synthesis of glucose from noncarbohydrate precursors. The body is not able to convert fatty acids to glucose, but most of the amino acids released during the breakdown of tissue protein and the glycerol from fat can be converted, through the process of gluconeogenesis, to glucose which enters the blood stream. Calculations by Cahill and associates (1968) from measurements in fasting man, illustrated in Fig. 96, indicate that as much as 130 gm of glucose may be synthesized per day from these sources and from lactic acid. This capacity falls some during prolonged fasting because of the slow rate of breakdown of body proteins and is accompanied by an adaptive response of the central nervous system which enables it to utilize some quantity of ketone bodies as an energy source.

Another function of glucose is evident from these observations. Obviously if glucose is not supplied in the diet, amino acids become an important precursor of the glucose that must be synthesized by the body. Since there is no reserve of amino acids they can be provided only through breakdown of tissue proteins. From comparisons of nitrogen excretion during starvation and during the ingestion of a protein-free diet, it is known that nitrogen loss is decreased below the starvation value when a protein-free diet is consumed and that this nitrogen-sparing action of a protein-free diet is largely attributable to the carbohydrate it contains (Munro 1964). Thus, although it may not be evident ordinarily when a mixed diet is being consumed, there are

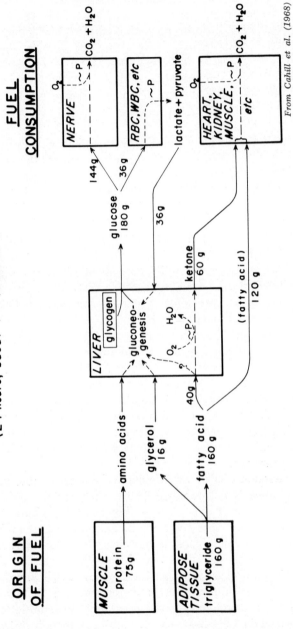

FIG. 96. QUANTITATIVE ESTIMATION OF HOW A SUPPLY OF GLUCOSE FOR THE CENTRAL NERVOUS SYSTEM IS MAIN-TAINED DURING STARVATION

conditions under which carbohydrate may reduce the wastage of amino acids.

Related to this is the role of carbohydrate in preventing ketosis (Krebs 1966). This is best illustrated by the diabetic state in which, even though glucose may be ingested, it is not utilized by most tissues because of the lack of insulin. This results essentially in glucose starvation and, although glucose concentration of the blood is high, the response of the body is to synthesize still more glucose. Hence, as in starvation, the body is largely dependent on fatty acid oxidation for energy; but at the same time, besides the withdrawal of glucose by the central nervous system, amino acids are being used to synthesize glucose which is subsequently excreted in the urine.

Both of these processes, fatty acid oxidation and glucose synthesis, are dependent upon an adequate supply of the 4-carbon dicarboxylic acid, oxaloacetate, which cannot be synthesized in the body from fatty acids, but only from amino acids and glucose. Hence, when glucose synthesis and fatty acid oxidation are both proceeding rapidly, the supply of oxaloacetate can become limiting. The basis for this is represented schematically in Fig. 97. Under these conditions a portion of

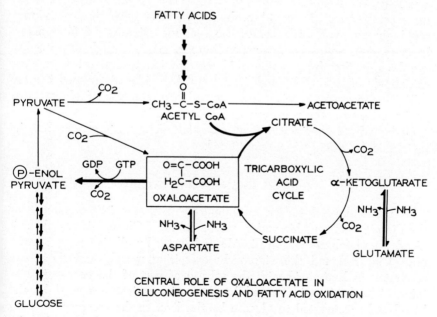

CENTRAL ROLE OF OXALOACETATE IN
GLUCONEOGENESIS AND FATTY ACID OXIDATION

Fig. 97. Schematic Representation of the Competing Demands for Oxalo-acetate by Gluconeogenesis and Fatty Acid Oxidation

the fatty acids, instead of being oxidized to carbon dioxide and water, are oxidized only to acetylcoenzyme A. This is converted to aceto-acetate which in turn may be reduced to β-hydroxybutyrate. These two substances, known as ketone bodies, accumulate, resulting in ke-tosis. In this condition, these organic acids are excreted in the urine in substantial quantities and sodium may be excreted with them, leading to a cation deficit and acidosis. If untreated, the glycosuria and aci-dosis progress, the condition worsens, and the end result is death. Insulin, which stimulates utilization of glucose by the tissues alleviates the condition.

A similar condition occurs in heavily lactating cows presumably for similar reasons. Carbohydrate production is high to provide the lactose in the milk secreted and the ruminant normally has a high dependence on fatty acid oxidation for energy. Both of these processes require oxaloacetate and unless the supply is adequate, ketosis develops just as in diabetic man. The condition is usually prevented or alleviated by provision of an ample quantity of glucose.

An old saying about these conditions is that fatty acids burn in the flame of carbohydrate (McHenry 1957). Although this is essentially true it tends, as Krebs (1966) has pointed out, to put the emphasis on the wrong place. It would probably be more accurate to say that rapid glucose synthesis removes the fuel for oxidation of fatty acids. In starvation, unless energy expenditure is great, enough oxaloacetate can be provided from amino acid degradation to support fatty acid oxidation. So again, although it may not be evident in normal healthy subjects consuming a mixed diet, an adequate supply of glucose can be important for the prevention of keto-acidosis.

Carbohydrate may not be an essential dietary component but its metabolic essentiality is unquestioned.

GLUCOSE AS AN ENERGY SOURCE

As was indicated earlier, carbohydrate is the major component of the diet of man and his major energy source. Glucose is readily oxi-dized completely to carbon dioxide and water by all tissues except the red blood cell and the vigorously exercising muscle and these can convert it to lactic acid with the release of part of the energy it con-tains. The lactic acid formed by these tissues is transported by the blood and reconverted to glucose by the liver so the energy remaining in the lactate is not lost. The oxidative process, including that by which the energy from the oxidative reactions is trapped in high energy phosphate compounds so that it is made available for synthetic reactions or for muscular or other work, has been elaborated in great detail. The classic work on this subject indicates that the body is well-equipped to use glucose efficiently as an energy source.

The series of reactions that yield energy are usually considered to be degradative reactions. Nevertheless, there are many intermediates along these pathways that serve as precursors for other substances. Thus, provided that nitrogen, as amino groups, is available, the dispensable amino acids can be synthesized from glucose. In fact the technique of feeding isotopically-labeled glucose has been used to determine which amino acids are dispensable, for only in the dispensable amino acids is radioactivity from glucose incorporated. These amino acids in turn are precursors of many nitrogen-containing substances such as glutathione, creatine, and heme, so carbohydrate provides carbon, hydrogen, and oxygen for the synthesis of such compounds. Carbohydrate also provides acetyl units which can serve as a precursor of the steroids, including steroid hormones. As well, glucose serves as a precursor of the pentoses and amino sugars found, respectively, in nucleic acids and in mucopolysaccharides. Although a dietary source of carbohydrate may not be an absolute essential because glucose can be synthesized from amino acids, it does, nevertheless, in practice, serve as a source of carbon skeletons for the synthesis of many substances in the body.

As was mentioned earlier, despite the high carbohydrate content of most diets the human body contains less carbohydrate than the average man eats in a day and less than one day's energy requirement (Table 25). Nevertheless, the amount of energy available in the body will permit survival of a well-nourished man during starvation for about 80 days. Obviously then, the energy consumed mainly as carbohydrate and not oxidized immediately is not stored as carbohydrate but rather

TABLE 25

APPROXIMATE ENERGY STORES OF 70 KG MAN

	Amounts of Energy Sources		
	Total in Body Gm	Available in Starvation Gm	Exhaustion Time Days
Glycogen-liver	60	200	2
-muscle	150		
Fat	12,000	10,000	
Protein	12,600	4,000	80

as fat. And even much of the carbohydrate ingested in a meal by a person who is in energy balance is not utilized directly as such, for usually half, and frequently well over half, of the daily energy requirement is ingested in a single meal so must be stored temporarily, probably mainly as fat.

Carbohydrate is very efficiently converted to fat as any good farmer who has raised pigs on corn can testify from experience. The efficiency of the process may approach 90% (Ball 1965). With efficiency of this order, the loss involved in storing energy consumed as carbohydrate in the form of fat is low, and even the portion not stored is presumably used to maintain body temperature. Energy stored as fat cannot be used for the resynthesis of carbohydrate nor for the synthesis of such amino acids as the body can synthesize. It is channeled directly into the energy-yielding reactions of the tricarboxylic acid cycle—except for the small part that is used for the synthesis of compounds, such as steroids, for which acetylcoenzyme A is a precursor.

Fat has decided advantages as a storage material. It contains 2.25 times as much energy per gram as carbohydrate. It can be stored in depots such as the adipose tissue which have a low water content. It has a low specific gravity. Hence, it provides a compact means of energy storage. It does not however, as is so often stated, serve as a superior source of metabolic water. Oxidation of one gram of fat does yield much more water than oxidation of one gram of carbohydrate, but the amount of water per calorie is actually less for fat.

Fats are more reduced compounds than are carbohydrates, so besides carbon skeletons, reducing substances are needed for the synthesis of fats. The biological precursors of fats are acetyl units, readily derived from the pyruvate formed through partial oxidation of glucose via glycolysis, and reduced coenzymes, readily formed during the oxidation of glucose via the pentose cycle. The process of fat synthesis, or lipogenesis, is shown schematically in Fig. 98. Most tissues have some capacity to synthesize fatty acids, but liver and adipose tissue are particularly important sites of fat synthesis.

Of particular interest in relation to the synthesis of fat from carbohydrate are some of the effects of a high carbohydrate intake on the activities of enzymes involved in lipogenesis. This is most clearly illustrated in animals kept without food for a period of time and then fed a high carbohydrate diet. During a period of fasting the body depends largely on the oxidation of fat for energy and lipogenesis is essentially shut down. Many of the enzymes of lipogenesis fall to very low levels. When the body is subsequently flooded with carbohydrate, readjustments occur that shift various metabolic processes in favor of fat synthesis.

In some of the early experiments that opened the way for the extensive investigations that have followed, Tepperman and associates (1943) noted that the respiratory quotient of animals that were fed only once a day and therefore had to store the energy consumed in a single meal for the subsequent 23-hr period was well above 1—indicating that fat was being synthesized. Subsequently Tepperman and Tepper-

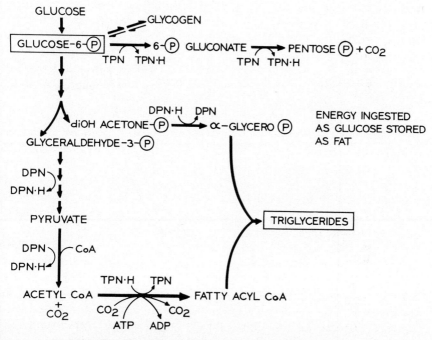

FIG. 98. SCHEMATIC REPRESENTATION OF LIPOGENESIS

man (1964) found that the activities of the TPN · H-generating enzymes of the pentose cycle in such animals were greatly elevated (Fig. 99). Such animals were also shown to synthesize fat at a rapid rate.

A high carbohydrate intake also stimulates the production of insulin which enhances the uptake of glucose by peripheral tissues, especially by adipose tissue. The activities of the enzymes that produce TPN · H, the reduced coenzyme required for fat synthesis, rise as Tepperman *et al.* (1943) had shown. This occurs in both liver and in adipose tissue—the two major sites of fatty acid synthesis. The limiting enzyme of the initial step in fatty acid synthesis—acetyl CoA carboxylase also rises (Gibson *et al.* 1966). The abundant supply of glucose provides a ready source of glycerol and acetyl units—the precursors of triglycerides—and fat synthesis in greatly enhanced.

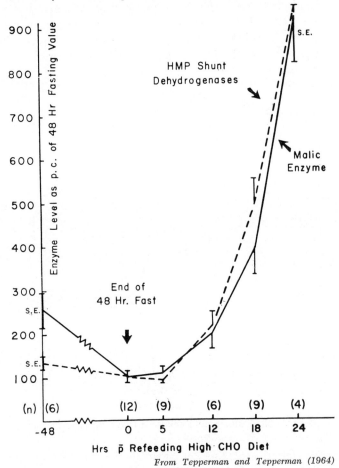

From *Tepperman and Tepperman (1964)*

FIG. 99. RESPONSE OF ENZYMES IN STARVED RATS TO A HIGH CARBOHYDRATE INTAKE

Animals, and presumably also man, respond to a high carbohydrate intake by metabolic adaptations that facilitate greatly the storage of the energy consumed, as fat in the fat depots (Tepperman and Tepperman 1965; Gibson *et al.* 1966). Some of the adaptive responses observed are summarized in Fig. 100. Although the observations have

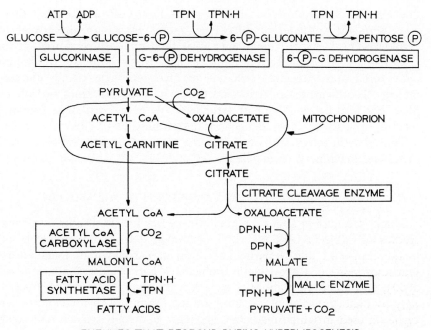

ENZYMES THAT RESPOND DURING HYPERLIPOGENESIS

Based on Gibson et al. (1966) and Tepperman and Tepperman (1965)

FIG. 100. SCHEMATIC REPRESENTATION OF ENZYMATIC RESPONSES IN THE RAT TO CONDITIONS THAT RESULT IN ACCELERATED LIPOGENESIS

not been discussed in detail this is perhaps enough to indicate that a change in the proportion of carbohydrate in the diet can result in striking metabolic changes within the body.

Besides enzymatic responses to a high carbohydrate diet, particularly after a period without food, there are, as well, enzymatic responses, some of them reciprocal, to the absence of carbohydrate or to a low carbohydrate intake. Noteworthy among these are marked drops in the activities of enzymes of lipogenesis. Also, several of the enzymes involved in gluconeogenesis—the synthesis of glucose from noncarbohydrate precursors—increase greatly in activity when carbohydrate intake is low.

An interesting point here is that a high intake of fructose results in responses of enzymes of gluconeogenesis similar to those observed when large amounts of noncarbohydrate precursors of glucose, such as protein, are consumed (Ashida 1963). These responses are presumably related to the indirect pathway followed by fructose during its conversion to glucose. It is initially phosphorylated to fructose-1-phosphate, which is subsequently cleaved to dihydroxyacetone phosphate and glyceraldehyde. The 3-carbon compounds must then undergo condensation to give rise to glucose. Interestingly a high fructose intake results in more rapid and extensive formation of glycogen than a high glucose intake. The 3-carbon units formed during cleavage of fructose-1-phosphate are also precursors of glycerol which is of interest since fructose feeding appears to favor lipogenesis (McGandy *et al.* 1967).

Thus the carbohydrate content of the diet can greatly influence the metabolic machinery of the body.

METABOLIC DEFECTS OF CARBOHYDRATE METABOLISM

Much information about metabolic processes has been obtained from studies of subjects with inherited metabolic defects—inborn errors of metabolism—and this is particularly true of carbohydrate metabolism. Our knowledge of these genetic problems is increasing (Stanbury *et al.* 1960), and a certain number of such diseases can at present be treated only by dietary means. In looking toward the future this would appear to be an area in which medicine, nutrition, and food science will meet more regularly than they have in the past.

The earliest recognized and best known metabolic defect of carbohydrate metabolism in man is diabetes mellitus (Renold and Cahill 1960), a widespread disease with an incidence estimated to be 1.7% of the population and possibly higher. A predisposition to develop the disease is inherited, but the actual development appears to be influenced by nongenetic factors.

Diabetes mellitus results from inability of the pancreas to produce an adequate amount of insulin but this is somewhat of an oversimplification because there are many modifying factors in the development of the disease and a question as to whether there may not be more than one form. The primary defect in diabetes mellitus, caused by lack of insulin, is reduced ability of many tissues to utilize glucose. This appears to be due mainly to the low ability of tissues to transport glucose across the cell membrane in the absence of insulin, true particularly of muscle and adipose tissue; or inadequate ability to convert glucose to glucose-6-phosphate, primarily in liver. The enzyme required for this is low in liver in the diabetic animal but responds quickly after insulin is administered. Brain appears not to be affected by insulin deficiency, and kidney and intestinal mucosa probably not either.

There is evidence that insulin deficiency also affects various aspects of lipid, amino acid, and protein metabolism, but here we are concerned primarily with carbohydrates.

The inability of major tissues to utilize glucose results in accumulation of glucose in body fluids, and although this may increase the penetration of glucose into the cells, the concentration of glucose soon exceeds the ability of the kidney to conserve it and glucose is lost in the urine. As well, the liver produces more glucose in response to the low tissue utilization. When diabetes is severe and glucose loss is high the body must depend on oxidation of fatty acids for energy and uses amino acids to synthesize more glucose. As was mentioned earlier, this leads to excessive production of ketone bodies. These are excreted together with cations, sodium and potassium. If the condition continues untreated acidosis becomes severe and coma and death result.

There are various degrees of severity of diabetes mellitus. Obviously in the most severe condition, with inability to use the major dietary source of energy, insulin must be supplied. In less severe cases, glucose utilization may be only moderately impaired so that dietary control of the condition can be achieved by careful adjustment of the intakes of carbohydrate, fat, and protein.

Less severe diabetes is frequently associated with obesity and develops during adult life. A question has been raised as to whether this may be the result of an increased requirement for insulin or inadequacy of insulin production owing to continuous increased insulin secretion stimulated by a high carbohydrate intake. There is some suggestive evidence that a prolonged high carbohydrate intake may contribute to the development of the disease in those with a predisposition for it. Despite the intensive study of this disease and the accumulation of knowledge about it, the role of dietary factors in its development is not clearly understood.

Other much rarer diseases resulting directly from genetic defects have been studied in considerable detail; two of these will be discussed briefly. One, characterized as glycogen-storage disease (Field 1960) is in reality a complex of diseases with distinctly different biochemical bases. In 5 of 6 known forms of the disease, glycogen accumulation has been shown to result from deficiencies of single enzymes involved in glucose or glycogen metabolism. Lack of or low activity of glucose-6-phosphatase, an enzyme necessary for glucose to be released into the blood during glycogen breakdown, results in the accumulation of large amounts of normal glycogen in liver and kidney; lack of or low activity of phosphorylase, the initial enzyme required for glycogen breakdown, results in the accumulation of large amounts of normal glycogen in liver or muscle, depending upon which tissue the enzyme is missing from; lack of debranching enzyme results in accumulation of a limit dextrin-type of glycogen in muscle, liver, and heart owing to inability of tissues to degrade glycogen beyond the branch points; and amylopectinosis is more or less the reverse, in which the enzyme for forming a branched structure is missing or low in activity so that a glycosan with very long chains and only slightly soluble in water, accumulates. This is a very rare disease. There is another, as yet uncharacterized, form of the disease that is usually fatal within the first year of life, in which deposition of nomal glycogen throughout the body is observed.

These diseases, which result in inability to mobilize glucose stored as glycogen, lead to varying degrees of incapacity—from inability to perform vigorous exercise with muscle phosphorylase deficiency; to enlarged liver, convulsions due to hypoglycemia, and severe acidosis in untreated glucose-6-phosphate deficiency. The latter condition, although fatal if untreated, can be improved if it is recognized early and care is taken to prevent overloading the patient with glucose.

Another inherited metabolic disorder, galactosemia, is the result of an inability to metabolize galactose owing to a deficiency of the enzyme, galactose-1-phosphate uridyl transferase which is necessary for the conversion of galactose-1-phosphate to glucose-1-phosphate (Isselbacher 1960). In untreated subjects the liver and spleen are enlarged, cirrhosis develops, cataracts occur, mental development is impaired, and death ensues. The condition is readily treated if it is recognized early by removing galactose and other galactose-containing carbohydrates, such as lactose, from the diet and substituting glucose or sucrose for them. This disease is of particular nutritional interest because experience with prolonged treatment of it indicates that there is no specific requirement for galactose. Also it illustrates in man, what had been recognized earlier in animals, that prolonged high blood and tissue levels of galactose can result in a toxicity syndrome leading, amongst other things, to the development of cataracts.

CARBOYDRATE AND THE ALIMENTARY CANAL

Since the various carbohydrates utilized by man are normally converted efficiently to glucose within the body; and since the small amounts of carbohydrates other than glucose (such as galactose, fructose, amino sugars, and pentoses) which are components of various body substances, can be synthesized in the body; an obvious place in which to search for specific and unique effects of individual carbohydrates is the alimentary canal, the lumen of which can be considered as a part of the external environment. Gastrointestinal effects of carbohydrate nutrition are usually discussed in some detail in articles on the role of carbohydrates so are given only passing attention here (Krehl 1955; Day and Pigman 1957).

The oral cavity, the point of entry of foodstuffs, has received great attention in relation to the development of dental caries. It is clear that acid producing microorganisms are required for the development of dental caries; and that the development of carious lesions is influenced by the type of carbohydrate in the diet (Day and Pigman 1957). A high intake of more refined foods containing readily fermentable carbohydrates is associated with an increased incidence of caries. Sucrose in particular appears to be a prime offender. Sucrose is more cariogenic in solid form than in liquid form, presumably because it adheres to teeth and provides nutriment for bacteria for a longer time. There is some suggestion that refined sugars also favor the growth of microorganisms that synthesize glycosans and that these polysaccharides provide a good source of nutriment for bacteria between meals, thus favoring caries production (Hartles 1967). Although knowledge of the etiology of dental caries is still incomplete there can be no doubt but that it is influenced greatly by the type of carbohydrate consumed, and more so by the type than by the quantity.

In the stomach, which is an organ of dilution as well as digestion, carbohydrates exert osmotic pressure. The emptying of the stomach is influenced by osmoreceptors and food intake is influenced by stomach distension. Although evidence in man is limited, the results of animal experiments indicate that the greater osmotic pressure of low molecular weight carbohydrates, by causing increased stomach distension, may reduce the consumption of diets that are already somewhat inadequate to the point where growth is impaired (Wiener et al. 1963). This effect may also be partially due to temporary dehydration of tissues, for in animals not allowed water, depressed food intake is accompanied by normal moisture content of the gut contents as a result of withdrawal of water from the tissues (Lepkovsky et al. 1957).

Several effects of the type of dietary carbohydrate on intestinal function have received attention. One, of considerable interest some years ago, but now given relatively little attention, is the influence of the type of carbohydrate on vitamin requirements (Harper and Elvehjem 1957). In animals, requirements for several of the B-vitamins are reduced when the diet contains a relatively insoluble or slowly absorbed carbohydrate. These effects were found later to depend largely upon coprophagy—the consumption of feces (Barnes 1962). Thus, effects of less soluble carbohydrates, such as uncooked starches and lactose, on B-vitamin requirements of man are probably less than was once thought. Although some of the B-vitamins may be produced in sufficient quantity high enough in the small intestine to be absorbed directly in amounts that are beneficial, the probable significance of these effects was certainly overestimated originally.

The influence of dietary carbohydrates on intestinal pH has also received considerable attention (Anon. 1962). It is thought that lactose in the diet of the infant, which favors an intestinal flora that produces organic acids, is beneficial because the growth of less desirable putrefactive organisms is thereby decreased. There is, as well, evidence that the acid environment owing to lactose ingestion, exerts a favorable influence on calcium absorption. There is some suggestion that this may be a specific effect of lactose, independent of its effect on the intestinal microflora, because the effect is still seen in animals treated with antibacterial agents.

Indigestible carbohydrates that do not serve as a source of energy for species other than the ruminant also have a role in the intestine. By providing bulk, they facilitate evacuation of wastes from the intestine and thus help to maintain a favorable environment in this organ.

Finally in considering carbohydrate nutrition in relation to the intestine we return to the problem of enzyme deficiencies (Townley 1967). These have attracted considerable attention during the past decade because they are known to be responsible for certain carbohydrate intolerances.

Disaccharides (lactose, sucrose, and maltose), are not hydrolyzed mainly in the intestinal lumen, but rather after their entry into the intestinal mucosal cells. It is within these cells that the hydrolytic enzymes lactase, sucrase, and maltase are found. Maltase is normally predominant in man, being four times as active as sucrase which is in turn twice as active as lactase. Lactase, however, is high in infants and tends to decline with age. Ordinarily the activities of these enzymes are more than adequate to hydrolyze efficiently all of the disaccharides ingested. However, intolerance of certain disaccharides is observed in a number of individuals. This state is characterized by diarrhea which may be quite severe in the infant and lead to dehydration.

It has become evident from examination of intestinal tissues that sucrose intolerance is frequently a genetic disease in which the enzyme sucrase is missing from the intestinal cells. As a result of this, sucrose is not hydrolyzed and, when it is eaten, accumulates in the intestine, leading to the development of diarrhea. The condition is usually associated with a deficiency of isomaltase as well.

Lactose intolerance also occurs. This condition would ordinarily be fatal to infants raised solely on milk, for the sugar of milk is lactose. The condition is not as well characterized as sucrase deficiency but evidence suggests that, although not all cases of lactose intolerance are due solely to lactase deficiency, low enzyme activity is an important factor. Lactose intolerance, associated with low lactase activity, is observed in adults and is more common in some racial groups than others. Lactase tends to decrease in the adult, so whether the adult syndrome is strictly genetic is not clear.

These metabolic defects are simply treated, merely by removing the offending carbohydrate from the diet. The occurrence of the conditions, however, emphasizes that there are groups within the general population who are unable to utilize carbohydrates that ordinarily make up a considerable part of the diet.

CARBOHYDRATES AND CARDIOVASCULAR DISEASE

One final topic, relationships among types of dietary carbohydrate, lipogenesis, blood lipids, and cardiovascular disease, have been receiving considerable attention lately. Some mention should therefore be made of this subject in relation to human nutrition, but since a recent symposium on it (MacDonald 1967), and a critical review of it (McGandy et al. 1967) have recently been published there is little justification for an elaborate discussion.

Interest in possible influences of carbohydrate in the development of cardiovascular disease has been stimulated by reports of a close statistical association between sucrose intake and the incidence of coronary artery disease in various countries (Yudkin 1967), an association the significance of which has been questioned (Grande 1967; McGandy et al. 1967); and also by recognition of a carbohydrate-inducible type of hypertriglyceridemia, elevated blood lipid concentrations.

In healthy subjects with normal concentrations of serum lipids the type of carbohydrate in the diet has little effect. However, when men ingesting a high fat diet are shifted to a high carbohydrate diet, serum triglycerides increase, and the increase is greater if the dietary carbohydrate is sucrose than if it is starch. The question as to whether this may be only a transitory state has been raised (McGandy et al. 1967). This effect is seen particularly in hyperlipidemic subjects who also show enhanced rates of synthesis of triglycerides in adipose tissue and glycerol in liver (Kuo *et al.* 1967). The carbohydrate-inducible triglyceridemia is frequently associated with atherosclerosis so dietary control is recommended. Lowering carbohydrate intake is an effective control measure and substituting starch for fructose has also been found to lower serum triglycerides in these subjects.

Although there is evidence that a high intake of sucrose or fructose tends to result in elevation of serum lipids above values observed when starch or glucose is the dietary carbohydrate, McGandy and associates (1967) point out that these effects are much smaller than those observed when the intake of saturated fats is increased. While recognizing the unique problem of carbohydrate-induced hyperlipidemia, they consider the effects of the type of dietary carbohydrate to be a minor factor in relation to the general problem of atherosclerosis in man.

In summary then, although carbohydrate is an essential substrate for the central nervous system and the red blood cell, and although it ordinarily is the major component of the diet, the carbohydrates needed by man can be synthesized in the liver and kidney from noncarbohydrate precursors; it is therefore doubtful whether there is a specific dietary requirement for carbohydrate. Nevertheless, because of its role as a substrate for the central nervous system, because of its importance for the prevention of ketosis and the conservation of body protein, for maximum health and activity it seems advisable to recommend that at least 25 to 50% of the daily caloric intake be provided as carbohydrate.

Calories consumed in excess as carbohydrate are efficiently stored as fat, and the body undergoes metabolic adaptations which facilitate this. It also undergoes metabolic adaptations in the absence of carbohydrate which facilitate carbohydrate synthesis.

There are a variety of defects of carbohydrate metabolism which range from inability to hydrolyze certain disaccharides in the intestine to inability to utilize glucose by major tissues, as in diabetes mellitus.

The type of dietary carbohydrate can influence the development of dental caries, the condition of the gastrointestinal tract, and the concentrations of blood lipids.

Essential or not, carbohydrate is of great significance in the diet of man.

DISCUSSION

L. M. Massey.—Would you comment about methods of establishing essential dietary requirements for man in view of our inability on moral and ethical grounds to subject human beings to many of the experimental procedures used in studies of animals.

A. E. Harper.—There are, and have been throughout history, volunteers who have willingly agreed to undergo a variety of experiments in which nutrients were withdrawn from the diet until the time that some deficiency sign actually developed. These are still being done under very carefully controlled conditions with medical supervision, frequently in metabolic wards in hospitals, and when any untoward signs or symptoms develop the experiment is curtailed or stopped entirely. Much knowledge has also been gained from the study of naturally occurring deficiencies such as scurvy, ricketts, pellagra, anemia, goiter, and beriberi. There are a few nutrients for which highly specific information about requirements have not been obtained in man; however, studies of the metabolism of tissues have shown that many reactions which depend on an adequate supply of specific nutrients in animals are identical in man. This provides indirect evidence that man requires the nutrients and estimates can be made by extrapolating from animal requirements.

D. French.—You mentioned intolerance of various disaccharides. I have heard the expression "starch intolerance." I do not know what it is or what causes it. Can you enlighten me on this?

A. E. Harper.—I am sorry, I do not know very much about it either. There is some evidence of deficiencies of glucosidases, such as oligo-1,6-glucosidase. This is quite rare and usually occurs in association with sucrase deficiency. Many intolerances are the result of a more general malabsorption syndrome.

D. French.—What would be the symptoms of an intolerance due to oligo-1,6-glucosidase deficiency?

A. E. Harper.—There would be loss of a fraction of the starch in the feces. One would not expect severe diarrhea because the molecular size of the unit excreted is relatively large. Low molecular weight substances exert high osmotic pressure and this causes severe diarrhea. In combined sucrase-isomaltase deficiency the signs of sucrase deficiency greatly predominate.

D. French.—If there are defects in the enzymes of digestion, would you expect that this would promote the development of a new distinct bacterial flora which would depend on the high concentration of these oligosaccharides? This might lead to extensive breakdown of these disaccharides into bacterial metabolic products.

A. E. Harper.—The main enzyme deficiencies cause severe diarrhea so must be corrected early if the infant is to survive. Adults (30–35% of some racial groups show evidence of lactase deficiency) tend to recognize their own lactose intolerance and avoid dairy products. The treatment thus tends to remove the opportunity for new bacterial forms to arise. It would, however, seem logical to expect changes in the intestinal flora of untreated subjects.

320   CARBOHYDRATES AND THEIR ROLES

BIBLIOGRAPHY

ANON. 1962. The nutritional significance of lactose. Dairy Council Dig. 33, 1–4.

ASHIDA, K. 1963. Diets and tissue enzymes. In Newer Methods of Nutritional Biochemistry, A. A. Albanese (Editor). Academic Press, Inc., New York.

BALL, E. G. 1965. Some energy relationships in adipose tissue. Ann. N.Y. Acad. Sci. 131, 225–234.

BARNES, R. H. 1962. Nutritional implications of coprophagy. Nutr. Rev. 20, 289–291.

BOLLMAN, J. L., MANN, R. E., and MAGATH, T. B. 1924. Studies on the physiology of the liver. VIII. Effect of total removal of the liver on the formation of urea. Am. J. Physiol. 69, 371–392.

CAHILL, G. F., JR., OWEN, O. E., and FELIG, P. 1968. Insulin and fuel homeostasis. The Physiologist 11, 97–102.

CATHCART, E. P., and MURRAY, A. M. T. 1931. A study in nutrition. Med. Res. Council, Spec. Rept. Ser. 151, London.

DAY, H. G., and PIGMAN, W. 1957. Carbohydrates in nutrition. In The Carbohydrates, W. Pigman (Editor). Academic Press, New York.

DOLE, V. P. 1965. Energy storage. In Handbook of Physiology Sec. 5 Adipose Tissue. Am. Physiological Society, Washington, D. C.

FIELD, R. A. 1960. Glycogen deposition diseases. In The Metabolic Basis of Inherited Disease, Stanbury et al. (Editors). McGraw-Hill, New York.

GIBSON, D. M., HICKS, S. E., and ALLMAN, D. W. 1966. Adaptive enzyme formation during hyperlipogenesis. Adv. Enz. Regulation 4, 239–246.

GRANDE, F. 1967. Dietary carbohydrates and serum cholesterol. Am. J. Clin. Nutr. 20, 176–184.

HARPER, A. E., and ELVEHJEM, C. A. 1957. A review of the effects of different carbohydrates on vitamin and amino acid requirements. Agr. Food Chem. 5, 754–758.

HARTLES, R. L. 1967. Carbohydrate consumption and dental caries. Am. J. Clin. Nutr. 20, 152–156.

ISSELBACHER, K. J. 1960. Galactosemia. In The Metabolic Basis of Inherited Disease, Stanbury et al. (Editors). McGraw-Hill, New York.

KREBS, H. A. 1966. The regulation of the release of ketone bodies by the liver. Adv. Enz. Regulation 4, 339–353.

KREHL, W. A. 1955. The nutritional significance of carbohydrates. Borden's Rev. Nutr. Res. 16, 85–99.

KUO, P. T. et al. 1967. Dietary carbohydrates in hyperlipemia (hypertriglyceridemia); hepatic and adipose tissue lipogenic activities. Am. J. Clin. Nutr. 20, 116–125.

LEPKOVSKY, S. et al. 1957. Gastrointestinal regulation of water and its effect on food intake and rate of digestion. Am. J. Physiol. 188, 327–331.

MACDONALD, I. (Editor). 1967. Symposium on dietary carbohydrates in man. Am. J. Clin. Nutr. 20, 65–208.

MCCLELLAN, W. S., and DuBois, E. F. 1930. Clinical calorimetry XLV, prolonged meat diets with a study of kidney function and ketosis. J. Biol. Chem. 87, 651–668.

McGANDY, R. B., HEGSTED, D. M., and STARE, F. J. 1967. Dietary fats, carbohydrates and atherosclerotic disease. New Engl. J. Med. *277*, 417–419, 469–471.

McHENRY, E. W. 1957. Basic Nutrition. Lippincott, Philadelphia.

MUNRO, H. N. 1964. General aspects of the regulation of protein metabolism by diet and hormones. *In* Mammalian Protein Metabolism, H. N. Munro, and J. B. Allison (Editors). Academic Press, New York.

OSBORNE, T. B., and MENDEL, L. B. 1921. Does growth require preformed carbohydrate in the diet? Proc. Soc. Exptl. Biol. Med. *18*, 136–137.

PÉRISSÉ, J. 1968. The nutritional approach in food policy planning. Nutr. Newsletter (FAO, Rome) *6*, 30–45.

RENOLD, A. E., and CAHILL, G. F., JR. 1960. Diabetes mellitus. *In* The Metabolic Basis of Inherited Disease, Stanbury *et al.* (Editors). McGraw-Hill, New York.

STANBURY, J. B., WYNGAARDEN, J. B., and FREDRICKSON, D. S. (Editors). 1960. The Metabolic Basis of Inherited Disease. McGraw-Hill, New York.

TEPPERMAN, J., BROBECK, J. R., and LONG, C. N. H. 1943. The effects of hypothalamic hyperphagia and of alterations in feeding habits on the metabolism of the albino rat. Yale J. Biol. Med. *15*, 855–874.

TEPPERMAN, H. M., and TEPPERMAN, J. 1964. Patterns of dietary and hormonal induction of certain NADP-linked liver enzymes. Am. J. Physiol. *206*, 357–361.

TEPPERMAN, J., and TEPPERMAN, H. M. 1965. Adaptive hyperlipogenesis—late 1964 model. Ann. N.Y. Acad. Sci. *131*, 404–411.

TOWNLEY, R. R. W. 1967. Disordered disaccharide digestion and absorption and its relation to other malabsorptive disorders of childhood. Borden's Rev. Nutr. Res. *28*, 33–48.

WIENER, R. P., YOSHIDA, M., and HARPER, A. E. 1963. Influence of various carbohydrates on the utilization of low protein rations by the white rat. V. Relationships among protein intake, calorie intake, growth and liver fat content. J. Nutr. *80*, 279–290.

YUDKIN, J. 1967. Evolution and historical changes in dietary carbohydrates. Am. J. Clin. Nutr. *20*, 108–115.

CHAPTER 16

Edgar S. Gordon

The Metabolic Importance of Carbohydrate in Obesity

INTRODUCTION

Obesity represents a state of excessive storage of chemical energy. The ultimate source of these energy containing compounds obviously must be the diet, and hence obesity is much commoner under conditions of affluence and plenty, and it disappears during periods of famine and privation. With a few exceptions related to genetic mutations, this condition does not occur in nature under normal circumstances unless it serves some specific beneficial purpose for the survival of the species, as for example, in the premigratory obesity of birds and the prehibernating obesity of certain animals. In considering obesity of many domestic animals, it is important to remember that this represents an artificial situation, planned and executed by the human animal for his own benefit.

Of the three categories of energy containing foodstuffs, proteins may be largely excluded as a significant source for stored energy for two specific reasons. The first of these is that protein has such a high satiety value that it becomes virtually impossible for humans to ingest large enough amounts at any one time or in frequently recurring episodes to induce a sustained positive caloric balance. In other words, hyperphagia of protein food is naturally self-limiting since loss of appetite does not permit it to continue. This is probably not true in carnivores. The second reason is that protein has a high specific dynamic action which means that it costs calories to the animal ingesting it, hence, making protein a poor source of energy for storage.

Fat is the most concentrated packet of energy that nature has devised. It provides nine calories per gram and it does not involve the accompanying storage of water to add extra weight so that fat is the logical form of storage for the energy requirements of animals that must maintain mobility. Contribution of dietary fat to the excessive storage of energy in obesity, however, is somewhat limited by the unpalatability, for the human at least, of diets exceedingly high in fat. This may be a cultural phenomenon, but more likely it is physiological. In any case, most people feel unable to consume a diet in which up to 75 or 80% of the calories are derived from fat. Unquestionably fat does make a significant contribution to the caloric excess that produces human obesity, but it is probably smaller than might be anticipated because of the unpalatability of high fat diets. If it were possible to label all the carbon atoms ingested, the source of the major part of the fat stored in obesity would be found to be dietary carbohydrate, rather than fat.

From the standpoint of acceptance as a foodstuff, carbohydrate is probably the best tolerated by humans. The largest component in the supply of energy containing food for the human race, unquestionably comes from carbohydrate of cereal grains and there is evidence that the carbohydrate intake in a great many civilized nations has increased steadily over the decades since the first observations were made. In the United States the typical "American diet" is well-known to be exceedingly high in carbohydrate and perhaps of equal importance is the statistically demonstrable fact that the consumption of refined carbohydrate in this country has also risen steadily as indicated by the per captia consumption of cane sugar. The incidence of obesity is high in the United States as it is also in a great many other western nations, with their highly industrialized culture and a great abundance of all the necessities of life including food. From the economic standpoint, carbohydrate food is cheap and therefore most easily available to people in all economic strata. Since these carbohydrate foods are abundant, cheap, attractive in acceptability, and well-tolerated for long sustained use, it is perhaps no surprise that their consumption, ranging all the way from the cereal grains to confections, accounts for the major energy source in the American as well as many other diets.

The capacity of animals to store energy, emancipates them from the necessity for monitoring food intake to their changing energy needs from minute to minute and hour to hour in the course of daily living. It is noteworthy that animals (in contrast to plants) are unable to store significant amounts of carbohydrate. The maximum amount of glucose available in storage form to the average adult human does not exceed about 75 to 100 gm. This reserve supply is located as glycogen in the liver which never exceeds about 4 or 6% of the wet weight of the organ, and also as glycogen in muscles. Extracellular fluids in the body contain approximately 80 to 100% of glucose, but the total amount available, based upon the energy contribution of 4 cal per gram of glucose, does not exceed about 300 to 400 cal. This limitation, in itself, emphasizes the extreme importance of energy storage in another form, which in animals obviously is fat. The conversion of carbohydrate to fat is well-known and has been carefully studied. It is an efficient process that can be turned on and off in the physiological economy of an animal by the homeostatic mechanisms that control the minute to minute flux of energy containing substrates during the ingestion of food and the performance of biological work. The adipose tissue organ, therefore, where the excess energy is stored, becomes of enormous importance in the entire energy exchange arrangements of the organism. In addition to an efficient method for lipogenesis and storage, it is obviously of equal importance that the stored lipid be available quickly and continuously through lipolysis and the discharge of fatty acids into the blood stream where they may be circulated to tissues that require a supply of fuel.

It is clear, therefore, that under normal circumstances carbohydrate and fat will both be used by the intact organism in supplying energy for biological processes. Although we have some understanding of the physiological circumstances that lead to the expenditure of either carbohydrate or fat for these functions, we are still woefully ignorant concerning the exact mechanism by which the signal is transmitted from the total metabolic state of the organism to the storage areas for glucose and fatty acid in order to trigger the release of one or the other according to the specific demands at the moment. Even more intriguing is the automatic homeostatic mechanism that determines whether glucose or fatty acid is to be used. In obesity, many of these controlling relationships are changed in one way or another, due presumably to the presence of large amounts of adipose tissue. In order to understand the functional integration of these processes, it is necessary to review some of the basic physiological facts about energy metabolism in reference to both carbohydrate and fat. To deal with this very complex subject in depth is beyond the scope of this discussion, but a number of excellent reviews have been published in recent years (Tepperman 1968; Jeanrenaud 1961; Renold and Cahill 1965; Whipple 1965).

ENERGY METABOLISM

Common experience confirms the practical reality of the ease of conversion of sugar into fat. This is a one way street, however, and it is a routinely accepted belief that in humans, as well as in most other animals, fat cannot be reconverted to glucose. Fuel supply for those tissues that have a specific dependence upon carbohydrate, therefore, must of necessity be derived either from further ingestion of carbohydrate food or from the conversion of other materials into glucose. This process is known as gluconeogenesis and the compounds employed for this function under normal physiological conditions are: (1) glycerol, a by-product of the breakdown of storage fat into its constituent fatty acids and glycerol; (2) lactic acid, which in more or less chemical equilibrium with pyruvic acid, is a product of glycolysis (the partial utilization of glucose); and (3) amino acids, derived from the breakdown of protein.

The storage of fat in the form of triglyceride (TG) occurs as a result of the chemical combination of free fatty acids (FFA) and alpha glycerophosphate (activated glycerol), in a molar ratio of 3:1. The former may originate from dietary fat or from the conversion of carbohydrate food into fatty acid (in the human this occurs largely in the liver); the latter is derived *only* from glucose entering fat cells where it is metabolized to glycerophosphate. Insulin is required for this process.

Mobilization of fat is dependent largely upon an enzyme in fat cells, adipose tissue lipase, which hydrolyzes the ester linkages of TG and frees the fatty acids and (inactive) glycerol, in a ratio that approximates 3:1. Both of these products are discharged from adipose tissue into the circulation. The FFA are immediately attached to serum albumin for

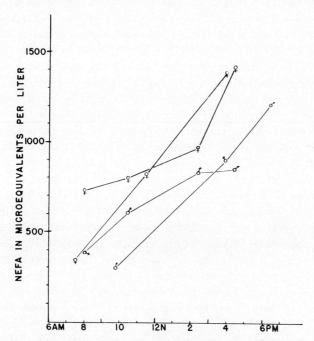

Fig. 101. FFA Changes in the Blood of Two Normal Men and Two Normal Women, During the Last 12 Hr of a 24-Hr Fast

The steady rise indicates increasing FFA mobilization as glucose reserves are depleted from fasting.

transport purposes and the glycerol is carried in the blood back to the liver where it serves as a substrate for resynthesis of glucose. Insulin inhibits this lipase, and hence lipolysis is suppressed whenever circulating insulin increases. Nearly all the FFA in the blood is derived from storage fat and is minimally affected by dietary fat. The normal blood concentration after the usual overnight fast is about 350 to 500 mEq/l and the largest component in the FFA pool is palmitate. In general, the spectrum of fatty acids in this pool reflects their relative abundance in body storage fat. TG is also present in circulating blood in a concentration of 50 to 150 mg%. For several hours after a fat containing meal, blood plasma normally is turbid (lactescent) because of the fat that has been absorbed from the gastrointestinal tract. This milky appearance is due to tiny droplets of fat, called chylomicrons, consisting almost entirely of TG with as little as 1 to 2% protein. These chylomicrons are cleared from the blood by the liver in 4 to 5 hr. Under fasting conditions, the blood TG fraction consists entirely of lipoproteins of variable chemical constitution. The lipid components contain triglyceride, phospholipid, and cholesterol, and their source is the liver. Circulating lipoproteins may be regarded as fat in transit from one anatomical site to another and they are always endogenous in origin. It is generally believed that (excepting chylomicrons) all TG found in blood has entered from the liver, since storage fat in adipose tissue can be mobilized for metabolic use, only through discharge of FFA.

An excellent perspective can be obtained of the entire breadth of energy metabolism, by a consideration of the famous and classical experiment of Benedict, performed in 1917 in which was carried out the most complete study ever performed up to that time, of the metabolic changes in a starving human. His subject, Mr. Levanzin, from the Island of Malta, was a professional devotee of starvation as a means of preventing sickness and disease. Taken from Benedict's original data, the acompanying figure depicts the changing sources of metabolic energy from day to day during this 31-day period of fasting. Several points are worthy of special comment: (1) The total caloric expenditure declined slowly from nearly 1,800 cal daily at the start until about the 16th day when it reached a plateau of less than 1,400 cal. (2) The contribution made by carbohydrate throughout the entire fast is amazingly small, and not of any real quantitative significance after about five days. It was not known at that time, that gluconeogenesis continues to supply carbohydrate for the obligatory requirements of some tissues, and hence the diagram erroneously shows no carbohydrate being metabolized during the entire last half of the experiment.

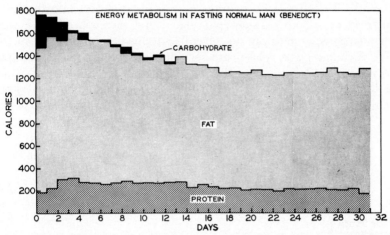

FIG. 102. THE EXPERIMENT OF BENEDICT DEPICTING THE ENERGY SOURCES THAT SUPPORT METABOLISM DURING A PROLONGED FAST

(3) The major, and by far the most important energy source throughout the 31 days was fat. (4) Protein is catabolized at a near constant rate of about 50 gm (200 cal) daily, chiefly used as a substrate for gluconeogenesis and accounting for excretion of about 8 gm daily of nitrogen. Thus, the basic pattern of degradation of body constituents is evident. It involves a major dependence upon fat as a source of fuel as might be anticipated from the normal reserve stores of lipid in all animals. Even without imposing the stress of a fast, this same pattern is seen in animals whenever a large acute or sustained demand for energy occurs, and accordingly suitable homeostatic control mechanisms have evolved to guard the integrity of nonexpendable tissues and substrates.

Only a few years ago adipose tissue was considered to be only a storehouse for inert "blubber," metabolically inactive and uninteresting. About 1937, however, experiments of Schoenheimer and Rittenberg (1935, 1937) provided the first solid evidence of the dynamic state of body constituents, and subsequent research in hundreds of laboratories has demonstrated that adipose tissue is far from the dull, lifeless protoplasm it was thought to be, but is instead, in a constant, highly dynamic state with a continuous rapid turnover of fatty acids and glycerol and with a rapid utilization of glucose. Fat cells are now recognized as the most sensitive cells in the body to the action of insulin, which is required as a regulatory agent for both the anabolic (lipogenesis) and the catabolic (lipolysis) activities of adipose tissue. Glucose arriving

328 CARBOHYDRATES AND THEIR ROLES

at the adipose cell surface requires insulin for transport into the intracellular space. Here it is phosphorylated and metabolized by a well-known enzymatic pathway to several intermediate products, one of which is alpha glycerophosphate, or "activated" glycerol. This high energy compound is then capable of esterifying acyl fatty acids which are abundant in fat cells into TG, the familiar storage form of lipid. Glycogen is present in adipose tissue in the fed state and tends to dis-

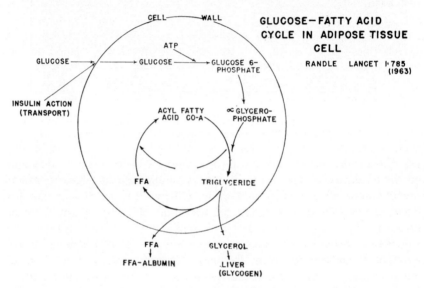

FIG. 103. GLUCOSE ENTERS THE FAT CELL WHEN INSULIN IS PRESENT, AND IS THEN PHOSPHORYLATED AND METABOLIZED TO GLYCEROPHOSPHATE AT WHICH POINT IT MAY COMBINE WITH ACYL FATTY ACIDS TO FORM TRIGLYCERIDE FOR STORAGE

In the absence of insulin and catalyzed by lipolytic hormones, adipose tissue lipase hydrolyzes the glyceride to fatty acids and glycerol, both of which products leave the cell to produce a sharp rise in blood concentration of both compounds. The availability of glucose and insulin therefore control fat metabolism.

appear during fasting. It probably represents an intermediate metabolite between entrance of glucose into the cell and its enzymatic utilization in lipogenesis. This entire process is literally under the control of insulin which insures the availability of carbohydrate carbon atoms for glycerophosphate synthesis. In the absence of adequate amounts of glucose and / or insulin, lipogenesis ceases and release of FFA occurs instead, with the simultaneous release of glycerol in a low energy (inactive) state. It was believed for many years that human adipose tissue does not contain the glycerol activating enzyme, glycerokinase.

In 1967, however, Robinson and Newsholme (1967) demonstrated its presence, so that, theoretically, at least, glycerol can be reutilized for production of more triglyceride when lipolysis ends. As a matter of practical fact, however, the concurrent rise in blood concentration of both FFA and glycerol in response to a strong lipolytic stimulus such as epinephrine indicates that the glycerol is being discharged from the cell and therefore not used again for reesterification. These two products of lipolysis are good indices of the magnitude of the process, but the rise in glycerol is regarded as more reliable than the change of FFA. It is possible, therefore, that the physiological significance of glycerokinase is not as great as its discovery originally seemed to suggest. The absolute requirement of glycerophosphate for the immobilization of FFA in the form of TG, and its storage in adipose tissue cells places special importance upon the availability of glucose. Although theoretically a small amount might be contributed by gluconeogenesis from lactate and protein, in practical terms, dietary carbohydrate is the only source of quantitative importance. It is, therefore, justifiable to state categorically that it is not possible to store fat (or in other words, to become obese) without ingestion of large amounts of carbohydrate. As a corollary, it is no less true that storage of fat cannot take place without an adequate supply of insulin with which to metabolize that glucose.

Lipolysis results from the action of adipose tissue lipase on stored triglyceride, with the release of free fatty acids (FFA) and glycerol, both of which are discharged from the cell into the blood stream where they are carried away, the FFA to cells that require a source of energy or to the liver where it may again be stored temporarily as triglyceride, and the glycerol to the liver where it is converted back to glucose. Adipose tissue lipase is exquisitely sensitive to insulin, and hence the presence of the latter, as under those conditions that stimulate lipogenesis, lipolysis will naturally be strongly inhibited. The lipolytic agents, on the other hand, activate this lipase and bring about a large flux of fatty acids from adipose tissue into the blood—a flux that can be blocked by a variety of antilipolytic compounds, including nicotinic acid, propranolol, sympatholytic drugs, and salicylates. The fat mobilizing or adipokinetic factors include epinephrine, norepinephrine, ACTH, lactogenic hormone, pituitary growth hormone, thyrotropic hormone, thyroxin, triiodothyronine, and probably cortisol (Rudman et al. 1963). In addition, exposure to cold and stimulation of sympathetic nervous activity also produce the same result. Thus, the parallel but opposing functions of fat storage and fat mobilization are both highly efficient and extremely labile processes, so that the storage of triglyceride fat in widely scattered adipose tissue sites is a remarkably dynamic process with the stream of fatty acid carbon atoms flowing in widely fluctuating amounts, first in one direction and then the other in a finely adjusted minute by minute response to the fuel requirements of energy metabolism of the whole organism. The general scheme of this fuel supply places glucose as the highest priority fuel and in the well-fed state, when carbohydrate stores are abundant, all tissues may use it freely; as these reserves begin to dwindle, however, fatty acid is mobilized in increasing amounts as glucose oxidation declines.

In obesity, fat reserves are extremely large and the concentration of fatty acids in blood is increased, often to almost twice the normal.

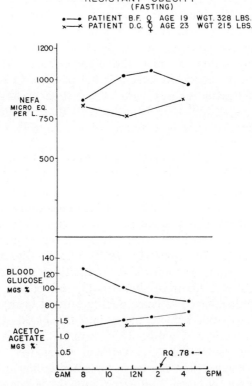

FIG. 104. THE FFA CONCENTRATIONS IN TWO
OBESE SUBJECTS DURING A DAY OF FASTING

The initial levels are about twice those seen in normal subjects and there is no tendency to rise as the fast continues. (Compare with Fig. 101).

On the basis of the foregoing brief resume of the main features of the metabolic behavior of fat, it should be apparent that the storage of lipid or its release into the "metabolic furnace" and the degree to which it participates as a component of the fuel mixture is actually under the control of glucose and insulin, and the site of this control is peripheral, at the fat cell. Of equal importance and interest in the composite picture is the mode of control of carbohydrate participation in the supply of energy.

With extracellular body fluids well supplied with fluctuating concentrations of both carbohydrate (glucose) and lipid (FFA), it is obviously of paramount importance to understand the mechanism by which the chemical constitution of the fuel mixture is determined from minute to minute for a variety of cell and organ types under widely variable environmental and physiological conditions. In 1963, Randle and his co-workers described a metabolic relationship which they named the Glucose-Fatty Acid Cycle (Randle, *et al.* 1963; Garland, *et al.* 1964;

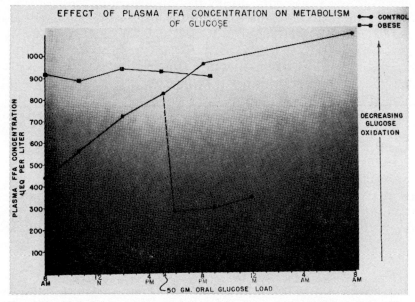

FIG. 105. THE SHADING FROM BLACK AT THE BOTTOM TO WHITE AT THE TOP INDICATES DECREASING GLUCOSE UTILIZATION AS BLOOD FREE FATTY ACID CONCENTRATION RISES

The solid dots demonstrate the rising titer during a 24-hr fast in a normal subject. The dotted line indicates the precipitous fall after administration of oral glucose, and the solid square dots represent the fatty acid levels in an obese subject which are elevated high enough to impair glucose utilization.

Schalch and Kipnis 1964). In the course of both *in vivo* and *in vitro* studies they had been able to demonstrate that a rising tide of circulating FFA consistently inhibits glucose utilization by a large number of tissues, most notably striated muscle. This suppression of glycolysis occurs through enzymatic inhibition at several specific sites, namely, glucokinase which permits the original phosphorylation of glucose, phosphofructokinase which catalyzes the addition of a second phosphate group to fructose, and pyruvate dehydrogenase which converts pyruvate to acetate and acetyl CoA. Thus lipolysis which is controlled in adipose tissue will suppress glucose utilization in a large number of tissues, simply by raising the extracellular fluid concentration of FFA. Notable exceptions include the central nervous system, the liver, the renal medulla, and erythrocytes (the red cell mass). All of these have been considered to have obligatory dependence upon glucose for all energy needs (Kety 1956). Nevertheless, despite a very substantial glucose requirement by these special tissues, the total effect of the fatty acid inhibition is to reduce very sharply the expenditure of glycogen stores, by the substitution of fat—an abundant and efficient fuel for unrestricted use. Since the brain and spinal cord under normal conditions are glucose dependent, and since the functional integrity of these tissues is necessary for life and survival, the entire mechanism for conservation of carbohydrate appears to be an evolutionary expedient designed primarily to guarantee the nutritional requirements of the nervous system.

In consideration of the fuel requirement of the brain and spinal cord (about 140 to 150 gm of glucose per 24 hr), some question and skepticism has arisen over the years concerning the capacity of hepatic gluconeogenesis to produce this amount of glucose from noncarbohydrate sources under conditions of fasting when no intake of new carbohydrate is taking place. Despite some doubt about this matter, adequate physiological methods for final resolution of these uncertainties have not been available until relatively recently. Within the last year, Owen *et al.* (1967), in the course of study of some starving volunteer subjects, were able to set up a balance sheet of glucose produced versus glucose utilized with the surprising discovery that under conditions of maximal stimulation of gluconeogenesis, adequate supplies of carbohydrate were not available to supply all those tissues that were believed to have an obligatory dependence upon this fuel for metabolic energy. Hence, there must be something wrong with these previously acceptable calculations. Either conversion of fat into carbohydrate must be taking place under starvation conditions (a concept which has not been demonstrated and which remains unacceptable to biochemists) or some of these tissues previously believed dependent upon glucose are, in fact, able to use other fuels for survival and function when the need arises. In experimental studies involving direct measurements on blood withdrawn from the jugular bulb and from an artery simultaneously, these investigators were able to demonstrate a marked utilization of ketone bodies by the brain. There was no evidence of extraction of free fatty acids as might be anticipated since these compounds do not cross the blood brain barrier, but the utilization of ketones was quantitatively large enough to account for a major part of cerebral energy metabolism by the complete oxidation of beta-hydroxy butyric and acetoacetic acids. Thus under conditions of extreme stress, as in total starvation, it now appears that the capacity of the central nervous system to oxidize substrates other than glucose for its energy needs may explain the discrepancy that has been noted in the energy balance sheet of human subjects in terms of glucose requirement as a metabolic fuel.

If these observations are adequately confirmed it is obvious that they will have a broad and profound physiological significance. This capacity to metabolize ketone bodies should be regarded as an adaptive change brought about by necessity for the purpose of survival, and it may be looked upon as "hypertrophy" of an enzyme pathway already present, probably through synthesis of increased amounts of enzyme proteins. However, since the body is believed never to synthesize a completely new protein, since it is able to produce only those that are available through the genetic code, the implication of these studies would mean that these enzyme pathways that become quantitatively functional under extreme conditions must be present at relatively insignificant levels of activity during normal conditions when dietary carbohydrate is abundant.

A further extension of this reasoning suggests that inasmuch as obese human subjects are utilizing fatty acids for almost all of their energy requirements, is it possible that they have in addition a "fat adapted brain?" There is known to be a resistance in obese subjects to the development of fasting ketosis despite the large flux of ketone bodies that must pour out of the liver, as a consequence of the heavy traffic of fatty acids that enters that organ. It is possible, therefore, that the slower rate of accumulation of ketone bodies in obese subjects under starvation conditions could be accounted for by the increased utilization by the brain, of ketone bodies in these subjects for its energy needs. This question obviously needs study and at present no answer is available.

ADAPTIVE LIPOGENESIS

Another important aspect of energy metabolism in obesity relates to the storage of energy substrates when food is ingested in excess of immediate caloric needs. The inability of humans, as with all animals, to store large amounts of carbohydrate has already been mentioned. It is quite apparent, therefore, that the ingestion of a large amount of carbohydrate or even protein food must entail the conversion of large amounts of these foods into a storage form which must, of necessity, be the conversion to fat. The process is known as lipogenesis and the process must obviously be markedly increased in magnitude whenever a huge meal is ingested (Tepperman and Tepperman 1964). Before this conversion takes place, the liver can be stored full of glycogen (which never exceeds about 6% of wet weight) and the muscles also can be packed full of the same fuel.

The desirability to animals of transporting extra fuel supplies in a concentrated, compact form, free of the surplus weight of additional water has already been mentioned. Conversion of all potential energy foods to fat, however, exacts a price from the animal. According to calculations of Flatt and Ball (1964) conversion of glucose to fatty acid costs approximately 10% of the energy content of the original carbohydrate, whereas conversion to glycogen costs only about half as much. Both processes actually are very efficient and animals must attribute their capacity to survive even brief periods of 4 to 6 hr of fasting, to their ability to carry out these metabolic functions.

The conversion to fatty acids and triglycerides takes place largely in the liver in humans and the process becomes exceedingly active following the consumption of a large meal, whereas the constant nibbling of small amounts of food over a long time span reduces to a minimum this necessity for conversion since energy supplies more nearly parallel energy requirements. In experiments with rats, it has been found that the substitution of 1 large meal per day leads quite rapidly to as much as a 25-fold increase in the magnitude of lipogenesis. This is especially true if the meal is taken in the morning when the animals are approaching a period of many hours of relative inactivity for rats are nocturnal creatures, and their ordinary eating habits classifies them as "nibblers." This dramatic increase in the magnitude of lipogenesis occurs as a necessity because of the need for a large and abnormal storage of energy in the form of fat (Hollifield and Parsons 1962; Cohn and Joseph 1960). This conversion entails the hypertrophy of all the enzymatic pathways involved in lipogenesis and the storage of fat may become so great as to produce a state of experimental obesity in the animals if they are given free access to food during the brief period when it is presented to them. If their food intake is restricted, however, without actually gaining weight, they still may manifest increased lipogenesis and actually sacrifice protein tissues in order to make excessive amounts of fat. This condition has been called "nonobese obesity." In these same experimental animals, a return to their normal pattern of eating by constant nibbling results in a gradual return to normal of these greatly expanded pathways of lipogenesis.

It seems probable and almost inevitable that a similar sequence of events must take place in humans whenever excessively large meals are consumed. Attempts in the author's laboratory to demonstrate these augmented enzyme pathways has met with failure, both in terms of increased tissue enzyme concentrations with heavy feeding and diminished concentrations from starvation. Presumably this inability to demonstrate these adaptive changes may be attributed to the slower adjustments that take place in a large animal, such as the human, in contrast to the rapid adaptive responses that occur in a small laboratory animal, such as the rat, with a much higher metabolic rate.

Despite these considerations, considerable interest has centered about the possibility that lipogenesis in humans might be significantly reduced in magnitude if eating were performed by the "nibbling" pattern (Gordon et al. 1963). Such a procedure would tend to provide smaller amounts of energy foodstuffs over a prolonged period of time and deliver this energy more or less metered according to need. The necessity for conversion into fat is thereby circumvented and presumably the enzyme pathways would adjust accordingly. Since excess adipose tissue cannot be stored unless the total caloric intake exceeds the total caloric output the nibbling pattern of eating would probably not have anything to do with the development of obesity, although it is possible that in terms of total body composition a "nonobese obese" state as described in rats could be prevented. Actual testing of human subjects in prolonged feeding experiments on a metabolic ward have failed to show any difference in energy utilization or any particular practical advantage in imposing a 6 or 8 meal per day feeding pattern upon obese subjects. Weight loss or weight maintenance in these experiments invariably depends solely upon the energy balance sheet as it always does under more normal conditions. The multiple feeding plan does, however, have one significant advantage when used in a clinical weight reducing program. If the diet is made reasonably high in protein content, and protein is administered in each one of the 6 or 8 feedings during the day, the patient is automatically obliged to ingest and assimilate protein food almost constantly throughout his waking hours. The high satiety value of protein, in turn, prevents the patient from becoming hungry and hence such diet plans are more easily followed by obese subjects who might otherwise suffer from constant hunger. It is of great importance, however, that on the basis of the calorie balance sheet, there is no known advantage either theoretical or practical in the ingestion of many small meals in preference to 1 or 2 large ones. Body weight will rise or fall or remain unchanged dependent only upon whether in a given unit of time, usually a day, there is an excess or a deficit of caloric intake over output.

THE IMPORTANCE OF INSULIN

The participation of and requirements for insulin in the process of fat storage deserves detailed comment. Insulin is the most powerful anabolic agent known in animal physiology. Without its presence, the concurrent and synchronous action of pituitary growth hormone in inducing somatic growth in young animals is known to be impossible. It is customary at present to regard growth hormone as the basic hormonal agent responsible for this phenomenon, but it is well recognized that the presence of adequate amounts of insulin in body fluids and tissues is an absolute requirement for growth hormone to express its growth potential. There are some dissenting voices expressing the opinion that insulin is truly the major growth agent with growth hormone participating as an accessory factor. This detail, however, is of academic interest only. The fact that uncontrolled juvenile diabetics with a seriously deficient production of insulin by the beta cells in the pancreatic islets, do not grow is eloquent testimony to the importance of insulin in controlling and promoting the growth process. Simple diabetic control with the administration of adequate amounts of insulin will then result in a resumption of a normal growth rate in such children. Insulin is known to be an active agent in effecting the transport of amino acids from the extracellular to the intracellular compartment of cells everywhere in order to permit the synthesis of new body protein. This includes, of course, the synthesis of enzymatic protein which is a part of the cellular machinery responsible for normal metabolic functions. In the absence of adequate amounts of insulin, protein synthesis slows down or stops due to an inadequate intracellular supply of amino acids which are the building blocks for tissue protein. Since all body constituents are in a dynamic state with a constant turnover in accord with physiological requirements, it is apparent that in adults as well as in growing children the capacity for new protein synthesis is of the utmost importance for normal function.

In regard to the carbohydrate storage of the body it is obvious that insulin is required for the normal operation of this process and the effect of insulin upon utilization of sugar in general is perhaps the best known of its actions since the human disease, diabetes, has always been considered to revolve chiefly about a deficiency of glucose metabolism. Glucose ingestion or injection serves as one of the major physiological stimuli for the production and release of insulin by the beta cells of the pancreatic islets into the blood stream. The normal subsequent metabolism of this glucose as it enters the blood stream depends upon the constant and universal availability of insulin in body fluids with the hormone serving probably as a transport mechanism to allow entry of glucose from the circulation into functioning cells. There is some disagreement as to whether insulin has an additional function in the intracellular space through the stimulation or control either direct or indirect, of anabolic biochemical processes. Since insulin is a protein, any hypothesis attempting to explain an intracellular function for the hormone must, of necessity, deal with the problem of transport of the protein molecule through the cell membrane which ordinarily does not admit molecules as large as insulin. At the present time, this protein molecule appears to be a polymer made up of aggregates of a basic unit that probably has a molecular weight of 3,000 to 6,000. Regardless of the manner in which this question is ultimately resolved, the fact still remains that energy metabolism of all animals depends upon the presence of this hormone in order to carry out the anabolic functions that are required to balance the catabolic effects of the "wear and tear of living."

Insulin is likewise an absolute requirement for the manufacture and storage of fat. In a preceding discussion the importance of glycerophosphate as the esterifying polyhydroxy-alcohol in the synthesis of triglyceride fat was discussed. Inasmuch as glucose is the only significant source for this compound it is apparent that not only a continuous supply of glucose but also of insulin for the transport and utilization of that glucose is a basic requirement. Therefore, in the insulin deficient diabetic child, synthesis of fat becomes utterly impossible and the inhibition by insulin of fat breakdown (lipolysis) is also lacking. For this reason, the adipose tissue of such children literally falls to pieces chemically with the release of a torrent of free fatty acids which, when metabolized in the liver and to a lesser extent in the kidney, produce such a flood of ketone bodies that their formation exceeds the maximum rate of their metabolism and utilization by peripheral tissues. Under these conditions they are excreted in the urine as sodium, potassium, calcium, and ammonium salts. The ensuing heavy loss of these cations (with the exception of ammonia) leads to a depletion of body electrolytes with the production of a metabolic acidosis. This is the familiar diabetic acidosis which leads to coma and death if it is not corrected by the administration of insulin which then permits the utilization of glucose to make adequate amounts of glycerophosphate and to inhibit further lipolysis.

The corollary to this interesting picture is in the obese subject in which a very large part of the carbon atoms that are stored as fat are derived ultimately from dietary carbohydrate. The process of lipogenesis which includes not only the esterification of fatty acids into triglyceride but also the conversion of carbohydrate to fatty acid requires enormous amounts of insulin. The enzymatic pathways involved in these processes are beyond the scope of this discussion, but they involve the supply of all metabolic components for the process of fat anabolism. This includes the cofactors such as NADP (TPN) the nucleotide which in reduced form (NADPH) provides the hydrogen necessary for the reductive steps in lipogenesis..

It seems highly logical for these reasons that the obese individual, must not only have adequate amounts of insulin in order to become obese but logically he might be expected to have excessive amounts of insulin. Ever since techniques became available for the accurate quantitation of circulating insulin it has become apparent that this speculative possibility is true. In response to an insulin releasing stimulus, such as the ingestion of glucose, the pancreas of the obese individual overreacts and supplies sometimes as much as 8 or 10 times more insulin for the same magnitude of stimulus as the pancreas of normal subjects (Bagdade et al. 1967). For this reason, it is possible to look upon the obese individual as one who is suffering from a form of hyperinsulinism. This is equally true of the obese diabetic who has, rather than an insulin deficiency as occurs in children, an insulin excess which accompanies the obesity. In further attempts to untangle the complicated abnormal metabolism of obesity, it is important to recognize that the "overshoot" of the pancreas in releasing insulin in this condition is distinctly a secondary phenomenon. It appears to depend upon a resistance to the action of the hormone on the part of all tissues, but apparently chiefly adipose tissue cells (Salans et al. 1968). For this reason, it may be justifiable to look upon the abnormal pancreatic response as an attempt at compensation for the ineffectiveness of insulin in normal amounts.   This resistance appears to focus on the adipose tissue cell itself because of which weight reduction in an obese individual with disappearance or shrinking of fat cells will restore insulin sensitivity and eliminate the overreaction of the insulin releasing mechanism to the normal stimulus of glucose administration.   The

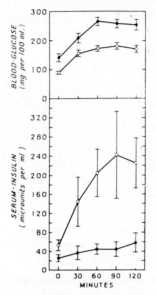

Fig. 106. Mean Blood-Glucose and Serum-Insulin Levels During Standard Oral Glucose-Tolerance Tests in Untreated Maturity-Onset Diabetics

These comprised 11 obese patients (3 males and 8 females) and 14 nonobese patients (8 males and 6 females). Open and closed circles, and bars represent mean ± standard error.

hyperinsulinism, therefore, must be recognized as a consequence of obesity rather than one of its causes, but the development and continued presence of high blood and tissue concentrations of insulin could exert an effect of perpetuating or even increasing the obesity in a vicious, accelerating spiral.   Investigators working in this field are still searching for the primary metabolic defect that is transmitted genetically which accounts at the basic cellular level for the original genesis of the obese state. For even though this disease is always due to overeating, it is very apparent that accessory factors are of great importance in determining who will become obese and who will remain normal.

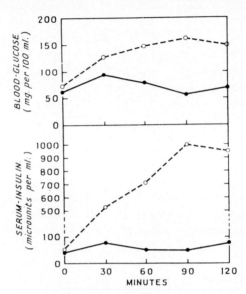

FIG. 107. EFFECT OF REDUCING FROM 20% OVER TO 5% UNDER IDEAL WEIGHT UPON BLOOD-GLUCOSE AND SERUM-INSULIN LEVELS DURING GLUCOSE TOLERANCE TESTS IN A MATURITY-ONSET DIABETIC

Levels shown as broken lines before, and solid lines after, reducing.

THE PRACTICAL PROBLEM OF OBESITY

In attempting to meet the practical problem of helping overweight human subjects to control their problem, the physician is constantly confronted by the troublesome genetic influence of the patient's own "natural weight." This factor, at the present time, cannot be adequately defined in scientific terms since it is poorly understood. Ordinarily when comparing two overweight subjects, living for weeks on the same metabolic hospital ward, eating precisely the same food and subjected to the same program of physical activity with one patient losing weight steadily and successfully and the other losing only very slowly or perhaps not at all, 1 of 3 explanations is usually invoked: (1) The unsuccessful patient is sneaking additional food and cheating on his diet; (2) The successful patient is somehow, perhaps surreptitiously getting more physical activity; (3) There is an invisible or insensible difference in energy expenditure of the two subjects, the successful one engaging in minute nearly undetectable increments of physical exertion, such as increased muscle tone or tiny, barely visible contractions of skeletal muscles, while he appears to be at rest, so far as total body activity is concerned. All of these obvious explanations must be recognized and evaluated, but it is also possible that there is an additional factor involved, namely, differences in basic chemical efficiency of the biological engine. It seems self-evident that if any subject, obese or thin, could somehow, perhaps through medication, increase the efficiency of his engine by as little as 1% if all other factors in activity and metabolism remained the same, a gain of weight would follow, simply because even that tiny increment of energy saved, operating constantly day and night year in and year out, would result in a very substantial gain in weight. Measurement of the efficiency of biological mechanisms has never been successfully achieved, but a variety of attempts have been made. Undoubtedly, some investigator in the future will devise a method for accomplishing this objective, but the task of measuring as tiny an increment as 1% is indeed a tough assignment. Recently, some experiments at the National Institutes of Health have come closer to providing a reasonable answer to this than has ever been attained before (Pool et al. 1968). Description of these studies is beyond the scope of this discussion but in view of the fact that differences in the efficiency of biological energy-producing systems do seem to occur, it is tempting to speculate that this factor, genetically determined, may provide the missing explanation for the "natural" body weight of humans, which may vary so widely from one individual to another. In view of the enormous numbers of ways in which human individuals differ one from another, such as height, color of hair, color of eyes, natural talents, susceptibility or resistance to disease, emotional stability, etc., it would indeed be strange if we all were provided with precise carbon copies of the energy transducing mechanism that supplies us with biological energy. It does not seem unreasonable therefore, to propose that the constitutionally obese individual inherited this trait from his parents so that he was predestined at the instant of conception to be a fat man. In exactly the same manner, the mesomorphic body structure of the "muscle man" with a superabundant endowment of bone, muscle, and connective tissue, could equally well be explained according to the now widely accepted concept of somatotype.

If all of this speculation should turn out to be true, it would be necessary to regard obesity as determined basically by genetic factors which cannot be altered until such time in the future as medical scientists become clever enough to "monkey" with human genes so that the resulting animal can be created on the basis of a blueprint that has been drawn in advance. Such a state of genetic science seems (fortunately) far in the future, and I think most of us would prefer to leave it that way. If we accept this proposition which admittedly is hypothetical, then obesity falls in the same category as diabetes, a disease about which we are coming to understand a great deal more, but which still remains incurable. Fortunately, we have ways of dealing with this condition, by methods that are none too satisfactory, but gradually growing better. We are still unable to cure it, however, in the same sense that we can cure pneumonia, and it is doubtful if, in the foreseeable future, the situation will change. Obesity, in the same way, is now treatable by methods that are none too satisfactory and which have hardly changed over the centuries, since obesity was recognized as an important human disease. If biological scientists in the more immediate future should learn to identify the precise biochemical reaction or sequence of reactions upon which thermodynamic efficiency depends, then there might be some hope that overweight people everywhere would have a happier and easier future, and the medical profession could then consider that obesity is truly treatable. At the present time we are obliged to be satisfied with simply starving these unhappy people in order to allow them to fit more comfortably into the range of acceptable body dimensions that we have established as "normal" in our culture.

DISCUSSION

R. W. Longley.—What is the role of muscle glycogen in maintaining the circulating glucose pool?

E. S. Gordon.—Actually muscle glycogen is not involved in maintaining the circulating glucose pool, but liver glycogen is the chief, and only important source for this glucose. Once glycogen is formed inside a muscle cell, it can disappear only as a result of glycolysis with release of lactic acid into the circulation. Muscle glycogen of course is used in the contraction of muscle . . . usually glycolysis, but it is also a precursor for triglyceride and it's interesting that there is an appreciable amount of triglyceride inside all muscle cells in well nourished individuals. This is a residual source of energy supply for the muscle cell, in addition to the circulating free fatty acids which gain access to the cell very easily. The glycerol portion of the triglyceride can come from the intracellular glycogen that is present in the muscle cell.

W. A. Rock.—What happens when a normal person ingests triglyceride or free fatty acids instead of glucose after this fasting period when the free fatty acid in the blood has risen? Does the ingestion of fatty acids cause any striking change?

E. S. Gordon.—I don't know of any instance where this has been studied by feeding fatty acid specifically but ingestion of triglyceride fat which of course is the common form in the diet, doesn't produce any appreciable change in the circulating free fatty acid level. I should add that triglyceride which is circulating in the blood as far as we know all comes from the liver. It cannot get out of fat cells except in the form of free fatty acid.

M. Bennion.—Would you comment on the idea that someone has suggested that there is glycerokinase in adipose tissue.

E. S. Gordon.—For a long time glycerokinase was not considered to be present in human fat cells. Just about a year or two ago it was discovered that this was an artifact due to the fact that the chemical methods had all been carried out at the wrong pH. When this was corrected it turned out that there is glycerokinase present. This is the enzyme (for those of you who don't know) that reactivates glycerol to make α-glycerophosphate. The glycerol which is not activated, of course, is of no use to the cell at all and this is the reason it is released. So there is the theoretical possibility that a glycerol molecule released from hydrolysis of triglyceride could be reused. The fact is that fatty acids and glycerol appear in the blood in a 3:1 molar ratio and everyone to this day regards the flux of glycerol coming out of adipose tissue as more reliable as an index of the amount of lipolysis that is taking place, than even a measure of the fatty acids. So the fact that the glycerol does come out indicates for the most part it is not being reused despite the fact that it has this enzyme available.

A. E. Harper.—You said quite a bit about the relationship between insulin and obesity. The other side of the coin is epinephrine. Can you say anything about this in relation to the obesity?

E. S. Gordon.—Insulin and epinephrine have mutually antagonistic effects. If a person develops hypoglycemia from too much insulin, one of the best ways to stop it is to give epinephrine because it not only causes glycogenolysis in the liver and raises the blood sugar but it also antagonizes the effect of insulin and puts a stop to the whole episode. So when it's a case of the two together acting on a fat cell, usually there is some lipolysis taking place because of the powerful effect of stimulating adipose tissue lipase that epinephrine has. But it is very much attenuated and blunted by the presence of the extra insulin since insulin has almost as much inhibitory effect as epinephrine does upon stimulating it. Now when it comes to the epinephrine in an ordinary fat person I don't think they have any more or any less than anyone else.

P. C. Markakis.—In a reducing diet would you say that the first things to remove would be glucose and other glucose carbohydrates rather than fats?

E. S. Gordon.—We devised a diet, which I guess most people know about now, on the basis of several principles, some of which we ourselves have proven to be not entirely true. Since carbohydrate is intensely lipogenic we decided to cut the glucose carbohydrate in the diet to a low level and instead of substituting fat, which has so often been done, we substituted protein and this also was done for three reasons: first, it is a good precursor for glucose formation, second, it has high satiety value, and third it has high specific dynamic action so that the patient does not get the benefit of all the calories it contains. We also divided it into six feedings a day and I wish time permitted discussing this because Dr. Harper has shown what happens when you eat meals as compared to when you nibble all the time. We thought that we'd like to prove that this is true in humans as it is in rats. So we biopsied adipose tissue in humans, both thin and fat, and then we starved them for as long as seven days and then did another biopsy. We ran enzyme quantitation on these biopsy specimens and we were surprised to find that most of the enzymes did not change in concentration at all. We couldn't even find some of the lipogenic enzymes considered necessary. For example, there is no citrate cleavage enzyme in human adipose tissue. This is believed to be the source of the acetyl CoA that is used in the soluble portion of the cell for the synthesis of fatty acids. We think that in the human, lipogenesis does not take place in adipose tissue but in the liver and is then transported to adipose tissue where it is reesterified by locally produced glycerophosphate. Now, what was it that you asked me? I forgot all about it.

P. C. Markakis.—Whether glucose should be the first thing to be withheld in a reducing diet?

F. S. Gordon.—We reduced it but not so far as to produce ketosis in anybody. Anybody can use these diets and will get no ketosis at all.

A. Mustafa.—Why is it that some people who want to either get fatter or thinner, no matter what they eat never seem to change? Can you comment on this for other than genetic reasons?

E. S. Gordon.—This is quite a problem. The explanation is actually genetic, but the manner in which the genetic constitution expresses itself in either fatness or leanness, is still poorly understood. Usually, the wastage of energy is evident even to casual observation, in the form of constant muscular movement, or intense nervous activity. Unless this energy expenditure is balanced by a comparable increase in caloric intake, extreme leanness will be the result. There is almost certainly another more subtle way in which energy can be wasted, and this involves the basic physiological concept of thermodynamic efficiency. People undoubtedly vary in this efficiency, possibly by as much as several percent—although it has never been possible to measure this factor experimentally in living animals. Unfortunately, I cannot answer the question beyond this brief comment.

BIBLIOGRAPHY

BAGDADE, J. D., BIERMAN, E. L., and PORTE, D. 1967. The significance of basal insulin levels in the evaluation of the insulin response to glucose in diabetic and non-diabetic subjects. J. Clin. Invest. 46, 1549–1577.

COHN, C., and JOSEPH, D. 1960. Effects on metabolism produced by rate of ingestion of diet. Am. J. Clin. Nutr. 8, 682–690.

FLATT, J. P., and BALL, E. 1964. Studies on the metabolism of adipose tissue. An evaluation of the major pathways of glucose catabolism as influenced by insulin and epinephrine. J. Biol. Chem. 239, 675–685.

GARLAND, P. B., NEWSHOLME, E. A., and RANDLE, P. J. 1964. Effects of fatty acids and ketone bodies, and of alloxan diabetes and starvation, on pyruvate metabolism and on lactate/pyruvate and L-glycerol-3-phosphate/dihydroxyacetone phosphate concentration ratios in rat heart and diaphragm muscles. Biochem. J. 93, 665–678.

GORDON, E. S., GOLDBERG, M., and CHOSY, G. J. 1963. A new concept in the treatment of obesity. J. Am. Med. Assoc. 186, 50–60.

HOLLIFIELD, G., and PARSONS, W. 1962. Metabolic adaptations to "stuff and starve" program. J. Clin. Invest. 41, 245–249.

JEANRENAUD, B. 1961. Dynamic aspects of adipose tissue metabolism. A Review. Metabolism 10, 535–581.

KETY, S. S. 1956. The general metabolism of the brain in vivo. In Metabolism of the Nervous System, D. P. Richter (Editor). Pergamon Press, London.

OWEN, O. E. et al. 1967. Brain metabolism during fasting. J. Clin. Invest. 46, 1589–1595.

POOL, P. E., SKELTON, C. L., SEAGREN, S. C., and BRAUNWALD, E. 1968. Chemical energetics of cardiac muscle in hyperthyroidism. J. Clin. Invest. 47, 80. (Abstr.)

RANDLE, P. J., GARLAND, P. B., NEWSHOLME, E. A., and HALES, C. N. 1963. The glucose fatty acid cycle. Lancet 1, 785–789.

RENOLD, A. E., and CAHILL, G. F. 1965. Handbook of Physiology: Sect. 5. Adipose Tissue. Am. Physiol. Soc., Bethesda, Md.

ROBINSON, J., and NEWSHOLME, E. A. 1967. Glycerol kinase activities in rat heart and adipose tissue. Biochem. J. 104, 2c–4c.

RUDMAN, D., BROWN, S. J., and MALKIN, M. F. 1963. Adipokinetic actions of adrenocorticotrophin, thyroid stimulating hormones, fraction H, epinephrine and norepinephrine in the rabbit, guinea pig, hamster, rat, pig and dog. Endocrinology 72, 527–543.

SALANS, L. B., KNITTLE, J. L., and HIRSCH, J. 1968. The role of adipose tissue cell size and adipose tissue insulin sensitivity in carbohydrate intolerance of human obesity. J. Clin. Invest. 47, 153–165.

SCHALCH, D. S., and KIPNIS, D. M. 1964. The impairment of carbohydrate tolerance by elevated plasma free fatty acids. J. Clin. Invest. 43, 1283–1284.

SCHOENHEIMER, R., and RITTENBERG, D. 1935. Deuterum as an indicator in the study of intermediary metabolism. J. Biol. Chem. 111, 163–168.

SCHOENHEIMER, R., and RITTENBERG, D. 1937. Deuterum as an indicator in the study of intermediary metabolism. J. Biol. Chem. 121, 235–253.

TEPPERMAN, J. 1968. Energy balance. In Metabolic and Endocrine Physiology. Yearbook Med. Publishers, Chicago, Ill.

TEPPERMAN, H. M., and TEPPERMAN, J. 1964. Adoptive hyperlipogenesis. Ann. N.Y. Acad. Sci. 131, (1), 404–411.

WHIPPLE, H. E. 1965. Adipose tissue metabolism and obesity. Ann. N.Y. Acad. Sci. 131.

CHAPTER 17

W. Werner Zorbach | **Cardiac Glycosides**

INTRODUCTION

It is, perhaps, a fair statement to say that, of all the classes of biologically active glycosides, the cardiac glycosides are the most noteworthy in that the sugar or sugars glycosidically bound contribute *significantly* to the physiological activity. The glycosides exert a specific action on the heart and are of inestimable value in cases involving decompensation, where the heart is beating in a deficient condition. Auricular fibrillation, a serious cardiac disorder, responds in a dramatic manner to treatment with the drugs, and daily doses of 0.05–0.1 mg of digitoxin *per os* generally suffice to maintain the heart in good condition.

Qualitatively, the cardiotonic activity of the glycosides resides in the steroidal aglycon. Partly because they are nearly insoluble in aqueous media, the aglycons have a poor physiological "distribution," and are not effective except at relatively high concentrations, under which conditions they are also toxic. In sharp contrast, the glycosides are therapeutically effective at very low dose levels, and it follows, therefore, that the sugars glycosidically bound enhance (often markedly) the activity inherent in the aglycon. The role of a sugar or sugars in this instance is especially noteworthy, because, by themselves, they are physiologically inactive.

The enhancement of activity observed in converting a cardiac aglycon to a cardiac glycoside may be ascribed, logically, to an "increment of polarity" provided by the additional hydroxyl function carried by the carbohydrate residue, in which the sugar residues influence the absorption and distribution of the drugs. It is further suggested that the carbohydrate component contributes to an adsorptive process that promotes the affinity of cardiac tissue for the cardiac glycoside.

The results and observations that follow are based solely on digitalis-type glycosides (because of the ready availability of digitoxigenin and strophanthidin for synthetic studies); aglycons derived from toad poisons and squill glycosides, although closely related structurally to the digitalis-type aglycons, will not be considered.

STRUCTURE OF THE AGLYCON(S)

When compared with other classes of saturated steroids, for example, the bile acids and corticosteroid metabolites, the cardiac aglycons are structurally unique. A true appreciation of their structural features may be best generated with a conformational representation of the molecules, as shown by the two, widely studied aglycons, digitoxigenin

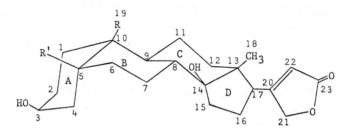

Digitoxigenin (R = CH₃, R' = H)

Strophanthidin (R = CHO, R' = OH)

and strophanthidin. Digitoxigenin is the steroidal component of digi-toxin (1 of the 3 major secondary glycosides obtained from *Digitalis* spp.) and is, perhaps, the most readily available cardiac aglycon com-mercially. Interestingly enough, it possesses the minimal structural requirements for cardiotonic activity. Both rings A/B and C/D have the *cis* fusion which imparts to the molecule a claw-like structure. An α,β-butenolide ring is attached at C-17, and both the β-oriented hy-droxyl groups at C-3 and C-14 are axial substituents. Minor changes in, or departures from, these structural features result in a sharp dimi-nution in cardiotonic activity; noteworthy, in this respect, is a reversal of configuration about C-5, which gives for rings A and B the *trans* arrangement, and a flattened character to the entire molecule.

STRUCTURE OF THE GLYCOSIDES

For cardiac glycosides, glycoside formation invariably takes place through the C-3 hydroxyl group of the aglycon. The sugars found in the naturally occurring glycosides are, for the most part, rare D- or L-hexoses of the deoxy type. For example, digitoxose (2,6-dideoxy-D-*ribo-*hexose) occurs in nature only as the carbohydrate component of cardiac glycosides obtained from *Digitalis* spp. D-Glucose is occasionally found in combination with other sugars, but rarely where it is joined directly to the aglycon (Mauli and Tamm 1957). Klyne (1950), in a compre-hensive study with steroid glycosides, showed that the molecular rotation is approximately the sum of the molecular rotation of the steroidal aglycon and of the methyl α- or β-glycoside corresponding configurationally to the sugar residue in the glycoside. On this basis, he has demonstrated that cardiac glycosides of natural origin containing D-sugars are β-D anomers, whereas those that contain L-sugars have the same absolute, or α-L configuration. Moreover, as shown, natural cardiac glycosides are generally 1,2-*trans* isomers, and, theoretically, it would

α-L̲ Isomer   β-D̲ Isomer

never be expected that glycosides containing either β-L-glucose or α-D-mannose would be found in nature. Although we shall deal here with hexosides ("monosides"), it is to be emphasized that, with many cardiac glycosides, the carbohydrate component may be composed of di-, tri-, or even tetrasaccharides, where the monosaccharide units may or may not be the same.

SOME MILESTONES IN THE SYNTHESIS OF CARDIAC GLYCOSIDES

The first synthetic cardiac glycosides were reported in 1943 by Uhle and Elderfield, who obtained, through a Koenigs-Knorr synthesis, strophanthidin pyranosides containing β-D-xylose, α-L-arabinose, β-D-glucose, and 2,3,4,6-tetra-O-acetyl-β-D-galactose residues, respectively. In each case, the new glycosides had the "natural" (either α-L or β-D) glycosidic linkage, in which only the C-3 hydroxyl group of the aglycon was involved, the same as that with cardiac glycosides of natural origin. Following this work, Reichstein and co-workers (Reyle *et al.* 1950) brought about the first partial synthesis of a naturally occurring glycoside, convallatoxin (α-L-rhamnopyranoside of strophanthidin), having the "natural," or α-L configuration. Convallatoxin is considered to be the most potent of all the known, naturally occurring cardiac glycosides of the digitalis type.

Convallatoxin

By 1957, the partial synthesis of cardiac glycosides was becoming commonplace; however, none containing a 2-deoxy sugar had been reported. This was surprising, especially because digitoxose (a 2,6-dideoxyhexose) was the carbohydrate component of digitalis glycosides. At this time, the writer began an investigation with a view, not only to substantiate the structure of the naturally occurring evatromonoside (β-digitoxoside of digitoxigenin) (Tschesche *et al.* 1955; Kaiser *et al.* 1957), but also to develop a general method of synthesis of cardiac glycosides containing 2-deoxy sugars, as well as 2-deoxyglycosides in general.

Completion of the first phase of this study (Zorbach and Payne 1958) resulted in the preparation of 2,6-dideoxy-3,4-di-*O*-*p*-nitrobenzoyl-β-D-*ribo*-hexosyl chloride (and bromide), the first reported stable, *crystalline O*-acylglycosyl halides of a 2-deoxy sugar. Subsequently, the chloride was reacted with digitoxigenin in the *absence* of an acid acceptor, giving, after saponification of the reaction products, not the desired evatromonoside, but the α-digitoxopyranoside, instead (Zorbach and Payne 1960). It is to be noted that the new glycoside has the "unnatural" (α-D) configuration, and intravenous assay disclosed that it has a cardiotonic activity less than that of its aglycon!

The preparation of the α-digitoxoside constitutes, simultaneously, the first synthesis of a cardiac glycoside containing a 2-deoxy sugar and the first recorded instance in which a biologically important 2-deoxyglycoside was synthesized, using a crystalline *O*-acyl-2-deoxyglycosyl halide.

The α-digitoxoside and evatromonoside are the first known anomer pair of cardiac glycosides. The surprisingly low potency of the α-digitoxoside could not have been predicted because, prior to this work, only α-L or β-D cardiac glycosides were known.

To determine whether the low potency of the synthetic glycoside was a special case, or whether this would hold generally for other "unnatural" glycosides, two additional α-D-glycosides were prepared. The hitherto unreported 2,3,4-tri-O-benzoyl-6-deoxy-α-D-mannosyl bromide was prepared and treated with digitoxigenin, in the presence of an acid acceptor, to give, after saponification, the α-D-rhamnoside of digitoxigenin (Zorbach et al. 1962). When strophanthidin was treated with the new halide in a similar manner, the α-D-rhamnoside of strophanthidin resulted (Zorbach et al. 1963). Each of the new glycosides has the "unnatural" anomeric configuration and, as predicted, showed an unusually low order of cardiotonic activity.

In an extension of our discovery that p-nitrobenzoic esters of 2-deoxy sugars may lead to stable, crystalline O-acylglycosyl halides, the preparation of crystalline halides of 2-deoxy-D-arabino-hexopyranose (Zorbach and Pietsch 1962) and 2-deoxy-D-ribo-hexopyranose (Zorbach and Bühler 1963) was accomplished. Each was employed successfully in the preparation of digitoxigenin glycosides containing 2-deoxy-β-D-arabino-hexopyranose (2-deoxy-β-D-glucopyranose) and 2-deoxy-β-D-ribo-hexopyranose (2-deoxy-β-D-allopyranose), respectively. When measured intravenously in cats, each was less potent than the fully hydroxylated β-D-glucoside (Elderfield et al. 1947). Relative to these studies, we sought to "oxygenate" the carbohydrate portion of the naturally occurring convallatoxin; this was accomplished by condensing the hitherto unknown 2,3,4,6-tetra-O-acetyl-α-L-mannosyl bromide with strophanthidin to afford, after saponification, the α-L-mannopyranoside of strophanthidin, alternatively named 6'-hydroxyconvallatoxin (Zorbach et al. 1963). The new glycoside shows the highest potency for any glycoside of the digitalis type, when assayed intravenously in cats.

As demonstrated in the foregoing, the presence of an "unnatural," α-D, glycosidic linkage has a deleterious effect on cardiotonic activity, despite the fact that the synthetic glycosides (vide supra) are 1,2-trans glycosides. It was of interest, in this connection, to investigate the preparation of some 1,2-cis glycosides with the "natural," β-D configuration. This is a difficult problem, and it has been adequately demonstrated (Reyle and Reichstein 1952; Zorbach et al. 1962; 1963) that 1,2-trans O-acylglycosyl halides, in a Koenigs-Knorr synthesis, give 1,2-trans cardiac glycosides, even in the presence of silver carbonate. Accordingly, a specially constituted D-rhamnosyl halide (1,2-trans) was prepared, which contained the "nonparticipating" 2,3-O-carbonyl group (Zorbach and Gilligan 1965). The new halide was coupled with digi-

4-O-Benzoyl-2,3-O-carbonyl-6-
deoxy-α-D-mannosyl bromide

toxigenin and with strophanthidin, resulting, in each case, in partial conversion into the corresponding β-D-rhamnosides. The two new glycosides are the first known β-D cardiac glycosides that have the *cis* relationship between the glycosidic oxygen atom and the hydroxyl group at C-2 of the carbohydrate component.

STRUCTURE VERSUS ACTIVITY: SOME OBSERVATIONS

Nearly all the known cardiac glycosides, natural and synthetic, have been assayed intravenously in the cat, by K. K. Chen and his co-workers (1962; 1965), under conditions as uniform as possible. The value of this assay is that the potency figures for cats are applicable to man, when the drugs are administered intravenously. In Table 26 is found a selected listing of hexosides of digitoxigenin and strophanthidin and their potency figures; these are given as the reciprocals of their lethal doses in micromoles per kilogram of cat to allow for a *direct* relationship between numerical value and potency.

For hexosides of digitoxigenin, the fully hydroxylated β-D-glucoside has the highest potency. The effect of deoxygenating C-2 of the carbohydrate moiety can be seen from inspection of the 2-deoxy-β-D-glucoside (2-deoxy-β-D-*arabino*-hexoside), in which the value has been lowered 40%. For the 2-deoxy-β-D-alloside (2-deoxy-β-D-*ribo*-hexoside), which differs from the 2-deoxy-β-D-glucoside only by a reversal of configuration about C-3 of the sugar residue, the figure is the same, and it appears that a configurational inversion at this position is an unimportant one. The β-digitoxoside (a 2,6-dideoxy-D-hexoside) is less potent than either the 2-deoxy-β-D-glucoside or 2-deoxy-β-D-alloside, showing that the effect of deoxygenation is cumulative, but not additive. The lower potencies of the β-D-rhamnoside and the α-L-rhamnoside suggest, again, that deoxygenation is unfavorable for cardiotonic activity; however, the β-D-mannoside is not yet available and, therefore, a direct comparison cannot be made.

With strophanthidin hexosides (Table 26), a similar trend can be noted. The synthetic 6'-hydroxyconvallatoxin (α-L-mannoside of strophanthidin) shows the highest potency recorded for any cardiac glycoside of the digitalis type. Both the β-D- and α-L-rhamnosides (6-deoxymannosides) show a significant diminution in potency when compared

TABLE 26

LETHAL DOSES OF SOME HEXOSIDES OF DIGITOXIGENIN AND STROPHANTHIDIN
AS MEASURED INTRAVENOUSLY IN CATS
In Lethal Doses Per Micromole

Hexoside	Glycosidic Linkage[2]	Lethal Dose
Digitoxigenin		0.8
β-D-glucoside	n	4.3
2-deoxy-β-D-glucoside[1]	n	2.7
2-deoxy-β-D-alloside[1]	n	2.7
β-digitoxoside (2,6-dideoxy-β-D-alloside)[1]	n	2.3
α-digitoxoside (2,6-dideoxy-α-D-alloside)[1]	u	0.9
α-D-rhamnoside[1] (6-deoxy-α-D-mannoside)	u	0.8
β-D-rhamnoside[1] (6-deoxy-β-D-mannoside)	n	1.5
α-L-rhamnoside[1] (6-deoxy-α-L-mannoside)	n	1.9
tetrahydropyranyl ether	n	0.2
Strophanthidin		1.2
α-L-mannoside	n	8.2
α-D-mannoside	u	2.2
α-D-rhamnoside[1] (6-deoxy-α-D-mannoside)	u	4.0
β-D-rhamnoside[1] (6-deoxy-β-D-mannoside)	n	5.6
α-L-rhamnoside[1] (6-deoxy-α-L-mannoside)	n	7.0
tetrahydropyranyl ether	n	0.9

[1] Trivial names have been used for ready identification.
[2] n refers to the "natural," and u, the "unnatural" linkage.

with the mannoside. In this case, a direct comparison is possible, leaving little doubt that deoxygenation is unfavorable. Perhaps the best evidence may be had by comparing the tetrahydropyranyl ether derivatives (completely deoxygenated pyranosides) of both digitoxigenin and strophanthidin, not only with the various corresponding hexosides, but also with the corresponding aglycons, themselves. In each case, the tetrahydropyranyl derivative has a potency substantially less than that of the corresponding aglycon, and it is suggested that the pyranoid ring makes no contribution of its own, but merely acts as a vehicle for carrying hydroxyl functions. It appears, therefore, that the unsubstituted tetrahydropyranyl ring serves only to "dilute" the cardiotonic activity of the aglycon (Zorbach et al. 1965).

In Table 26 are given, also, potency figures for four "unnatural" cardiac glycosides. The α-digitoxoside has a figure less than ½ its anomeric form (β-digitoxoside) and the same as that for its aglycon. With the α-anomer, therefore, it appears that the sugar makes no contribution to the cardiotonic activity whatever. The "unnatural" β-D-rhamnoside, when compared with both the α-D-rhamnoside and the α-L-rhamnoside, shows a sharp drop in activity, having a value the same as that for digitoxigenin.

In strophanthidin, there is aldehyde function at C-10 and a hydroxyl group at C-5, giving rise to new, potential interactions between the aglycon and the sugar residue, as compared with digitoxigenin. Consequently, the relatively straightforward relationships observed with the digitoxigenin hexosides may or may not obtain. The "unnatural" α-D-mannoside, when compared with the "natural" α-L-mannoside shows a significant drop in potency, yet it is about two times more active than strophanthidin (aglycon). Surprisingly, the α-D-rhamnoside is only slightly less active than the β-D-rhamnoside, and more than ½ as active as the α-L-rhamnoside (convallatoxin), and it is more than 3 times more active than its aglycon.

Most intriguing is the fact that the "natural" α-L and β-D cardiac glycosides have potencies that fall within a rather narrow range, especially when enantiomeric pairs are considered (compare the α-L and β-D rhamnosides of digitoxigenin and strophanthidin in Table 26). It has been stated that α-L and β-D cardiac glycosides have the same absolute configuration at the anomeric center, but this is barely an explanation to account for this phenomenon, because the conformations of the sugar residues are not specified by the statement.

In the usual terms of the conformational rules for carbohydrates, specifying that most hexoses exist in that chair form in which the hydroxymethyl group attached at C-5 is an equatorial substituent, α-L and β-D cardiac glycosides would contain pyranosyl residues in alternative chair forms. This would connote distinctly different conformations for α-L vis-a-vis β-D cardiac glycosides, which, reasonably, should result in a significant disparity in the potencies of the two.

In the absence of alternative explanations, we have advanced the hypothesis that, with glycosides having bulky aglycon residues, the usual rules regarding conformation for carbohydrates are not applicable, and that, instead of being governed by the hydroxymethyl group at C-5, the pyranoid residues assume that chair form which accommodates the aglycon residue as an *equatorial* substituent. Accordingly, the α-L- and β-D-rhamnosides of digitoxigenin (Table 26), for example, could be shown below as the conformers, with which a correspondence obtains, in that the aglycon components are both equatorially disposed, both pyranoid rings have the *same* chair form, and the *order* of substituents on each ring is the same. The only difference between the conformers

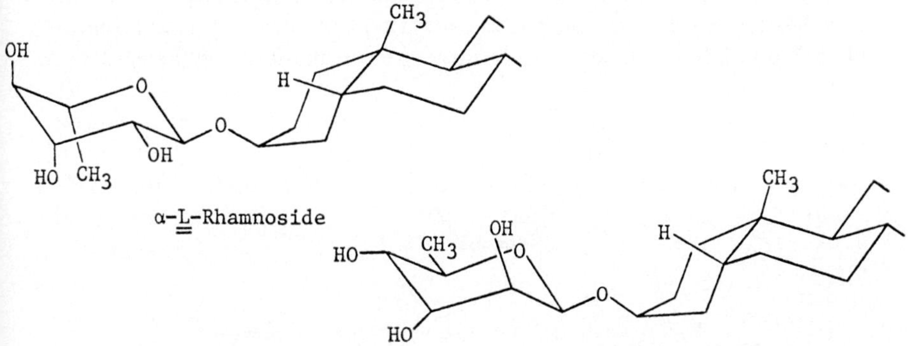

α-L-Rhamnoside

β-D-Rhamnoside

is a reversal of configuration of substituents on one sugar ring with respect to the other, amounting only to small spatial displacements of the respective substituents. However, because such a small difference (in the structures shown) does exist, it is not unreasonable to ascribe the slightly greater potencies of α-L cardiac glycosides, as compared with those of their β-D isomers, to such structural variation.

PROSPECTS FOR THE SYNTHESIS OF CARDIAC GLYCOSIDES CONTAINING 2-DEOXYALDOHEXOSE RESIDUES IN THEIR FURANOID FORMS

Cardiac glycosides that contain sugars in their furanoid form do not occur naturally. Such glycosides might show unexpected biological activity and, because of this possibility, their synthesis is warranted. Although fully hydroxylated aldoses may give highly potent pyranosyl glycosides, our current efforts have been directed toward the preparation of O-acyl-2-deoxyglycosyl halides, primarily because of the occurrence of 2-deoxy sugars (*e.g.*, digitoxose) as the carbohydrate constituents of naturally occurring cardiac glycosides. Recently, we have been successful in the preparation of crystalline, relatively stable O-acylglyco-furanosyl halides of two 2-deoxyhexoses; with these, there has now been presented the possibility of preparing, by a direct method, furanoid cardiac glycosides containing the subject sugars.

We have shown (Bhat and Zorbach 1965, 1968) that the direct methyl glycosidation of 2-deoxy-D-glucose (2-deoxy-D-*arabino*-hexose) results in 30–35% of crystalline methyl 2-deoxy-α-D-*arabino*-hexofuranoside, which underwent the following conversion: methyl furanoside → methyl 5,6-O-carbonyl-2-deoxy-α-D-*arabino*-hexofuranoside → methyl 5,6-O-carbonyl-2-deoxy-3-O-*p*-nitrobenzoyl-α-D-*arabino*-hexoside → crystalline 5,6-O-carbonyl-2-deoxy-3-O-*p*-nitrobenzoyl-α-D-*arabino*-hexosyl bromide.

```
5,6-O-Carbonyl-2-deoxy-3-O-p-nitrobenzoyl-
α-D-arabino-hexosyl bromide
```

The methyl glycosidation of 2-deoxy-D-*ribo*-hexose has recently been shown (Bhat *et al.* 1968A) to result in 100% conversion into a 1 : 1 anomer mixture of crystalline methyl 2-deoxy-D-*ribo*-hexofuranoside, which was converted, in good yield, into methyl 3,5,6-tri-O-*p*-nitrobenzoyl-D-*ribo*-hexoside. The latter was readily converted into crystalline 3,5,6-tri-O-*p*-nitrobenzoyl-D-*ribo*-hexosyl bromide (Bhat *et al.* 1968B). Both of the new halides have been shown to have utility in glycosidation reactions.

```
2-Deoxy-3,5,6-tri-O-p-nitrobenzoyl-
D-ribo-hexosyl bromide
```

DISCUSSION

A. Neuberger.—Do you have any other evidence other than that mentioned for the belief that the large aglycon is always equatorial? That's very interesting and a rather exciting sort of suggestion.

W. W. Zorbach.—This is only a hypothesis, and, although I don't propose to fight too hard for it, no one has offered a better explanation.

If you take a series of cardiac glycosides comprised of α-L and β-D isomeric pairs, you find a remarkable correspondence in potencies. As a result of the work that we did with the unnatural, α-D glycosides, one finds a great disparity in the potencies, when these are compared with either the corresponding α-L or β-D isomer, and the hypothesis presented is the only way in which I could think to explain the situation. I can anticipate another question, namely, "why don't you do some nmr studies?" First of all, this would require data on a respectable number of compounds, but this isn't very easy, as it may take 6 or 8 months to make enough of a cardiac glycoside with a rare sugar residue. You must synthesize your sugar from available carbohydrate starting materials and, by the time you do all the characterization of intermediates, do your couplings (which are tedious and costly), and repeat these on a larger scale, 6 or 8 months may go by. We did, however, study the α-L-, α-D-, and β-D-rhamnosides of digitoxigenin with a 60-MHz spectrometer, and I can tell you that nothing definitive whatever could be gleaned from the spectra.

D. R. Lineback.—In relation to this, Dr. Zorbach, I wonder if it's really so unusual that the conformation has flipped, because, in this case, the function you have at C-5 is methyl (rather than hydroxymethyl), which, of course, would be smaller. Also, I think that someone in Lemieux's group a few years ago has shown, with some mannosyl substituted products (some N-glycosyl derivatives, I believe), that, when the aglycon was sufficiently larger, the preferred conformation of the mannopyranosyl ring was the "flip" of the normal one.

W. W. Zorbach.—We appreciate that point also, especially with rhamnose, with its methyl group at C-5, in which case, such a "flip" would not be so difficult. As far as that goes, you don't have to assume an ideal, conformational inversion and that, even with a partial inversion, you're going to get something more comfortable conformationally.

BIBLIOGRAPHY

BHAT, K. V., and ZORBACH, W. W. 1965. 5,6-O-Carbonyl-2-deoxy-3-O-p-nitrobenzoyl-D-*arabino*-hexosyl bromide. A stable, crystalline O-acylglycofuranosyl halide of a 2-deoxyhexose. Carbohydrate Res. 1, 93–95.

BHAT, K. V., and ZORBACH, W. W. 1968. Pyrimidine nucleosides derived from 2-deoxy-D-*arabino*-hexofuranose. Carbohydrate Res. 6, 63–74.

BHAT, C. C., BHAT, K. V., and ZORBACH, W. W. 1968A. The direct methyl glycosidation of 2-deoxy-D-*ribo*-hexose. Chem. Commun., 1968, 808–809.

BHAT, C. C., BHAT, K. V., and ZORBACH, W. W. 1968B. The four isomeric methyl glycosides of 2-deoxy-D-*ribo*-hexose. Abstr., CARB 16, 156th Meeting Am. Chem. Soc., Atlantic City, N.J.

CHEN, K. K. 1962. Possibilities of further developments in the glycoside field by modifying the glycoside structure. Proc. Intern. Pharmacol. Meeting, 1st, Stockholm, 1961. 3, 27–45. (For a compilation of Chen assay values prior to 1961, see Hoch, J. H. 1961. A Survey of Cardiac Glycosides and Genins. University of South Carolina Press, Columbia).

CHEN, K. K., and HENDERSON, F. G. 1965. Digitalis-like substances of *Antiaris*. J. Pharmacol. Exptl. Therap. *150*, 53–56.

ELDERFIELD, R. C., UHLE, F. C., and FRIED, J. 1947. Synthesis of glucosides of digitoxigenin, digoxigenin, and periplogenin. J. Am. Chem. Soc. *69*, 2235–2236.

HENDERSON, F. G., and CHEN, K. K. 1962. Cardiac activity of newer digitalis glycosides and aglycones. J. Med. Pharm. Chem. *5*, 988–995.

HENDERSON, F. G., and CHEN, K. K. 1965. Cardiac glycosides and aglycones by synthesis and microbiological conversion. J. Med. Chem. *8*, 577–579.

KAISER, F., HAACK, E., and SPINGLER, H. 1957. Concerning the mono- and bis-digitoxosides of digitoxigenin, gitoxigenin, and gitaloxigenin. Justus Liebigs Ann. Chem. *603*, 75–88.

KLYNE, W. 1950. The configuration of the anomeric carbon atoms in some cardiac glycosides. Biochem. J. *47*, xli.

MAULI, R., and TAMM, C. 1957. Glycosides of *Periploca nigrescens*. Helv. Chim. Acta *40*, 299–305.

REYLE, K., MEYER, K., and REICHSTEIN, T. 1950. Partial synthesis of convallatoxin. Helv. Chim. Acta *33*, 1541–1546.

REYLE, K., and REICHSTEIN, T. 1952. Partial synthesis of a digitoxigenin β-D-thevetoside, its identification with honghelin, and an attempt at partial synthesis of a digitoxigenin L-thevetoside. Helv. Chim. Acta *35*, 195–214.

TSCHESCHE, R., WIRTZ, S., and SNATZKE, G. 1955. Cardioactive glycosides from roots of *Euonymus atropurpurea*. Ber. *88*, 1619–1624.

UHLE, F. C., and ELDERFIELD, R. C. 1943. Synthetic glycosides of strophanthidin. J. Org. Chem. *8*, 162–167.

ZORBACH, W. W., and BÜHLER, W. 1963. Digitoxigenin 2-deoxy-β-D-allopyranoside. Justus Leibigs Ann. Chem. *670*, 116–121.

ZORBACH, W. W., and GILLIGAN, W. H. 1965. The partial synthesis of two 1,2-*cis* cardenolides. Carbohydrate Res. *1*, 274–283.

ZORBACH, W. W., and PAYNE, T. A. 1958. 3,4-Di-*O*-*p*-nitrobenzoyl-1-chloro (and 1-bromo)-1,2,6-trideoxy-D-*ribo*-hexose. Two crystalline 2-deoxy-acylglycosyl halides. J. Am. Chem. Soc. *80*, 5564–5568.

ZORBACH, W. W., and PAYNE, T. A. 1960. 3β-(2,6-Dideoxy-α-D-*ribo*-hexopyranosyl)-14β-hydroxy-5β-card-20(22)-enolide. A direct method of synthesis of 2-deoxyglycosides involving a crystalline 2-deoxy-acylglycosyl halide. J. Am. Chem. Soc. *82*, 4979–4983.

ZORBACH, W. W., and PIETSCH, G. 1962. Partial synthesis of the anomeric 2-deoxyglucopyranosides of digitoxigenin. Justus Leibigs Ann. Chem. *655*, 26–35.

ZORBACH, W. W., BÜHLER, W., and SAEKI, S. 1965. Tetrahydropyranyl ether derivatives of digitoxigenin and strophanthidin. Chem. Pharm. Bull. (Tokyo) *13*, 735–736.

ZORBACH, W. W., SAEKI, S., and BÜHLER, W. 1963. Strophanthidin cardenolides containing hexoses of the mannose series. J. Med. Chem. *6*, 298–301.

ZORBACH, W. W., VALIAVEEDAN, G. D., and KASHELIKAR, D. V. 1962. Partial synthesis of evomonoside. J. Org. Chem. *27*, 1766–1769.

CHAPTER 18

Glenn E. Mortimore

Diabetes Mellitus: Hormonal Regulation of Hepatic Carbohydrate Metabolism

INTRODUCTION

A major feature of the insulin deficient state is the excessive rate with which energy stores are broken down. Regardless of whether the deficiency arises spontaneously, as in human diabetes mellitus, or is produced experimentally in animals, many of the basic manifestations are the same: Protein breakdown is accelerated, blood free fatty acid levels rise, and glucose production is enhanced. The latter, coupled with reduced glucose utilization by such tissues as skeletal muscle, heart, and adipose tissue, results in the accumulation of glucose in the blood and urine. In health, energy storage is carefully regulated so that the supply of intermediates from these stores is able to keep pace with the varying demands of the body. In situations where supply exceeds demand, as during feeding, the balance is tipped away from degradation and energy stores are renewed. As a result of this homeostatic activity, the concentration of glucose in blood, for example, remains remarkably constant and is not greatly altered by such metabolic stresses as exercise and fasting.

It would appear that insulin plays an important role in this mechanism. In this chapter I should like to single out one aspect and discuss the role of insulin on hepatic glucose production. Evidence will be presented to show that the restraining influence of insulin on this process can be modified profoundly by hormones such as glucagon and epinephrine, which accelerate glucose formation and release.

THE LIVER AS A SOURCE OF GLUCOSE

Virtually all of the glucose which enters the circulating glucose pool during the postabsorptive state is derived from the liver. The importance of this organ as a glucose source was convincingly demonstrated by Mann who was one of the pioneers in the development of the technique of hepatectomy (Mann 1927). He showed that fatal hypoglycemia invariably followed the completion of this surgical procedure in the dog. However, if glucose was injected in amounts sufficient to equal rates of glucose utilization by the peripheral tissues, life could be sustained for several hours. The glucose which is released from the liver is derived either from the breakdown of glycogen or synthesized from pyruvate, amino acids or glycerol by a process known as gluconeogenesis.

The synthesis of glucose from pyruvate incorporates carbon from amino acids as well as circulating lactate. The initial reaction involves the carboxylation of pyruvate to form oxalacetate. Oxalacetate then is converted to phosphopyruvate, the precursor of triose phosphate used in the generation of glucose. Glycerol enters the gluconeogenic pathway at the triose phosphate level and, hence, bypasses steps between pyruvate and phosphopyruvate. These steps are known to include an important hormonally-sensitive site for the control of gluconeogenesis (Exton and Park 1968).

Glycogen is synthesized from glucose-1-phosphate by steps involving glucose-1-phosphate and uridine diphosphate glucose, and is degraded by the action of phosphorylase yielding glucose-1-phosphate. Since glycogen is both the product and a source of glucose phosphate, it represents a quantity of preformed glucose which may serve as a physiological buffer against sudden changes in the circulating pool of glucose. The amount of glycogen contained in the normal liver, if utilized as the sole source of glucose during the postabsorptive period, would not maintain adequate levels of blood glucose for more than a few hours. For this reason the process of gluconeogenesis assumes major importance in providing glucose over long periods of fasting. Most of the glucose that is lost from the body by excretion in the urine in severe diabetes is formed by this process.

From metabolic studies in the normal, fasting individual, Cahill has calculated that for a basal hepatic glucose output of 180 gm per day 20% of the glucose came from circulating lactate, 10% from glycerol, and 70% from amino acids (Cahill et al. 1968). In these human studies, as opposed to rat experiments, liver glycogen was stable and no significant quantity of glucose came from this source. The glycerol that was utilized was released largely from adipose tissue during the hydrolysis of triglyceride. The source of the amino acids is less certain, but both skeletal muscle and liver are thought to release substantial quantities of amino acids (Cahill et al. 1968; Miller 1961; Mortimore and Mondon 1968). These tissues may thus serve as important reservoirs of gluconeogenic carbon. The net contribution to hepatic glucose release that was derived from lactate represents the amount of glucose that was utilized by glycolysis in the red cell and other tissues.

EFFECT OF INSULIN ON HEPATIC GLUCOSE RELEASE *IN VIVO*

Many attempts have been made to show that the liver is directly responsive to the administration of insulin *in vivo*. Until comparatively recently, however, the results were either small in magnitude or lacking in consistency. Sherlock and her co-workers were the first to report that the injection of insulin into human subjects might cause an abrupt and significant reduction in hepatic glucose release (Bearn et al. 1952). The effect was obtained in both normal and diabetic subjects and, in some instances, an appreciable uptake of glucose was noted. In these studies the hepatic vein was catheterized and transhepatic glucose differences were estimated from glucose concentrations in hepatic and arterial blood. Her findings were criticized on the basis that the glucose differences reflected changes across the entire splanchnic bed and might therefore have included effects in tissues other than liver. Mesenteric fat was one such possibility. In further experiments of this general type, Mahler and collaborators failed to observe any decrease in the transhepatic gradient in dogs after insulin administration (Mahler et al. 1959).

Madison reported that the infusion of very small amounts of insulin into the portal circulation of dogs would evoke a highly significant reduction in hepatic glucose output (Madison et al. 1960). The effect was diminished or abolished when the amount of insulin given was increased to the point where a strong uptake of glucose by the peripheral tissue was elicited. The failure to obtain an hepatic effect under these conditions was attributed to the fact that the additional insulin had caused a more marked hypoglycemia and had thereby stimulated the liver to release glucose, thus overriding the inhibitory effect. This interpretation presupposes that the concentration of glucose itself may be an important factor in the net transfer of glucose between liver and the circulating glucose pool. Such a possibility was put forth earlier by Soskin who emphasized that the basic regulation of blood glucose is due to an intrinsic homeostatic activity on the part of the liver (Soskin and Levine 1952). Thus, when the glucose level rises, the liver responds by diminishing the output of glucose and, conversely, when the level falls, the output increases. As Soskin viewed this activity, hormones such as insulin might modify the "setting" or the range of glucose concentration over which it operates, but they would not affect its basic mode of operation.

Isotope dilution methods of various kinds have been employed to provide indirect assessments of hepatic glucose output (deBodo et al. 1957; Dunn et al. 1957; Searle et al. 1959). Inherent in all of the methods is the assumption that the specific activity of plasma glucose decays as a function of the rate with which unlabeled glucose leaves the liver and dilutes the isotope in the circulating glucose pool. The injection of insulin into human subjects or dogs thus labeled has been reported to evoke a reduction in the rate of decay, thus substantiating effects observed with the use of direct, but far more cumbersome methods.

The effects of acute insulin withdrawal in rats have been demonstrated in rats with the use of antiinsulin serum, which effectively binds free insulin. In one study the administration of antiinsulin was followed by a doubling of blood glucose and a significant reduction in liver glycogen (Jefferson et al. 1968). The increase in the circulating glucose pool corresponded closely to the net change in liver glycogen, suggesting that glycogenolysis had been accelerated. From the foregoing, such a response to insulin withdrawal would be expected on the supposition that hepatic glucose release was normally under restraint by insulin.

STUDIES WITH THE ISOLATED, PERFUSED RAT LIVER

The result of attempts to demonstrate effects of insulin in isolated liver preparations has generally paralleled experience with the use of intact animals in the sense that consistent, workable responses were established only after years of effort. The reason for this is not altogether clear, but it may be related in part to methods of tissue preparation. Investigators have become aware that the liver is extraordinarily sensitive to hypoxia and other injuries which are associated with the preparation of slices and minces and organ perfusion has largely supplanted these methods, particularly for metabolic and endocrinologic work. This change in methodology undoubtedly accounts for much of the recent success in this field.

Figure 108 depicts a recent model of an organ perfusion apparatus for metabolic studies with the isolated rat liver that was developed by the author at the National Institutes of Health (Mondon and Mortimer 1967). It consists of a temperature-regulated box which houses four independently operating perfusion units. With such an apparatus we are able to operate four separate perfusion experiments in parallel under closely controlled conditions. The perfusion medium is pumped out of the rotating, spherical oxygenating flask and into the portal vein of the liver at a constant, predetermined rate of flow. The medium then returns to the flask through tubing inserted into the inferior vena cava between the liver and the heart. Despite the apparent complexity of the apparatus as seen in the figure, its operation is relatively simple and direct. By perfusing the liver in situ rather than after it has been removed from

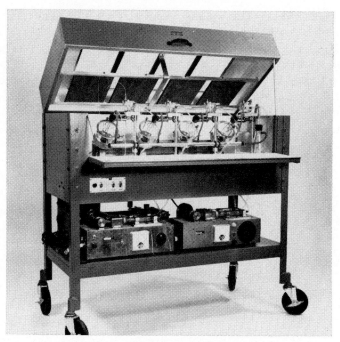

FIG. 108.   ORGAN PERFUSION APPARATUS
See text for description.

the carcass, unnecessary trauma from hypoxia and manipulation of the tissue is avoided.

Effect of Perfusate Glucose on Hepatic Glucose Release

It may be seen in Figure 109 that glucose in the medium strongly influences the release of glucose from the liver during perfusion. Within the limits studied, the relationship between glucose concentration and release is both linear and reciprocal, and it illustrates the type of hepatic response that was mentioned previously in connection with glucose homeostasis *in vivo*. It should be emphasized at this point that the sensitivity of these particular preparations to glucose was far from ideal. For example, it was difficult to obtain a net glucose uptake even at very high glucose levels, and the livers appeared to become less responsive to glucose with increased duration of perfusion.

In further experiments it was found that the sensitivity of glucose could be modified by altering the red cell concentration of the perfusion medium (Glinsman and Mortimore unpublished). When perfusate having a packed red cell volume simulating that of whole blood was used in place of the regular medium (defibrinated rat blood, diluted about 1 : 1, v/v, with isotonic bicarbonate buffer), rates of glucose release were much lower for a given initial glucose level. Furthermore, significant rates of glucose uptake could be obtained at glucose levels above 3.0 mg/ml of medium. When red cells were omitted from the medium, rates of glucose release were maximal at all glucose concentrations.

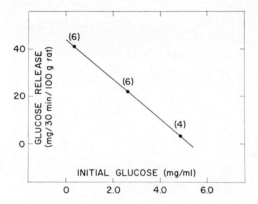

Fig. 109. The Relationship Between the Initial Concentration of Glucose in the Medium and Net Hepatic Glucose Release During the First 30 Min of Perfusion

Livers were obtained from normal, nonfasted rats. Number of experiments given in parentheses.

While the nature of this red cell effect is not presently understood, evidence to date suggests that a rate-limiting step in glucose uptake by liver may be adversely affected by oxygen lack. By analogy with other tissues, such a step might include transport and/or phosphorylation, since these are close to the external environment of the cell. In the case of liver, however, glucose is able to penetrate the cell membrane freely and there is yet no evidence for a sugar transport mechanism of the kind that has been demonstrated in heart, skeletal muscle, adipose tissue, and the red cell (Cahill et al. 1958). Moreover, hypoxia is known to accelerate glucose uptake in muscle by an action on transport (Morgan et al. 1959). Studies from Sols' laboratory have indicated that glucose uptake by the liver may be limited by the rate of phosphorylation (Vinuela et al. 1963). An enzyme, glucokinase, which is sufficiently active to account for most of glucose phosphorylation in liver, is highly specific for glucose and has a K_m that is close to the physiological level of glucose in blood. The activity of this enzyme is increased by insulin *in vivo*, but the time base over which significant changes can be measured is too long to account for the rapid inhibitory effects of insulin on net glucose release (Salas *et al.* 1963).

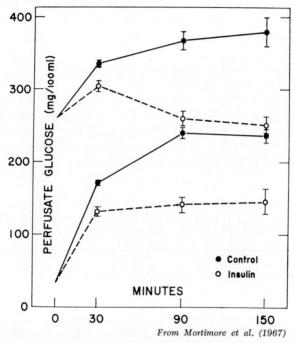

From Mortimore et al. (1967)

FIG. 110. EFFECT OF INSULIN ON PERFUSATE GLUCOSE
CONCENTRATION AT INITIAL GLUCOSE LEVELS OF 260
AND 33MG/100ML

Livers from 6 pairs of nonfasted rats comprised each group. Insulin was infused into the medium in amounts sufficient to yield maximal effects.

Effects of Insulin on Carbohydrate Metabolism

Representative effects of insulin on glucose release from perfused livers of normal, nonfasted rats are shown in Fig. 110 (Mortimore 1963). The addition of insulin at both high and low initial glucose levels significantly inhibited net release as judged by the time-courses of perfusate glucose.   In line with what was mentioned earlier, it may be seen that glucose itself was an effective inhibitor.   Increasing the initial concentration of glucose from 0.33 to 2.60 mg/ml reduced the rise in glucose concentration during perfusion by half.   It is of interest that the magnitude of the effects were similar in both sets of experiments and were not affected by the initial glucose level.   This suggests that insulin did not inhibit the net release of glucose by stimulating its uptake from the medium.   Such an action would likely have resulted in an augmentation of the response at the higher glucose level.   In this sense the insulin effect differed strikingly from the red cell effect, discussed above.

The possibility that net glycogenolysis was inhibited by insulin was investigated in a series of carbohydrate balance experiments (Mortimore *et al.* 1967).   From the results in Table 27, it is evident that the reduction by insulin in the accumulation of total carbohydrate in the medium (glucose + lactate) could be accounted for largely by a sparing of glycogen breakdown.   The inhibitory effect on the accumulation of perfusate carbohydrate was somewhat larger than the glycogen sparing

TABLE 27

EFFECTS OF INSULIN ON NET CARBOHYDRATE ALTERATIONS AFTER 2 HR OF RAT LIVER PERFUSION

	Perfusate		Liver Glycogen, Final- Initial			
	Glucose A	Lactate B	C	A + B	A + B + C	C/(A + B)
Control, 7 Experiments						
Mean	+13.4	+0.91	−19.3	+14.3	−5.0	
SE	0.79	0.25	0.62	0.95	0.96	
Insulin, 7 Experiments						
Mean	+ 6.2	+0.05	−13.4	+ 6.3	−7.1	
SE	0.43	0.09	0.97	0.46	1.09	
Δ (Insulin-control)						
Exp. No.						
10	− 9.5	−0.73	+ 8.6	−10.2	−1.6	0.84
11	− 6.6	−1.08	+ 6.0	− 7.7	−1.7	0.78
12	− 7.0	−0.52	+ 8.4	− 7.5	+0.9	1.12
13	− 4.8	−0.75	+ 3.7	− 5.6	−1.9	0.66
14	− 4.1	−0.29	+ 2.3	− 4.4	−2.1	0.52
15	−10.0	−2.22	+ 9.0	−12.2	−3.2	0.74
16	− 8.2	−0.39	+ 3.4	− 8.6	−5.2	0.40
Mean Δ	− 7.17	−0.85	+ 5.91	− 8.03	−2.11	0.723
SE	0.84	0.25	1.06	1.00	0.69	0.088
P	< .001	< .02	< .01	< .001	< .05	

Values are expressed as mg/gm liver. Mean initial glycogen was 40.8 ± 2.32 mg/gm.
Source: Mortimore *et al.* (1967).

effect, indicating that other pathways or reactions were involved.   The difference observed is consistent with a reduction in gluconeogenesis from amino acids, previously noted in the perfused rat liver (Mortimore 1963).   Insulin also reduced the accumulation of perfusate lactate.   The reason for this is not apparent, but there is evidence which suggests that the effect may be the result of diminished glycolysis, secondary to the lowering of glucose in these experiments (Mortimore et al. 1967).

Metabolic Effects of Glucagon and Cyclic AMP

The similarity between the effects of glucagon (or catecholamines) and those produced by the withdrawal of insulin has suggested the idea that insulin may act as a physiological inhibitor of these hormones. In the case of hepatic glucose regulation, for example, the increased glucose production that occurs after the administration of antiinsulin serum would then represent an unopposed stimulation of glucose release by these agents.   Since the diverse effects of glucagon or catecholamines in liver are thought to be mediated by increases in tissue levels of adenosine $3'5'$-monophosphate, referred to hereafter as cyclic AMP (Sutherland et al. 1966), the appeal of this hypothesis lies in its ability to explain the equally diverse effects of insulin by antagonism of a single chemical agent.

The isolated, perfused rat liver has proved to be an ideal system for studying the metabolic effects of both glucagon and cyclic AMP.   With the addition of comparatively small amounts of the cyclic nucleotide to the medium, all the major effects of glucagon have been reproduced (Glinsmann and Mortimore 1968).   In addition to its ability to stimulate glycogenolysis by the activation of phosphorylase, cyclic AMP is a potent stimulator of gluconeogenesis (Exton and Park 1968; Sutherland and Robison 1966).   Exton and co-workers have shown that a rate-limiting step (or steps) in the gluconeogenic pathway from pyruvate to glucose is involved in this effect (Exton and Park 1968).   Glucagon and cyclic AMP also increase urea formation (Glinsmann and Mortimore 1968; Miller 1961).   It is not clear at the moment whether this latter effect represents a direct enhancement of amino acid oxidation or is the result of other actions of cyclic AMP.   It is conceivable, for example, that stimulation of gluconeogenesis by glucagon, which is known to reduce pyruvate levels (Exton and Park 1968), might increase deamination by pulling reactions in the direction of the depleted keto acid pool.   Such an action would explain the rapid deamination of existing amino acids, but it clearly fails to provide for a steady increase in the supply of amino acids from protein. Inasmuch as the stimulus to urea formation and gluconeogenesis by glucagon is not a transient phenomenon but well sustained, proteolysis must also be directly stimulated.

Interaction between Effects of Insulin and Glucagon or Cyclic AMP

One test of a possible interaction between insulin and glucagon would be to see whether insulin is able to suppress metabolic responses to glucagon administration under conditions in which insulin alone exerts

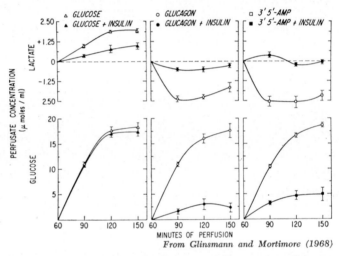

From Glinsmann and Mortimore (1968)

FIG. 111. TIME COURSE OF PERFUSATE CHANGES IN GLUCOSE AND LACTATE BETWEEN 60 AND 150 MIN OF PERFUSION

Livers were obtained from nonfasted rats, averaging 125 gm in weight. All additions were made by infusion into the medium after a 60 min period of equilibration at which time glucose and lactate concentrations were 9.8 and 2.6 μmoles/ml, respectively. The following were infused: Insulin, 4.72 μg/hr; glucagon, 0.053 μg/hr; cyclic AMP, 0.84 mg/hr; glucose, 13.1 μmoles/min between 60 and 120 min and 2.6 μmoles/min from 120 to 150 min.

little effect. The results of such a test are given in Fig. 111 (Glinsmann and Mortimore 1968). It may be seen in the left panel that when insulin was added after 60 min of perfusion (at a time when spontaneous glucose release had leveled off), no effect was obtained even though glucose levels were raised to moderately high values by the continuous infusion of glucose. Insulin, however, was able to inhibit strongly the release of glucose that was induced by the infusion of a small amount of glucagon (Fig. 111, center panel). This suppression could be demonstrated only when the amount of glucagon given produced less than a maximal effect. The response to maximal doses of glucagon could not be antagonized with 50-fold increases in the quantity of insulin. The sharp drop in perfusate lactate in Fig. 111 may be explained by the stimulation of gluconeogenesis by glucagon (Exton and Park 1968). Insulin antagonized this effect as well.

Jefferson and colleagues have recently reported that the addition of insulin to the perfused rat liver will reduce cyclic AMP, both under ordinary conditions of perfusion and after levels had been increased by the administration of antiinsulin serum *in vivo* or the addition of glucagon *in vitro* (Jefferson et al. 1968). These investigators suggested that the moment-to-moment output of glucose from the liver is regulated by a balance between glucagon and/or catecholamines, which elevate cyclic AMP, and insulin, which lowers it. Such a hypothesis provides a reasonable explanation for the glucagon-insulin antagonism in Fig. 111. If one supposes that levels of the cyclic nucleotide decrease during perfusion in the absence of stimulatory agents, then the failure of insulin alone to influence hepatic glucose release after 60 min of perfusion may be attributed to the fact that the level of nucleotide had fallen below the effective regulatory range. When cyclic AMP was increased by the administration of glucagon, then insulin was capable of exerting an effect.

The results shown in the right-hand panel of Fig. 111 are pertinent to this hypothesis. In these experiments cyclic AMP levels were increased by infusing the nucleotide into the medium rather than by stimulating its formation within the cell. It may be seen that insulin was just as effective in antagonizing the metabolic responses to added cyclic AMP as it was in antagonizing the effects of glucagon. Assuming that a reduction of tissue cyclic AMP was the basis of antagonism by insulin in these experiments, it seems unlikely that nucleotide synthesis was inhibited since cyclic AMP was supplied exogenously. The possibility that insulin had increased the degradation of cyclic AMP by an action on phosphodiesterase has not been conclusively ruled out. These data are also consistent with the idea that insulin had antagonized the effects of cyclic AMP without altering the effective concentration of the nucleotide. An inhibitor of cyclic AMP has been described (Murad 1965), but its physiological role is not known.

A major problem in the interpretation of changes in cyclic AMP is the fact that a sizeable quantity of the nucleotide in liver is bound to a particulate fraction of the cell (Jefferson et al. 1968). It is thus difficult to know to what extent the relatively small changes in tissue levels that were produced by insulin in the study cited actually reflected alterations in the concentration of the nucleotide at enzyme sites. Since glucagon is known to activate phosphorylase by increasing cyclic AMP at a biochemically defined locus, it was of interest to see whether an increase in the activity of this enzyme by glucagon during liver perfusion could be antagonized by insulin. As shown in Table 28, phosphorlyase activity was stimulated 62% by glucagon over the 30-min test period. Glycogenolysis was greatly enhanced. When the same

TABLE 28

EFFECTS OF INSULIN ON ALTERATIONS IN PERFUSATE CARBOHYDRATE, LIVER GLYCOGEN, AND PHOSPHORYLASE IN THE PRESENCE AND ABSENCE OF SMALL AMOUNTS OF GLUCAGON

Infusion	No.	Glucose	Glucose + 1/2 Lactate	Glycogen (glucose equiv)	Phospho- rylase
		μmoles/gm liter			μmoles ρ_i/gm per min
None	10	9.56 ± 1.17 (<.001)	8.83 ± 1.11 (<.01)	14.0 ± 4.7 (NS)	17.2 ± 1.0 (<.01)
Insulin	10	1.67 ± 1.50	1.44 ± 1.61	18.7 ± 4.8	13.3 ± 0.9
Glucagon	6	56.61 ± 4.67 (<.001)	54.06 ± 4.28 (<.001)	47.6 ± 3.9 (<.001)	27.9 ± 1.1 (NS)
Glucagon + insulin	6	20.89 ± 4.89	20.33 ± 5.17	16.5 ± 2.3	27.7 ± 1.3

Values are means ± SE. Net changes were calculated over a 30-min interval between 60 and 90 min of liver perfusion. Changes in glucose and lactate represent net increases occurring in the perfusate. Changes in glycogen represent net losses from the liver. Glucagon (0.048 μg/hr) and insulin (4.72 μg/hr) were infused into the perfusion reservoir beginning at 60 min. Liver phosphorylase was assayed at the conclusion of the experiments. Enzyme activity in unperfused controls was 22.9 ± 1.1 μmoles ρ_i/gm per min and following 30 min of maximal glucagon stimulation (10 μg hr) it was elevated to 38.1 ± 1.4 μmoles ρ_i/gm per min. P values in parentheses represent significance of mean effect of insulin; NS = not significant.
Source: Glinsmann and Mortimore (1968).

dose was given in the presence of insulin, the stimulus to glycogenolysis was sharply reduced without any apparent reduction in the activation of phosphorylase. If the inhibition of glycogenolysis was achieved by a decrease in cyclic AMP, then one should have noted a significant effect on enzyme activity. This would be so since the activation of phosphorylase was less than maximal and this step is generally considered to be rate-limiting in glycogen degradation. One can only speculate as to what other factors might be involved in the regulation of glycogenolysis and hepatic glucose release by insulin. Some of these have been recently discussed elsewhere (Glinsmann and Mortimore 1968).

DISCUSSION

V. Riddle.—What experiments have been done in which intermediates of glycolysis such as glycerol or dihydroxyacetone, Krebs cycle intermediates, or sorbitol have been utilized both in the amelioration of some of the effects of diabetes and in attempts to rationalize effects due to insulin on permeability of glucose? Also, how much work has been done on the glucose?

G. E. Mortimore.—I'm not aware of any studies in which intermediates, such as phosphate esters, have been used successfully. Glycerol is readily converted to glucose by the liver but I'm not aware that it is useful in modifying any particular responses in liver. Is part of your question directed towards the effects of fatty acid breakdown products on glucose transport?

V. Riddle.—Let me limit the question to amelioration of the diabetic effects on ketogenesis and so forth that occur in diabetes. Can these effects be minimized by the administration of Krebs cycle intermediates since one of the causes of acidosis is the impaired metabolism of glucose in diabetes?

G. E. Mortimore.—I personally have not studied ketogenesis, but it's my understanding that ketone body formation in diabetes is primarily the result of increased lipolysis, secondary to insulin lack.

V. Riddle.—In the work that you're describing, the effects on cyclic AMP and so forth, the interaction between insulin and glucagon, have you eliminated the effects of these hormones on the transport of glucose itself?

G. E. Mortimore.—There is no evidence for a glucose transport mechanism in liver. At all concentrations of glucose in the medium one finds almost the same concentration of glucose in liver cell water. Furthermore, in systems that do have a transport step there is no evidence that cyclic AMP is involved in regulating sugar transport.

BIBLIOGRAPHY

BEARN, A. G., BILLING, B. H., and SHERLOCK, S. 1952. The response of the liver to insulin in normal subjects and in diabetes mellitus: hepatic vein catheterization studies. Clin. Sci. *11*, 151–165.

CAHILL, G. F., JR., ASHMORE, J., EARLE, A. S., and ZOTTU, S. 1958. Glucose penetration into liver. Am. J. Physiol. *192*, 491–496.

CAHILL, G. F., JR., OWEN, O. E., and FELIG, P. 1968. Insulin and fuel homeostasis. The Physiologist *11*, 97–102.

DE BODO, R. C. *et al.* 1959. Effects of exogenous and endogenous insulin on glucose utilization and production. Ann. N.Y. Acad. Sci. *82*, 431–451.

DUNN, D. F. *et al.* 1957. Effect of insulin on blood glucose entry and removal rates in normal dogs. J. Biol. Chem. *225*, 225–237.

EXTON, J. H., and PARK, C. R. 1968. Control of gluconeogenesis in liver. II. Effects of glucagon, catecholamines, and adenosine $3',5'$-monophosphate on glucogenesis in the perfused rat liver. J. Biol. Chem. *243*, 4189–4196.

GLINSMANN, W. H., and MORTIMORE, G. E. 1968. Influence of glucagon and

372   CARBOHYDRATES AND THEIR ROLES

3′,5′-AMP on insulin responsiveness of the perfused rat liver. Am. J. Physiol. *215*, 553–559.

GLINSMANN, W. H., and MORTIMORE, G. E.   Unpublished observations.

JEFFERSON, L. S. *et al.*   1968.   Role of adenosine 3′,5′-monophosphate in the effects of insulin and anti-insulin serum on liver metabolism.   J. Biol. Chem. *243*, 1031–1038.

MADISON, L. L., COMBES, B., ADAMS, R., and STRICKLAND, W.   1960.   The physiological significance of the secretion of endogenous insulin into the portal circulation. III.   J. Clin. Invest. *39*, 507–522.

MAHLER, R., SHOEMAKER, W. C., and ASHMORE, J.   1959.   Hepatic action of insulin.   Ann. N.Y. Acad. Sci. *82*, 452–459.

MANN, F. C.   1927.   Effects of complete and partial removal of liver.   Medicine *6*, 419–511.

MILLER, L. L.   1961.   Some direct actions of insulin, glucagon and hydrocortisone on the isolated, perfused rat liver.   Recent Progr. Hormone Res. *17*, 539–568.

MONDON, C. E., and MORTIMORE, G. E.   1967.   Effect of insulin on amino acid release and urea formation in perfused rat liver.   Am. J. Physiol. *212*, 173–178.

MORGAN, H. E., RANDLE, P. J., and REGEN, D. M.   1959.   Regulation of glucose uptake by muscle. III.   Biochem. J. *73*, 573–579.

MORTIMORE, G. E.   1963.   Effect of insulin on release of glucose and urea by isolated rat liver.   Am. J. Physiol. *204*, 699–704.

MORTIMORE, G. E., KING, E., JR., MONDON, C. E., and GLINSMANN, W. H.   1967.   Effects of insulin on net carbohydrate alterations in perfused rat liver.   Am. J. Physiol. *212*, 179–183.

MORTIMORE, G. E., and MONDON, C. E.   1968.   Inhibition of proteolysis by insulin in perfused rat liver.   Federation Proc. *27*, 495.

MURAD, F.   1965.   An inhibitor of adenosine 3′,5′-phosphate (3,5-AMP) action in heart and liver extracts.   Federation Proc. *24*, 150.

SALAS, M., VINUELA, E., and SOLS, A.   1963.   Insulin-dependent synthesis of liver glucokinase in the rat.   J. Biol. Chem. *238*, 3535–3538.

SEARLE, G. L., MORTIMORE, G. E., BUCKLEY, R. E., and REILLY, W. A.   1959.   Plasma glucose turnover in humans as studied with ¹⁴C-glucose.   Diabetes *8*, 167–173.

SOSKIN, S., and LEVINE, R.   1952.   Carbohydrate Metabolism, 2nd Edition. University of Chicago Press, Chicago.

SUTHERLAND, E. W., and ROBISON, G. A.   1966.   The role of cyclic 3′,5′-AMP in responses to catecholamines and other hormones.   Pharmacol. Rev. *18*, 145–161.

VINUELA, E., SALAS, M., and SOLS, A.   1963.   Glucokinase and hexokinase in liver in relation to glycogen synthesis.   J. Biol. Chem. *238*, PC1175-PC1177.

Role of Carbohydrates in the Food Industry

CHAPTER 19

G. N. Bollenback | Sugars

When Prof. Cain started organizing this Symposium well over a year ago, I was completely sold on the philosophy—that of building a bridge of understanding between those engaged in basic scientific research and the scientists and technologists who are doing research more directly related to the food industry (Cain 1967).

Such a foundation for a Symposium is a firm one and has been throughout the life of this series; perhaps the industry-academic inter-face theme that is getting such considerable attention nowadays could be considered an alternate definition of this philosophy (Anon. 1968B; Nesty 1968; Hass 1968; Harris 1968). Frankly I believe research is research whether it is "basic" or being carried on in relation to the food industry. The "bridge of understanding" is needed not only between the academically oriented *scientist* and industrial practitioner or *technologist* but also between industrial technologists and research personnel within the same company or working out of a research association which may be industry sponsored (see, for instance, reports on some recent problems of the Research Association of the British Food Manufacturing Industries, Anon. 1966).

Perhaps here is the best spot to clarify the difference between food science and food technology. I like J. R. Blanchfield's (1966) definition[1]:

"Food science is a coherent and systematic body of knowledge and

---

[1]The Oregon State University definition is that "Food Technology is the application of science and engineering to food manufacturing," or more fully stated, "Food technology is the application of the sciences and engineering to the manufacturing, preservation, storage, transportation, and consumer use of food products."

understanding of the nature and composition of food materials, and their behaviour under the various conditions to which they may be subject.

"Food technology is the application of food science to the practical treatment of food materials so as to convert them into food products of such nature, quality and stability, and so packaged and distributed, as to meet the requirements of the consumer and of safe and sound practice."

The basic understanding needed, then, is one between scientists and technologists. As one who spent many years as a scientist and the past 5 or 6 yr in the technology area, I can verify the need for such an understanding. It is still disturbing to me to remember that the many facts I brought with me from lab to production/sales/marketing found a small audience and, indeed, had to be highly diluted in order to become palatable to those practicing the manufacture of food products. Such dilution was made with patience, gaining of confidences, persistence against ingrained habits and recognition that *opinions* run many processes.

I find, also, on looking over our roster of speakers, that I am among the few who may be considered as having acquired a severe tarnish of pragmatism. I will not speak in defense of industrial practices (good or bad), but I will attempt to point out to the many of you who are highly research oriented (1) what some of the *technical* problems are in the technology of sugar, including some that have been solved by use of scientific data as well as some that still need solving; (2) what some of the nontechnical problems are in the sugar industries; such frequently are more difficult to solve than the technical ones.

Before getting into specifics, we should recognize a few ground rules:

(1) When one refers to sugars in the food industries, it should be recognized that sucrose is the sugar used in largest amounts (Ballinger and Larkin 1964). Others of importance—and, perhaps for our purposes, *sweeteners* is a better term—include dextrose, corn syrups, synthetics. Of much lower volume are such products as lactose and sorbitol. Most of our discussion will revolve around sucrose, dextrose and corn syrups.

(2) You will have to adjust your horizons before we move on. In referring to sugars used in the food industries we are talking not in mg, gm, kg, not in lb or cwt. We are talking in terms of tons, thousands of tons and, indeed millions of tons; for instance, the current allocation of sugar for use in the United States in 1968 is 10.6 M tons (approximately 22 billion lb) (U.S. Dept. Agr. 1968B). To emphasize this point, data are given in Table 29 for the deliveries of major sweeteners to food

TABLE 29

SUGAR, DEXTROSE AND CORN SYRUP DELIVERED TO INDUSTRIAL FOOD PROCESSORS
IN THE UNITED STATES, 1949–1967 IN 1,000 TONS

Average for Period	Sugar	Dextrose	Corn Syrup	Total
1949–53	3132	301	559	3992
1954–58	3804	270	676	4750
1959–63	3654	327	893	4874
1964	4997	382	1137	6516
1965	5415	380	1152	6947
1966	5777	400	1190	7367
1967	5828	390	—	—

Source: Ballinger (1966) and U.S.D.A. (1967, 1968A).

processing industries for the period 1949–1967 (Ballinger 1966; U.S.
Dept. Agr. 1967, 1968A).

(3) There are certain trends in sugar usage that help define problem areas in the manufacture and use of sugars in the food industries. These are an increased use of sugars in bulk[2] form (liquid and dry) (Cook 1968) (Table 30), and an increasing use of sweetener blends (especially those of sugar with corn syrup) (Ballinger and Larkin 1964). Reasons given for both major trends are convenience, sanitation, and ease of handling.

TABLE 30

DELIVERIES OF SUGAR IN CONSUMER SIZE PACKAGES, INDUSTRIAL PACKAGES, BULK
GRANULATED AND LIQUID SUGAR AS PERCENTAGES OF TOTAL U. S. DELIVERIES
1958–1967

Year	Consumer Size Packages	Industrial Size Packages	Bulk Granulated	Liquid Sugar	Total Bulk Granulated and Liquid
			%		
1958	35.3	41.8	9.7	13.2	22.9
1959	34.2	39.3	11.4	15.1	26.5
1960	34.0	36.0	13.6	16.4	30.0
1961	33.2	33.8	15.5	17.5	33.0
1962	31.6	32.2	17.7	18.5	36.2
1963	30.1	32.3	19.3	18.3	37.6
1964	30.7	27.8	21.1	20.4	41.5
1965	29.0	27.4	22.5	21.1	43.6
1966	27.7	26.2	23.6	22.5	46.1
1967	26.9	25.3	24.7	23.1	47.8
10-Year average	31.3	32.2	17.9	18.6	36.5
Annual trend	−0.94	−1.83	+1.72	+1.05	+2.77

Source: Cook (1968).

[2] The trend toward bulk usage perhaps reflects Dr. Cantor's comment in his remarks opening this Symposium with regard to increase in convenience foods.

These ground rules are the framework for our discussion. Let us now examine some of the gross structures as well as certain fine details comprising this picture. However, before proceeding, I believe it would be beneficial to repeat Blanchfield's (1966) listing of the areas wherein information generated by scientists is useful in food fields— such use being directed toward causing "food materials to undergo desirable changes of nature and/or form, while inhibiting and if possible preventing undesirable changes of nature and/or form." Such areas include: a knowledge and understanding of the chemical composition of food materials; their physical, biological and biochemical nature and behavior; human nutritional requirements and nutritional factors in food materials; the nature and behavior of enzymes and of microorganisms and their action on foods; the interaction of food components and the effect on these of additives and contaminants; any pharmacological and toxicological considerations; the reactions of food materials with atmospheric oxygen and with substances with which they may come in contact during handling, processing, and packaging; the effects of various manufacturing operations, processes and storage conditions on all the foregoing; and the application of statistical methods for the design of experimental work and evaluation of these results.

Many of these areas have been discussed during this Symposium. Let me highlight some of them as defined by the trends (bulk, liquid, blends) described above, by sweetener usage in food industries, and by sweetener manufacture.

(1) We deal almost exclusively with systems that are not only nonideal but also impure from the scientist's point of view. Thus one must be wary of applying to foods systems physical and chemical data collected on solutions of pure sugars. Within the sugar refining industry this is recognized as witness publication on physical properties of "impure" sucrose solutions (Breitung 1956; Lees 1965; Mantovani and Indelli 1966; Thiele 1962). Because we are talking of sweeteners, perhaps this is a good time to mention *sweetness*. For many years a "sweetness scale" for sugars has been used as a guide for selecting a given sweetener for a particular application. Such a scale is based on solutions of pure sugars, frequently dilute. One wonders at the use of a scale so based on realization that sweeteners are rarely used alone and frequently in such high concentrations that it is difficult to evaluate their sweetness level within a narrow range of reproducibility. (see Nieman 1960 and, especially, Pangborn 1963; Wick 1963).

(2) We deal often with groups of systems which interact and make the attempt to predict properties of products by adding the

properties of members of the system a bit like Russian roulette. As an example of the danger involved Davis and Prince (1955) showed that when a corn syrup of viscosity of 90,000 cp was blended with an equal volume of liquid sugar of 235 cp viscosity, the actual viscosity of the blend was 2,600 cp. Arithmetic proportioning gave a value of 45,117 cp while logarithmic proportioning came closer (4,599 cp) but was still far removed from the actual value. Viscosity is a very important physical property both to the sugar refiner and user. There is a need for measurements of viscosities of such nonideal sweetener solutions and perhaps here is a scientist-technologist gap that could be easily bridged.

(3) We are frequently—and in practice many times unknowingly—more involved with *rates* of reactions or equilibrations rather than actual equilibrium values. *Rate* is part of the basic vocabulary of the scientist. Many times a food processor adopts the *equilibrium value* with little conception of the meaning and importance of the *rate* involved in arriving at that value. An example may be found in the crystallization of sugars. The problem exists for sellers of liquid sugars and blends of sugars. To help the industry, we published a series of curves showing limits of solubilities of sucrose-dextrose blends designed from data obtained at the National Bureau of Standards (Hoynak and Bollenback 1966). The point we failed to make was in not emphasizing the part the *rate* of crystallization plays at the various concentrations cited. Such is far more important to the refiner and user of liquid sugars than limits of solubility.

(4) Even with a basic understanding of *rates* on the part of the practitioner, trouble can be met in food products in the form of inter-action of systems. In this connection, I must call to your attention a very fine paper by McWeeny (1968). He gives a number of examples where the apparent rate of a particular reaction *decreases* when temperature is raised. This is contrary to the rule of thumb that the rate of a reaction doubles for each 10° rise in temperature. He concludes that "such an effect can usually be related to the fact that the change in rate of reaction is being measured in a system where conditions other than temperature (e.g., rates of concurrent reactions, concentrations of reactants, phase conditions) are being allowed to vary and the rate is being compared in different systems as well as at different temperatures." McWeeny's paper is a delightful example of not only bridging science and technology but also of putting a lot of lively traffic on that bridge.

(5) Microbiological problems in sweeteners industry are of great

importance especially when one considers the trends in use of bulk quantities of sugars and in use of sugars in liquid forms. The industry has considerable need for rapid methods of detection of viable micro-organisms—perhaps recent work on identification via gas chromatography will be part of an answer (O'Brien 1967). There is also a need for determining proliferation rates on sugar mixtures (blends) and sugar grades that are called "industrial." In addition, we can use information regarding the effect on flavor and odor of food products of microbial action on such mixtures and sugar grades.

(6) *Irradiation* is becoming more of a question mark, with the FDA casting doubt on the safety of preserving foods with γ-ray bombardments (Anon. 1968A; Anon. 1968C). Ultraviolet light is often used and promoted for control of microbial growth in liquid sugar systems (Ellner 1967; Gaddie *et al.* 1964; Robe 1965). Under conditions used— and to many the technique is not completely satisfactory—are important amounts of materials being formed that cause mutations of human cells? (cf. Shaw and Hayes 1966).

(7) *Synthetics* or nonnutritive sweeteners are undoubtedly of worldwide importance. Economically, they are important—there is some suggestion that their use has bitten into what might have been the normal trend of nutritive sweetener usage in the beverage industry (Ballinger 1967).[3] Nutritionally—or physiologically, there is still an open discussion going on (Greig 1968). One might wish in this as well as the irradiation area that humans could pass through as many generations as rapidly as do *Drosophila* so that mutations could result which would be adaptable to these new environs. Valid data and objective viewing are called for—and this should be a challenge both to scientists and technologists.

(8) In addition to these highly technical problems, the scientific community should be informed of some others the sweetener manufacturer and user runs into. These include:

a) government regulations (standards of identity)
b) ingrained habits—reluctance to change (e.g., processors insistence on a weight for weight usage when passing from one particular sweetener to another); NIH factors[4]

[3] Even if this conclusion is valid the reasons may be difficult to pin down. One possibility is the relation cited by both Dr. Cantor and Dr. Gordon that as affluence increases, carbohydrate consumption decreases.

[4] *Not Invented Here.* This attitude results in the refusal by a group or industry to accept any suggestion, discovery or data unless such are developed within the group or industry. Such an attitude seems to the author to be practiced with great proficiency within large segments of the sugar refining and related industries.

c) literature of the food processing industry—the hazards of articles written by public relations and advertising personnel. Actually, trade journal literature is like many other bodies of information—when you live with it long enough you learn which journals are the most objective, which authors the most reliable.

We have been somewhat general to this point although we have tried to throw in some examples illustrative of areas wherein information generated (or that could be generated) by scientists is useful in sweetener production and use. Let us now dip into the five recognized areas of the food processing industries and offer some selected illustrations of (1) the generation of scientific data and its use in technology as well as (2) existing problems in technologies that could use the assistance of scientific investigation. Throughout this sketch it should be remembered that the success in using scientific data in a technology and the design of scientific experiments to solve technological problems is greatly dependent on having proper liaison between the scientist and technologist.

Beverage Industry

(Soft drinks, malt, malt liquors, distilled liquors, flavorings) For the purpose of this discussion let us emphasize soft drinks and carbonated soft drinks in particular. To orient you economically, the distribution of sugar to this industry in 1967 accounted for 30% of all sugar shipped to industrial users (U.S. Dept. Agr. 1968A). The same holds for 9% of the dextrose and 3.5% of the corn syrup (1966 data) similarly shipped. The overall trend of usage of these sweeteners in the beverage industry over the past 10 yr is upwards. (Ballinger and Larkin 1963B; U.S. Dept. Agr. 1963, 1964, 1965, 1966, 1967, 1968A).

A recent and quite exciting development in our own labs shows what can be done when scientist and technologist are brought together under proper conditions. Up until about 2 yr ago, the evaluation of beverage formulations rested mainly on statistical data gathered from large volume production runs. This is the ultimate test of any formula; however, the absence of rigid control of variables—solids, degree of carbonation, acid content, etc., made interpretations of results somewhat more hazardous than they should have been. Dr. E. T. Bohm of our labs designed and built a pilot plant bottling unit (Fig. 112) such that one bottle of a given flavor/formulation could be produced under precisely controlled conditions. This has allowed rapid evaluation of carbonated beverages wherein the only variables present are under control of the operator and not subject to the vagaries of the manufacturing process.

By Permission of E. Bohm and P. Hoynak (1967)

FIG. 112A. PILOT PLANT BOTTLING AND CANNING UNIT

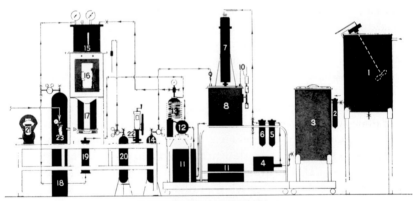

"Q-PLUS" PILOT BOTTLING AND CANNING UNIT
Designed and built by Refined Syrups & Sugars Division, Corn Products Co.

1. Water Treatment Tank	8. Water Storage Tank	15. Filler
2. Filter	9. CO2 Inlet	16. Bottle & Can Filling Cabinet
3. Holding Tank	10. CO2 Outlet	17. Cylinders
4. Pump	11. Heat Exchange Unit	18. High pressure CO2 Unit
5. Cotton Filter	12. Pump	19. CO2 Filter
6. Carbon Filter	13. Carbonator	20. Air Replacement Unit
7. De-aerator - Vacuum Pump	14. Low Pressure CO2 Unit	21. Canning Unit
22. Crowning Unit	23. Compressed Air Intake	

By Permission of E. Bohm and P. Hoynak (1967)

FIG. 112B. SCHEMATIC OF PILOT PLANT BOTTLING AND CANNING UNIT

Now, if we add to this machine a shelf-life interpreter (e.g., gas chromatography) as well as flavor ingredient control (also gas chromatography or other analytical method) we are well on our way to converting an art into a science. Certainly, a problem area today in this industry is satisfactory determination of shelf-life. If gas chromatography is a satisfactory technique then proper instruction of the technologist in the technique becomes the problem—includes head space analysis, sugars analysis, correlation of results with panel testing.

Flavor stability in another part of this industry presents yet another unsolved problem. A rule of thumb prevails that corn syrup cannot be used in carbonated soft drinks (although they can in still drinks and fountain syrups) because of flavor loss. Here is a project for someone to take on.

Another many faceted problem of great importance in soft drink technology is that of microbiology. Several troubles of microbial origin have been pinpointed through the diligent accumulation and solving of customer complaints by various flavor houses. (See for instance a particularly reliable and excellently put up brochure and trouble shooting manual (Naarden 1966; Sand 1966)). Some microbial contamination problems have also been solved by using the knowledge that high concentrations of sugar (e.g., 73–76° Brix invert sugar vs. 67° Brix sucrose solutions) discourage growth. Other problems do exist in the microbiological area—the need for a more rapid method of determining presence of microorganisms, the viability of microorganisms in blends as compared to solutions of pure sugars, more effective (and allowable) means of preventing infection.

Further to the question of beverage problems is the occasional occurrence in carbonated soft drinks of a precipitate called "floc." This material may be of microbiological origin or of sugar and/or sugar processing origin—no one knows for sure (at least with cane based sucrose). We understand that the bottling industry is considering establishing a fellowship directed toward elimination and identification of floc at the University of Georgia—so the scientist-technologist collaboration is in evidence—let's hope the liaison is effective.

Before leaving beverages I'd like once more to mention the characteristic of "sweetness." You will recall that we mentioned sweetness and its variation with concentration—one implication being that sweetness of different sweeteners can be compared in a more reproducible manner at lower solids concentrations. Thus, soft drinks might appear to be an ideal medium for proper evaluation of sweetness. We should, however, point out that there are additional ingredients (flavors, acid, carbonation, water) in soft drinks, and, to quote R. J. Wicker (1966): "Clearly the only satisfactory way of testing sweetness is by evaluation by a reasonably sized panel of a particular product under practical conditions. For example the necessary amount of sweetening agent for formulation of a soft drink will result from experiment and not by calculation from literature figures."

Sweetness necessarily will pop up time and again throughout our discussion. Right now it leads us into a second major segment of the food processing industries.

Baking Industry

The baking industry includes not only bread but also yeast raised goods (such as doughnuts) and nonyeast raised cakes, pies, snacks, etc. (Ballinger and Larkin 1963C). Sweetener usage for 1967 shows 22% of the sugar, 47% of the dextrose and 15% (1966 value) of the corn syrup distributed to industrial users went to the baking industry. (U.S. Dept. Agr. 1968A). Trends of these sweeteners in the baking industry implies an increased usage over the past 10 yr. There has, however, been a precipitous decrease in dextrose usage over the 1963–1967 interval (Ballinger and Larkin 1963C; U.S. Dept. Agr. 1963, 1964, 1965, 1966, 1967, 1968A).

So far as bread is concerned the primary purpose of sweeteners is as a leavening agent. Here we can show a very close knit association between the scientist and technologist. Indeed, the data show the sweeteners sucrose and dextrose as leavening agents do not differ significantly on a leavening basis (Bohn 1954; Bohn and Junk 1960; Lee and Geddes 1959; Piekarz 1963) if proper adjustment is made for weight gain on inversion of sucrose. However, what of the questions of residual sugars, crust color, and flavor development? The *determination* of residual sugars in breads was first made using paper chromatography (Koch *et al.* 1954; Griffith and Johnson 1954) and this represents an admirable collaboration between research and technology. The question is still open, I believe, as to the reliability of some of the values (%) recorded for residual sugars let alone a satisfactory interpretation of the meaning (organoleptic or otherwise) of their presence (Bohn and Junk 1960; Piekarz 1963).

Crust color and flavor development in baked goods reflect the impact of knowledge of the browning reaction on the baking industry (Lee and Chen 1966; Liau and Lee 1966; Salem *et al.* 1967). The variation of intensity in crust color (Anon. 1967) is used promotionally in the marketing of sweeteners with varying spectra of saccharides. Flavor in bread—again attributable much of the time to products of the browning reaction—has been the subject of much study in a number of laboratories (Hodge and Nelson 1961; Johnson 1963; Pence 1967; Thomas and Rothe 1956; Wiseblatt 1961). The problem of flavor cannot be considered as fully under control, for one frequently cited reason for the absence of a strong upward trend in bread consumption[5] is the lack of flavor due to processing techniques now in vogue. The English are working valiantly to understand what goes on in breadmaking (Axford et al. 1968; Collyer 1966, 1967). Again a proper conveyance of their results to the technologist is a must.

Where cakes and other nonyeast leavened baked goods are concerned we have what has been reported (Nagle 1968) as a major source of profits for some of the baking concerns. The type of sweetener used in this type of baked goods seems to vary depending on regional preferences in, e.g., texture. Some notable contributions of the scientist in baking of cakes, etc., might include nitrogen freezing (McIntyre 1965) and the use of microwave heating (Fetty 1966; Ward 1966). (The latter may find even more use in the future, especially for articles such as pound cakes which are generally heated at high temperatures for long periods of time. Use of microwave would allow use of reducing sugars as sweeteners. Such would generally contribute too much color in the current high temperature long bake procedure.)

I have given reference here to a literature that may not be known beyond the practitioners in the baking industry. This I have done to make the point that this body of literature (Proceedings of the American Society of Bakery Engineers) *should be* primary reference material. It is not recognized as such because of the absence of bibliographies—thus implying all reports cover original work. The omission of cross-references apparently came about due to a curious interpretation of a Society policy that resulted in the classification of bibliographies as "commercialism." Here, then, is an instance of the need for educating the technologist on the great value of one of the basic tools of well founded science—a thorough record of what has gone on before.

Dairy Industry

A third part of the food processing industries, the dairy industry, includes products such as ice cream, frozen desserts, sweetened con-

---

[5] The total bread produced in the United States amounted to 9,2225,000 lb in 1958, 10,248,000 lb in 1964, and 10,903,000 lb in 1967. Such a rate of increase—1 to 2% per year—may be considered unspectacular (Davis and Trempler 1965, 1968).

densed milk, and sweetened milk products. Again it has been shown
that in 1967 the dairy industry received 8% of sugar and 1% of the
dextrose shipped to industrial users. The 1966 value for corn syrups
was 11%. The trend, based on 1958–1967 figures is towards an increased use of both sugar and corn syrup but a very decided decline
in use of dextrose. (Ballinger and Larkin 1963A; U.S. Dept. Agr. 1963,
1964, 1965, 1966, 1967, 1968A).

At least two major alarms are sounded when considering use of high
concentrations of corn sweeteners in ice cream—flavor deterioration and
freezing point lowering (Crowe 1964; Keeney 1963, 1965). Let us not
spend too much time on these subjects, but let us pose one question
that has frequently been a bothersome one to some of us. What is the
meaning of freezing point depression where ice cream is concerned?
True, for a given weight of a sweetener, one can expect a greater freezing point lowering with the lower molecular weight monosaccharides
than with the disaccharides. The corollaries to the basic question are:
at the sweetener solids concentrations usually found in ice cream is the
freezing point significant? Is it measurably different? Does not the
manufacturer of ice cream have more than enough cooling capacity to
handle any such difference? If the difference finds itself emphasized
in the retailer's shop, does not his great variety of frozen desserts allow
sales of products whether soft, hard, grainy, etc.? Perhaps here—
freezing point in ice cream—is a project both scientist and technologist
could benefit from by giving an objective evaluation of the subject.

Canning Industry

A fourth food processing industry, the canning industry, puts us smack
in the midst of a subject called standardization. For anyone not yet
acquainted with the multitude of restrictions for various sweeteners
usage in the canning industry I should like to refer you to the annual
publication of the National Canners Association—The Almanac (1968).
This is a wonderful up-dater on standards among other items of interest
to the canning industry.

I mention the standards aspect because such controls cannot but be
reflected in the trends of usage in this industry. In 1967, 14% of the
sugar and 8% of the dextrose distributed to industrial users went to
the canning industry. The 1966 value for corn syrups was 14.5%.
The volume of all three sweeteners over 1958–1967 to the canning industry showed an overall increase and projected tendency to maintain
such an increase. (Ballinger and Larkin 1962; U.S. Dept. Agr. 1963,
1964, 1965, 1966, 1967, 1968A).

The canning industry by definition includes not only canned, bottled or frozen foods, but also jams, jellies, pickles, and preserves. One of the strong links between the scientist and technologist in this industry may be found in the preserving of, e.g., fruits, in syrups. One general technique involves treating fruits, for instance, with a light density (low sugar concentration) syrup then, after equilibration, with a syrup of higher density to arrive at the final, desired sugar concentration. The *rate* of penetration of sugar (or sweetener) determines whether or not the processor ends up with fruit of the desired texture (plumpness) or fruit that has shrunken because fruit has been treated with syrup of too high a concentration of sweetener (or of sweetener with a relatively higher molecular weight) (Anon. 1964; Feit 1964).

Crystallization of sweeteners can be a problem area in the processing of jams and jellies (Kossoy 1968; Rauch 1965). It can also be a problem in the case of a phenomenon known as a cap-lock (Ward *et al.* 1966). Here, once more, we are confronted not with a transferral of sweetener solubility data—nor even sweetener solubility data in the presence of other ingredients in a formulation. We are involved with a rate of crystallization and perhaps this area of canning deserves a review and objective viewing by the scientist on behalf of the technologist.

Let me give you one more example because it is a recurring one which we advise on from data collected on pure solutions. On the coating of, e.g., cherries with a sugar glaze, a particular problem of stickiness arises when the sugar syrup being used inverts (Grosso 1965). We usually grab our rate of inversion chart and estimate the life of a given syrup for the processor in trouble. What we really need is a rate of inversion of syrup under the actual use conditions and, again, perhaps here is an area ripe for the aid of the scientist to the technologist.

Confectionery Industry

Last of the five major food processing industries for us to consider includes candy, chewing gum, and chocolates. Again, in 1967 17% of the sugar and 9% of the dextrose shipped to industrial users went to the confectionery industry; the 1966 value for corn syrups was 30%. The general trend over 1958–1967 and predicted for the future is an increased volume usage of all three sweeteners. (Ballinger and Larkin 1963D; U.S. Dept. Agr. 1963, 1964, 1965, 1966, 1967, 1968A).

This is a fascinating industry from our viewpoint because there are many examples pointing up the science-technology collaboration as well as the need for continuing this close association.

There seems, for instance, to be a preference for dry sweeteners in this industry because of the difficulty in supplying high solids liquid sweeteners. (High solids liquids require high temperatures for storage leading to color development and perhaps sweetener degradation.) At least a partial solution to this problem is offered by the suggestion (Dutch Patent Applic. 1965) of using slurries of sweeteners. Again, a study of *rates* of crystallization, browning, sweetener degradation vs. solids and temperature and time seems a worthwhile project.

Color development is interesting in itself. In ion exchanged corn

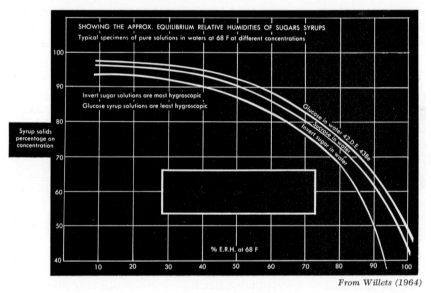

From Willets (1964)

FIG. 113. EQUILIBRIUM RELATIVE HUMIDITIES OF SOME SOLUTIONS OF CARBO-HYDRATES

syrups, industry has provided relatively color-stable products. Yet, when hard candy types of candies are made the absence of some kind of buffer (as in ion exchanged corn syrups) results in dropping of pH and causes in some cases an inversion of sucrose present to the detriment of the final product. And so, the thoroughly deashed ingredients are treated with various buffer salts (Suri 1967) to stabilize the final candy product.

Ash is also suspect in the formation of off-shape hard candies and surely here is a problem in technology that calls for the scientific touch.

ERH (equilibrium relative humidity) is a characteristic which has received much attention in the confectionery field from all sides. The

ERH of a product is the relative humidity of the atmosphere in equilibrium with it. Should the ERH of a product be higher than that of the atmosphere around it, the product will dry out; should the ERH be lower than the air around it, the product will absorb moisture and become tacky or sticky. An approximation of the ERH is given by the equation ERH $= 100/1 \times 0.27N$ were N = number of moles of solute in 100 gm of water (Money and Born 1951). (A more thermodynam-

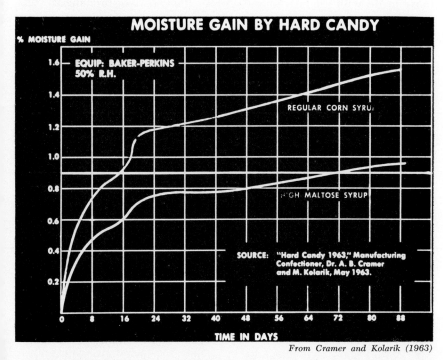

MOISTURE GAIN BY HARD CANDY

% MOISTURE GAIN

EQUIP: BAKER-PERKINS 50% R.H.

REGULAR CORN SYRU;

HIGH MALTOSE SYRUP

SOURCE: "Hard Candy 1963," Manufacturing Confectioner, Dr. A. B. Cramer and M. Kolarik, May 1963.

TIME IN DAYS

From Cramer and Kolarik (1963)

Fig. 114. Moisture Gain by Hard Candy Made with Different Corn Syrups

ically correct equation is that of Norrish (1964) but the approximate one will do for our purposes.) Willets (1964) has drawn curves (Fig. 113) showing approximate ERH of some sugar syrups. Note from the curves the dependence of ERH on molecular weight. Thus, for a given weight of a product, the lower the molecular weight, the higher the N and thus the lower the ERH.

The practical value of ERH in part of candy technology can be seen from Fig. 114. The curves shown in this figure are based on data obtained by Cramer and Kolarik (1963) on hard candies made from two different corn syrups of approximately the same molecular weight (Grosso 1963). The *rate* of uptake of moisture is the important observation here.

Perhaps to emphasize the necessity of continuing education via proper liaison between the scientific and technological communities, there is a recent publication (Pannell 1968) noting the improved moisture stability of hard candies made with high maltose type corn syrups over those made with "regular" corn syrups. *Rate* of moisture uptake was lost as the conclusion drawn was that such behavior is clearly out of line with ERH equation based on molecular weight.

Sweetness we have touched on before and it's rather difficult to slip over when considering the confectionery industry. Let me point out that I believe that sweetness is a problem but the problem may not be sweetness. We can determine the sweetness value of a compound through testing on protein receptors of an animal's tongue (Dastoli and Price 1968) or, possibly, by determining the extent of intramolecular hydrogen bonding (Shallenberger 1966). We can offer sweetness scales showing the sweetness of one compound relative to others and follow that with the precaution (Wick 1963) that in highly concentrated solutions it is very difficult to reproduce one's own evaluation of the degree of sweetness of a product. Yet we can make a chocolate bar (*non*standard) in which the only sweetener is dextrose and not be able to differentiate it from a chocolate bar based on sucrose (Hoynak and Bollenback 1963). We can show marketable products (Dextrogen) or, another (Sweetarts) which contain only dextrose as the sweetener (Fig. 115). Is not all this enough to make one wonder at the value and validity of our taste panels upon which many of our decisions of preference and to market are based? An article in a recent issue of the New York Times by Mitchell Wilson (1968) may emphasize the point. Mr. Wilson notes that advances in medical sciences have added years to the average span of human life. Many of these years have been preempted by the young such that many people of 25 today have more in common with those ten years younger than with those ten years older.[6]

What with over half our population at 25 or younger one might speculate as to whether our taste panels reflect the tastes and preferences of the proper end of the consumer spectrum.

Several selected examples of problems existing in the food processing

[6] The theme of Mr. Wilson's article is not reflected in the use of his observations as cited here. However, this point as well as others he makes fit not only his essay but seemed adaptable to the object at hand.

industries that have been solved by judicious use of scientifically generated data have been offered here. We've also pointed out several problems that seem to command similar treatment.

FIG. 115A. SOME MARKETABLE PRODUCTS CONTAINING DEXTROSE AS SWEETENER

Overall and basically though we consider there is a great need for continuing and improving the relationship between the scientific and technological societies. So far as technology is concerned we could suggest a more thorough and repeated indoctrination and education by industry of scientists designated as consultants. So far as scientific environs are concerned we could suggest a serious consideration for use as such contacts or bridges those who have had the industrial experience and now have the desire to leave industry for academia.

FIG. 115B. SOME MARKETABLE PRODUCTS CONTAINING DEXTROSE AS SWEETENER

DISCUSSION

D. Bills.—Do you have any data concerning consumer acceptance of corn syrup or corn syrup solids when used as a partial replacement for sucrose? Ice cream is an example. Flavors aside from sweetness are quite often associated with corn syrup; this is true of the syrup before it's heated in the food product. One would certainly also expect the amount of browning to be more considerable when a reducing sugar is used. Do you have any data which reflect what the consumer thinks of corn syrup vs. sucrose?

G. N. Bollenback.—I think our public relations department does. I do not have any really objective data. Contact Phil Keeney at Penn State University for a good answer to your question relative to ice cream.

BIBLIOGRAPHY

ANON. 1964. Better pickles with dextrose. Food Eng. *36*, No. 5, 66.

ANON. 1966. Shoestring research. Nature *21*, 553.

ANON. 1967. Crust colour of bread. Milling, August 11, 105, 113.

ANON. 1968A. Irradiated foods go sour. Bus. Week, May 25, 160.

ANON. 1968B. Conference held on industry-academia. Chem. Eng. News, May 27, 59.

ANON. 1968C. Food Chem. News *10*, No. 15, 19–20; No. 17, 29–31; No. 19, 3–6.

AXFORD, D. W. E., COLWELL, K. H., CORNFORD, S. J., and ELTON, G. A. H. 1968. Effect of loaf specific volume on the rate and extent of staling in bread. J. Sci. Food Agr. *19*, 95–101.

BALLINGER, R. A. 1966. Markets for sweeteners. U.S. Dept. Agr., Agr. Econ. Rept. *95*, 314.

BALLINGER, R. A. 1967. Noncaloric sweeteners: their position in the sweetener industry. U.S. Dept. Agr., Agr. Econ. Rept. *113*.

BALLINGER, R. A., and LARKIN, L. C. 1962. Sweeteners used by food processing industries in the United States: their competitive position in the canning industry. U.S. Dept. Agr., Agr. Econ. Rept. *20*.

BALLINGER, R. A., and LARKIN, L. C. 1963A. Sweeteners used by the dairy industry: their competitive position in the United States. U.S. Dept. Agr., Agr. Econ. Rept. *30*.

BALLINGER, R. A., and LARKIN, L. C. 1963B. Sweeteners used by the beverage industry: their competitive position in the United States. U.S. Dept. Agr., Agr. Econ. Rept. *31*.

BALLINGER, R. A., and LARKIN, L. C. 1963C. Sweeteners used by the baking industry: their competitive position in the United States. U.S. Dept. Agr., Agr. Econ. Rept. *32*.

BALLINGER, R. A., and LARKIN, L. C. 1963D. Sweeteners used by the confectionery industry: their competitive position in the United States. U.S. Dept. Agr., Agr. Econ. Rept. *37*.

BALLINGER, R. A., and LARKIN, L. C. 1964. Sweeteners used by food processing industries: their competitive position in the United States. U.S. Dept. Agr., Agr. Econ. Rept. *48*.

BLANCHFIELD, J. R. 1966. The evolution of the food technologist. J. Food Technol. *1*, 37P–42P.

BOHM, E., and HOYNAK, P. 1967. Q-plus: tool for soft drink research. Soft Drinks 85, No. 1118, 28–30.

BOHN, R. T. 1954. Sugar in bread breaking. Bull. Sugar Information, Inc., New York, N.Y.

BOHN, R. T., and JUNK, W. R. 1960. Sugars. In Bakery Technology and Engineering, S. A. Matz (Editor). Avi Publishing Co., Westport, Conn.

BREITUNG, H. 1956. The viscosity of technical sugar solutions. Z. Zuckerind. 6, 185–193. (German)

CAIN, R. F. 1967. Personal communication. Oregon State Univ., Corvallis, Oregon.

COLLYER, D. M. 1966. Fermentation products in bread flavor and aroma. J. Sci. Food Agr. 17, 440–445.

COLLYER, D. M. 1967. Sugar brews in bread improvement. J. Sci. Food Agr. 18, 428–439.

COOK, E. T. 1968. Deliveries of packaged and bulk (granulated and liquid) sugars 1958–1967. U.S. Dept. Agr., Sugar Rept. 191, 5–14.

CRAMER, A. B., and KOLARIK, M. 1963. Hard candy 1963. Proc. 17th Ann. Prod. Conf. Penna. Mfg. Confectioners Assoc., Sect. XII; Mfg. Confectioner 43, No. 5, 42–46.

CROWE, L. K. 1964. Aim of sweetening agents is to enhance flavor. Ice Cream World 72, No. 8, 22, 24.

DASTOLI, F. R., and PRICE, S. 1968. A sweet-sensitive protein from bovine taste buds. Purification and partial characterization. Biochemistry 7, 1160–1164.

DAVIS, P. R., and PRINCE, R. N. 1955. Liquid sugar in the food industry. In Uses of Sugars and Other Carbohydrates in the Food Industry. Advan. Chem. Ser. 12, 35–42.

DAVIS, R. E., and TREMPLER, M. 1965. How bakers fared in '64. Baking Ind. 123 No. 1560, 53–60.

DAVIS, R. E., and TREMPLER, M. 1968. Baking Industry magazines 31st annual survey. Baking Ind. 129, No. 1637, 39–57.

DUTCH PATENT APPLICATION NO. 283,583. 1965. Process for the preparation of a starting material for confectionery.

ELLNER, S. 1967. How plants use UV equipment to purify water, liquid sugar. Soft Drinks. 86, No. 1119, 38–40.

FEIT, T. 1964. Bulk volume for maraschino cherry packer. Canner/Packer 133, No. 9, 20–21.

FETTY, H. 1966. Microwave baking of partially baked products. Proc. 42nd Ann. Mtg. Am. Soc. Bakery Eng., 145–152.

GADDIE, R. S., WEST, R. R., and BENNISON, E. G. 1964. Thin film ultraviolet sterilization of liquid sugar, using the Aquafine sterilizer. J. Am. Soc. Sugar Beet Technologists 13, 214–217.

GREIG, W. S. 1968. Consumer images of sugar and synthetic sweeteners. Michigan State Univ. Agr. Expt. Sta. Bull.

GRIFFITH, T., and JOHNSON, J. A. 1954. Chromatographic analysis of sugars in bread. Cereal Chem. 31, 130–134.

GROSSO, A. L. 1963. E.R.H.—observations on the Money and Born equation. Confectionery Prod. 29, 757, 758, 804.

GROSSO, A. L. 1965. Candied and Glaced Fruit. Refinerias de Maiz S.A.I.C. Buenos Aires, (Argentina).

HARRIS, M. 1968. The education-industry interface. Res. Management 11, 159–166.

HASS, H. B. 1968. History never repeats. Res. Management 11, 153–158.

HODGE, J. E., and NELSON, E. C. 1961. Preparation and properties of galactosylisomaltol and isomaltol. Cereal Chem. 38, 207–221.

HOYNAK, P. X., and BOLLENBACK, G. N. 1963. Unpublished data.

HOYNAK, P. X., and BOLLENBACK, G. N. 1966. This is Liquid Sugar, 2nd Edition. Refined Syrups & Sugars, Yonkers, N.Y.

JOHNSON, J. A. 1963. Bread flavor factors and their control. Proc. 39th Ann. Mtg. Am. Soc. Bakery Eng., 78–84.

KEENEY, P. G. 1963. The sugar dilemma—comments of a technologist. Ice Cream Trade J., 59 No. 6, 50, 52, 103–107.

KEENEY, P. G. 1965. A survey of ice cream manufacturing plants with emphasis upon ingredient and processing variables affecting quality. Ice Cream Field 85 No. 6, 19, 20, 22, 58, 60, 62.

KOCH, R. B., SMITH, F., and GEDDES, W. F. 1954. The fate of sugar in bread doughs and synthetic nutrient solutions undergoing fermentation with baker's yeast. Cereal Chem. 31, 55–72.

KOSSOY, M. W. 1968. The function of corn syrups in jams, jellies and preserves. Food Products Develop. 2, No. 1, 34–37.

LEE, C. C., and CHEN, C.-H. 1966. Studies with radio active tracers. IX. The state of sucrose-^{14}C during bread making. Cereal Chem. 43, 695–705.

LEE, J. W., and GEDDES, W. F. 1959. Studies on the brew process of bread manufacture: the effect of sugar and other nutrients on baking quality and yeast properties. Cereal Chem. 36, 1–18.

LEES, R. 1965. Factors affecting crystallization in boiled sweets, fondants and other confectionery. The British Food Manufacturing Ind. Res. Assoc. Scientific and Technical Surveys No. 42, 15–17.

LIAU, Y. H., and LEE, C. C. 1966. Studies with radio active tracers. X. The effect of glycine-1-^{14}C during bread making. Cereal Chem. 43, 706–715.

MANTOVANI, G., and INDELLI, A. 1966. Densities of impure sucrose solutions. Intern. Sugar J. 68, 104–108.

McINTYRE, D. L. 1965. Liquid nitrogen freezing—comparative economics and practices with other freezing methods. Proc. 41st Ann. Mtg. Am. Soc. Bakery Eng., 130–138.

McWEENY, D. J. 1968. Reactions in food systems: negative temperature coefficients and other abnormal temperature effects. J. Food Technol. 3, 15–30.

MONEY, R. W., and BORN, R. 1951. Equilibrium humidity of sugar solutions. J. Sci. Food Agr. 2, 180–185.

NAARDEN. 1966. Some aspects of soft drinks manufacturing. Naarden-Flavorex Inc., Baltimore, Md.

NAGLE, J. J. 1968. Baking industry profit makes strong recovery. The New York Times, March 24.

NATIONAL CANNERS ASSOCIATION. 1968. The Almanac of the Canning, Freezing, Preserving Industries. 1968. E. E. Judge & Son, Westminster, Md.

NESTY, G. A. 1968. The need for industry/university understanding. Res. Management *11*, 149–152.

NIEMAN, C. 1960. Sweetness of glucose, dextrose and sucrose. Mfg. Confectioner *40* No. 8, 19–24, 43–46.

NORRISH, R. S. 1964. Equilibrium relative humidity of confectionery syrups—a nomogram. Confectionery Prod. *30*, 769, 771, 808.

O'BRIEN, R. T. 1967. Differentiation of bacteria by gas chromatographic analysis of products of glucose catabolism. Food Technol. *21*, 1130–1132.

PANGBORN, R. M. 1963. Relative taste intensities of selected sugars and organic acids. J. Food Sci. *28*, 726–733.

PANNELL, R. J. H. 1968. New advances in glucose syrups and their application to flour confectionery. Food Trade Rev. *38*, No. 3, 50–53.

PENCE, J. W. 1967. Factors affecting bread flavor. Proc. 43rd Ann. Mtg. Am. Soc. Bakery Eng., 122–129.

PIEKARZ, E. R. 1963. Evaluation of sugars in ferment systems. Proc. 39th Ann. Mtg. Am. Soc. Bakery Eng., 118–126.

RAUCH, G. H. 1965. Jam Manufacture. Leonard Hill Books, London.

ROBE, K. 1965. Continuous cold sterilization of liquid sugar in storage tanks markedly reduces cleaning frequency. Food Process. Marketing. *26*, No. 11, 178, 179, 182.

SALEM, A., ROONEY, L. W., and JOHNSON, J. A. 1967. Studies of the carbonyl compounds produced by sugar-amino acid reactions. II. In bread systems. Cereal Chem. *44*, 576–583.

SAND, F. E. M. J. 1966. Investigations on soft drink spoilage organisms. Intern. Bottler Packer *40*, 63, 64, 66, 68, 70, 72, 74.

SHALLENBERGER, R. S. 1966. The chemistry of sweetness. Symp. Proc., Frontiers in Food Res. Cornell Univ., 45–56.

SHAW, M. W., and HAYES, E. 1966. Effects of irradiated sucrose on the chromosomes of human lymphocytes *in vitro*. Nature *211*, 1254–1256.

SURI, B. R. 1967. Stabilization of pH of corn syrup for hard candy. Proc. 21st Ann. Prod. Conf. Penna. Mfg. Confectioners Assoc. 32–36; Mfg. Confectionery *47*, No. 7, 35–37.

THIELE, H. 1962. The influence of temperature on the density and volume of pure sucrose solutions and technical sugar solutions. Z. Zuckerind. *12*, 424–434. (German)

THOMAS, B., and ROTHE, M. 1956. The aroma content of bread. Brot Gebäck *10*, 157–162. (German)

U.S. DEPT. AGR. 1963. Sugar Rept. *131*. U.S. Agr. Stab. & Conserv. Serv.

U.S. DEPT. AGR. 1964. Sugar Rept. *143*. U.S. Agr. Stab. & Conserv. Serv.

U.S. DEPT. AGR. 1965. Sugar Rept. *154*. U.S. Agr. Stab. & Conserv. Serv.

U.S. DEPT. AGR. 1966. Sugar Rept. *166*. U.S. Agr. Stab. & Conserv. Serv.

U.S. DEPT. AGR. 1967. Sugar Rept. *178*. U.S. Agr. Stab. & Conserv. Serv.

U.S. DEPT. AGR. 1968A. Sugar Rept. *190*. U.S. Agr. Stab. & Conserv. Serv.

U.S. DEPT. AGR. 1968B. Administrative actions relating to 1968 sugar supplies. Sugar Rept. *193*, 6–8.

WARD, D. R., LATHROP, L. B., and LYNCH, M. J. 1966. Taste and cap-locking behavior of pharmaceutical syrups. Drug Cosmetic Ind. *98*, No. 1, 48–53, 157.

WARD, J. R. 1966. High frequency baking 1966. Food Trade Rev. *36*, No. 12, 50–55.

WICK, E. L. 1963. Sweetness in fondants. Proc. 17th Ann. Prod. Conf. Penna. Mfg. Confectioners Assoc. Sect. X, 1–14; Mfg. Confectioner *43*, No. 5, 59–63.

WICKER, R. J. 1966. Some thoughts on sweetening agents old and new. Chem. Ind. (London), 1708–1716.

WILLETTS, E. 1964. Modern glucoses. Confectionery Prod. *30*, 941–949.

WILSON, M. 1968. Topics: the fountain of eternal adolescence. The New York Times, Sat., July 6.

WISEBLATT, L. 1961. Bread flavor research. Baker's Dig. *35*, No. 5, 60.

CHAPTER 20

Thomas J. Schoch | Starches in Foods

This article will discuss the various physical properties which starch may impart to prepared foods, as well as the basic mechanochemistry involved in these uses. The term starch includes not only the refined and modified starches of commerce, but also the natural starch occurring in cereal flours and in such root sources as potato. Various types of starch are used to perform seven different functions in foods: (1) thickening agent for sauces, soups, and pie-fillings; (2) colloidal stabilizer for oil-in-water emulsions such as salad dressing; (3) moisture-retention agent, as in toppings and icings for cake; (4) gel-forming agent in such confections as gum drops; (5) binding agent for ice-cream cones, and for "breading" fish and chicken preparatory to frying; (6) coating and glazing agent for nut meats and candies; and (7) dry-dusting of bakery products and certain candies.

All these functions are essentially aesthetic, to make the food more pleasing to the eye, to impart better texture or mouth-feel, to prevent separation of ingredients, and to provide a carrier for flavors. While the first use as a thickener is by far the most important, the other functions will also be briefly considered.

But before discussing individual starch applications, two basic phenomena of starch behavior must be understood: (1) The composition of starch with respect to its linear and branched polysaccharide components (i.e., the amylose and amylopectin), and the influence of these components on such paste properties as congelation and opacity. (2) The mechanochemical nature of the starch granule, its swelling and solubilization when cooked in water, and the relation of granule behavior to paste viscosity.

Most starches contain two glucose polymers (Fig. 116). One of these, the amylose, is an extended linear chain of glucose units; this component comprises 17–27% of the common starches (e.g., corn, wheat, potato, tapioca). The major component or amylopectin is a highly branched or tree-like glucose polymer. Certain recessive genetic varieties of many cereal starches contain only the branched fraction. These include the so-called waxy or glutinous corn and sorghum starches, which are available commercially. Dry-milled waxy rice flour is marketed in limited quantity in the United States, and the refined starch is manufactured in Germany. At the other end of the spectrum are

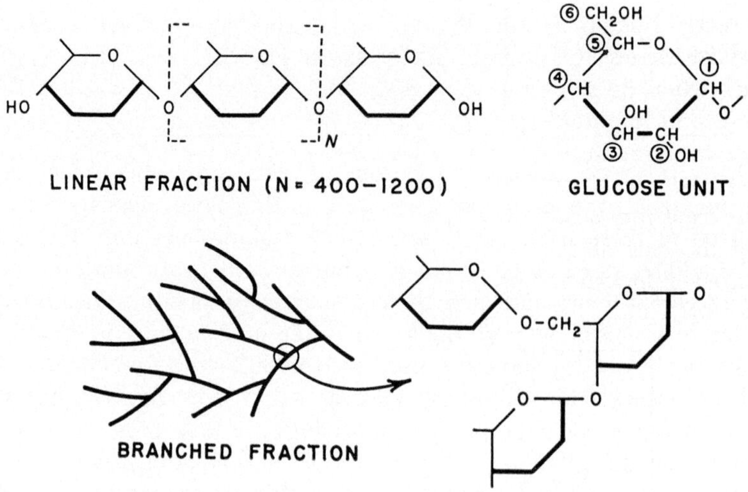

LINEAR FRACTION (N= 400—1200)

GLUCOSE UNIT

BRANCHED FRACTION

FIG. 116. STRUCTURE OF THE LINEAR AND BRANCHED STARCH FRACTIONS

the commercial high-amylose corn starches, containing 55–85% linear fraction.

Spontaneous insolubilization of starch from solution (termed "retrogradation") is highly important in various food applications. Like any highly polar linear polymer, the linear starch fraction shows strong hydrogen-bonding tendencies, which may act in two different ways (Fig. 117). A hot relatively concentrated solution will rapidly congeal to a rigid irreversible gel when cooled to room temperature. Or a more dilute solution will slowly become opaque and deposit an insoluble precipitate on standing. Since the strength of this associative bonding

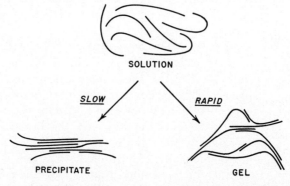

SOLUTION

SLOW

RAPID

PRECIPITATE

GEL

FIG. 117. MECHANISMS OF RETROGRADATION

is directly related to the degree of linearity, the branched fraction shows much less tendency to retrograde; indeed, insolubilization only occurs when its solutions are frozen and thawed.

A second important property of the linear fraction is its strong tendency to form insoluble complexes with higher fatty acids or monoglycerides. It is presumed that the normally extended linear chain winds itself into a helical coil, with the enclosed fat oriented lengthwise along the axis of the helix. The branched fraction shows no affinity whatsoever for such lipids.

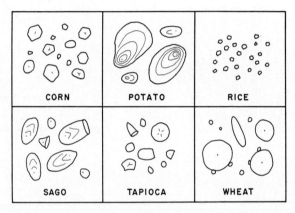

FIG. 118. MICROAPPEARANCE OF VARIOUS GRANULAR STARCHES

This same associative bonding is likewise responsible for the structure and pasting behavior of the starch granule. The granules are microscopic spherocrystals, whose size and shape are characteristic for each species of starch (Fig. 118). The granule is built up in the plant by the deposition of successive layers of starch substance on its exterior surface. If we magnify a segment of a granule layer, the intermingled linear and branched molecules appear to be oriented radially, i.e., toward the center of the granule (Fig. 119). Where linear chains or linear segments of the branched molecules run parallel, they associate by hydrogen bonding, and this maintains the granule as an intact spherocrystalline entity.

When starch is suspended in water and the slurry heated to progressively higher temperatures, nothing happens until a critical "gelatinization temperature" is reached. Then the heat energy begins to dissociate the more weakly bonded regions within the granule, and the latter commences to swell. This swelling continues as the temperature is raised. While some of the shorter linear molecules may actually dissolve and diffuse out of the swollen granule, the longer linear chains meander through the granule and act as reinforcement to prevent extensive solubilization. Only by going to an autoclave temperature of 120°C can the granule be completely dissolved. The old myths describing the pasted granule as a swollen sac which bursts or ruptures on cooking are totally false.

As the granules swell, they take up the free water and hence begin to jostle one another. The viscosity (or more properly, the consistency) of a cooked starch paste simply reflects the resistance to stirring of this

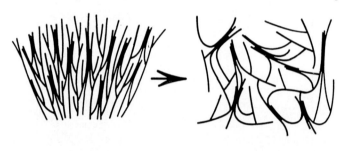

UNSWOLLEN   SWOLLEN
SEGMENT   SEGMENT

Fig. 119.   Mechanism of Swelling of Granular Starch

Regions of association shown as thickened sections.

mass of swollen gel particles. The greater the granule swelling, the higher will be the viscosity of the paste. However, high swelling generally gives a fragile granule which is readily fragmented by stirring, with substantial thinning of the paste. Hence the extent of swelling can be correlated with viscosity behavior.[1]

The swelling power of a starch can be readily determined as the weight of the swollen granules per gram of dry starch. If this value is plotted against the temperature of pasting over the entire cooking range, the curves shown in Fig. 120 are obtained. Corn starch shows

---

[1] An exception is shoti starch from a Pakistan tuber of the curcuma family, which shows high swelling very similar to tapioca, but has a stable Brabender viscosity comparable to that of a cross-bonded starch (Schoch and Maywald 1968). This anomalous behavior is attributed to two opposing characteristics of the starch, a relatively high content of linear fraction (35-40%) which inhibits swelling, and a very high content of natural ionic phosphate ester groups (0.2% P) which increases swelling.

only moderate swelling over the range of 65°–95°C, while the granules of potato and waxy sorghum show very high swelling. But if this waxy sorghum is first cross-bonded with trace amounts (e.g., 0.02%) of epichlorohydrin, its subsequent swelling is greatly restricted. Presumably, the introduction of a very few ether cross-linkages within the granule tightens up the molecular network, restricts granule swelling,

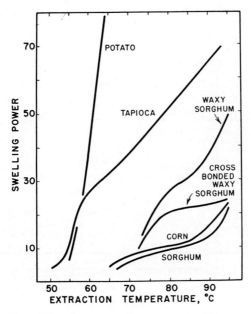

Fig. 120. Swelling Patterns of Various Starches

and hence stabilizes the viscosity against breakdown by agitation or by acidity (Kite *et al.* 1963).

The relationship between viscosity stability and extent of granule swelling is illustrated by stirring 5% pastes for 20 min at various speeds, and then determining the Brookfield viscosity (Fig. 121). The high-swelling starches (potato, tapioca, and waxy sorghum) show the greatest breakdown in viscosity, corn is intermediate, and the cross-bonded waxy starch shows maximum stability.

The consistency changes during pasting and cooking are best measured with the Brabender amylograph (Mazurs *et al.* 1957), which automatically charts the viscosity as the starch slurry is heated at a constant rate to 95°C, stirred at that temperature for an hour, then cooled to

50°C and held an hour (Fig. 122). Thus corn starch gives a relatively low peak viscosity (because the granules are only moderately swollen), followed by some breakdown during cooking as a result of granule fragmentation under shear, and finally the very substantial congelation during cooling due to association of linear molecules. Tapioca and waxy sorghum starches swell to a much greater extent, but the fragile overswollen granules break down and thin out during stirring. Note that waxy sorghum starch shows very little retrogradation on cooling,

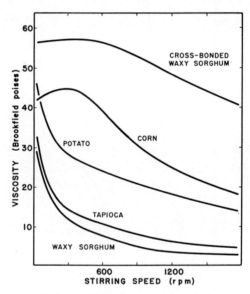

Fig. 121.   Breakdown of 5% Starch Pastes
Under Shear

due to the absence of a linear fraction. The cross-bonded waxy starch shows no breakdown at all and even a slight increase in consistency during cooking. It is also the most efficient starch from a viscosity standpoint, requiring only 27.5 gm of starch per 500 ml as compared with 35 gm for the other 3 starches.

The chemically cross-bonded products undoubtedly represent the greatest single advance in food starches during the past 30 yr. Cross-bonding imparts stability not only with respect to mechanical agitation, but likewise resistance against supertemperature cooking for sterilization and against thinning of the starch thickener by fruit acids. Thus Fig. 123 shows the final 50°C Brabender viscosities of various starches plotted as a function of the pH at which they were cooked; the cross-

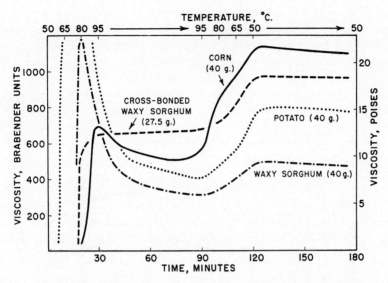

FIG. 122. BRABENDER VISCOSITIES OF VARIOUS STARCHES
Concentration given in grams dry starch per 500 ml.

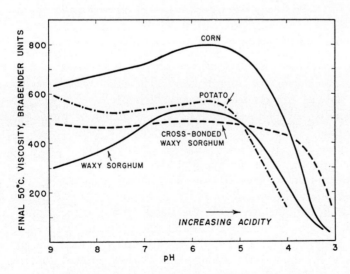

FIG. 123. EFFECT OF ACIDITY ON STARCH VISCOSITY

bonded waxy starch retained good consistency down as low as pH 3.5. Any starch can be stabilized by cross-bonding, but the major application has been to the waxy starches, to improve their granule stability and yet retain the high paste clarity typical of a branched-type starch. For example, a cherry pie filling thickened with a cross-bonded waxy starch enhances the bright clear color of the fruit, while a comparable pie made with corn starch is muddy and unappetizing in appearance. The types of food-permissible cross-bonding agents include epichlorohydrin to introduce ether linkages, and sodium trimetaphosphate or phos-

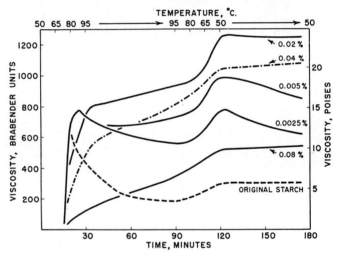

FIG. 124. BRABENDER VISCOSITY OF WAXY SORGHUM STARCH CROSSBONDED WITH VARIOUS AMOUNTS OF TRIMETAPHOSPHATE

phorus oxychloride for ester linking. The ether-linked product shows better stability at very low and very high pH levels. The amount of cross-bonding agent must be carefully regulated to provide optimum stabilization. For example, 0.02% trimetaphosphate gives maximum viscosity and paste stability; higher cross-bonding excessively limits granule swelling to the detriment of viscosity (Fig. 124).

Pregelatinized starches are used as thickening agents in many of the instant-type foods reconstituted by addition of water (Osman 1967). The starch manufacturer cooks the starch in water, then spray- or roll-dries the paste. Or a mixture of starch and water may be passed through a high-temperature extruder, the cooked mass flashed off and simultaneously dried, and the sponge-like product ground to size. Such pregelatinized starches never give as high a viscosity as a properly-cooked granular starch, nor is the paste as smooth and lump-free. Spray-dried and finely-ground roll-dried products tend to give clots when added to water. This can be minimized by slowing down the hydration rate, for example by spraying oil or melted monoglyceride onto the dry pregelatinized starch. Pastes prepared by reconstituting a coarsely-ground roll-dried product are generally rather grainy in texture. This can be advantageous in spaghetti sauces. In such instances, a pregelatinized cross-bonded corn starch may be used, to increase the grainy texture and to impart acid stability.

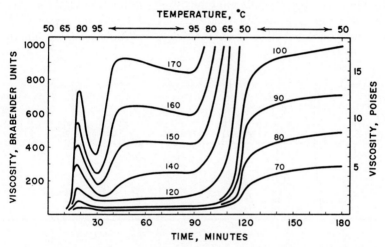

FIG. 125. BRABENDER VISCOSITY OF THIN-BOILING CORN STARCH (85-FLUIDITY)

Concentrations in grams per 500 ml.

However, in the great majority of food applications where starch is used as a thickening agent, a heavy-bodied or even salve-like consistency is desired, with no evidence of graininess and with no elastic gel structure. A few instances may be cited where congelation is desired, such as gum confections, the old-fashioned tough rigid corn starch pudding of our grandmothers' day (usually chocolate flavored), and certain synthetic "custards." It is of course the linear fraction which produces these gels. In order to develop maximum gel strength for gum confections, "thin-boiling" corn starch is commonly used, manufactured by pretreating the granular starch with warm dilute mineral acid. Glucosidic bonds within the granule network are thereby hydrolyzed, and the product may be cooked at 3 to 5 times the concentration of an unmodified starch (Fig. 125). Hence the congelation on cooling becomes very pronounced. However, it is still necessary to age the molded confection for 24–36 hr before sugaring and packaging. Increasing the amylose content of the starch gives gum confections which set up to rigid gels within 20–30 min, and thus can be packaged directly without lengthy conditioning. This is accomplished by joint use of thin-boiling corn starch and high-amylose corn starch; the latter ingredient is separately pressure-cooked in water to effect solubilization, and then combined with the main starch-syrup mix (Robinson and Brock 1965). The normal cereal starches contain 0.5–0.8% of natural fatty acid, which complexes with some of the linear fraction of these starches to impair their gel-forming quality. This fat can be removed by a practicable commercial process involving continuous supertemperature

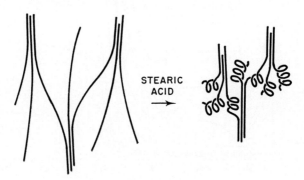

FIG. 126.   HELICAL CONFIGURATION OF LINEAR STARCH
SEGMENTS IN THE PRESENCE OF FATTY ACID

extraction with 85% methanol (Rist *et al.* 1952). Cooked pastes of this defatted starch set up to gels of unusually high strength immediately on cooling to room temperature (Schoch 1941; Schoch 1952).

The addition of polar lipids such as monoglyceride or fatty acid is an important way of modifying the paste properties of starchy foods. The lipid apparently forms a helical complex with segments of linear chains within the granule network (Fig. 126); this complexed linear material is insoluble and hence incapable of forming a gel. In addition, swelling is greatly restricted (Fig. 127), and the granule is stabilized against mechanical breakdown (Gray and Schoch 1962). The net result is a high and very stable Brabender viscosity (Fig. 128). Other similar food additives are sodium stearyl fumarate and sodium stearyl lactylate. Several outstanding examples of the application of these agents may be cited: (1) Probably half of the commercial white bread

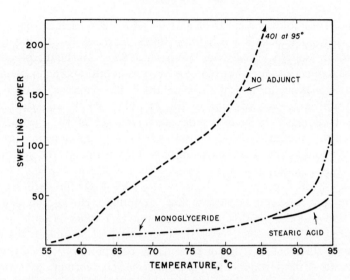

FIG. 127. EFFECT OF FATTY ADJUNCTS ON THE SWELL-
ING OF POTATO STARCH

baked in the United States contains monoglyceride added as a softening
agent. The action here is to prevent congelation of the linear fraction
of the wheat starch, to give a bread which is overly soft when freshly
baked. However, with large commercial bakeries, some 24–36 hr may
elapse before the bread is finally delivered to the ultimate consumer.
During this time, staling develops sufficient crumb structure so that

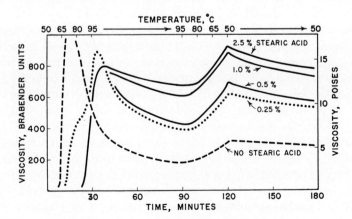

FIG. 128. EFFECT OF VARIOUS AMOUNTS OF FATTY ACID
ON THE BRABENDER VISCOSITY OF POTATO STARCH

the bread resembles normal fresh bread. However, such softened bread makes very poor toast, since heating totally reverses the staling to give an over-soft doughy consistency (Schoch 1965). It should be stressed that monoglyceride does not prevent or even retard staling. (2) In the process of making instant mashed potatoes, the potatoes may be over-cooked with excessive swelling of the granules. Such a product reconstitutes with water to give a slimy paste instead of the desired fluffy consistency. Similarly, where dehydrated mashed potatoes are prepared for institutional or restaurant serving, too long a time on the steam table will give the same pasty consistency. Also, if the user wishes to whip the reconstituted product to beat in air, this agitation will fragment and thin out the fragile overswollen granules. Hence the processor incorporates monoglyceride during cooking to restrict granule swelling and stabilize the viscosity. (3) In the cooking of spaghetti, the starch may easily overcook to give an excessively soft product. Also, sufficient soluble starch and fragmented granules may be formed to cause the spaghetti strands to stick together. Monoglyceride is added to the dough to correct both of these faults (Brokaw 1962). The resulting product has a firm consistency approaching that of high-protein pasta goods.

It will be noted that the effect of lipids is somewhat similar to that of chemical cross-bonding. However, there are two important differences: (1) the starch-monoglyceride complex has no resistance against acid thinning, and (2) it dissociates at temperatures in the range of 100°–110°C, and therefore cannot survive retorting. However, where the food product is not acidic and where sterilization is not necessary, the use of such adjuncts is strongly recommended.

As another example of this relationship between granule swelling and the character of the paste viscosity, starches which swell greatly like potato give pastes which are cohesive or weakly elastic. An elegant term for this behavior is "viscoelasticity;" a somewhat vulgar but much more expressive adjective used in the starch industry is "snotty." A good method of illustrating this behavior is to rule a tracer line on the surface of a 5% starch paste with an indelible pencil, then pass a spatula edgewise through the paste at right angles to the tracer line (Fig. 129). The width of distortion is then a measure of this paste elasticity. Potato and waxy sorghum are highly cohesive, corn is much less so, and cross-bonded waxy starch shows scarcely any elasticity at all. With starches containing a linear fraction, this elasticity can of course be prevented with monoglyceride. American tastes would be highly unfavorable toward any such elasticity in sauces or gravies.

However, a number of German clear soups have shown this characteristic, presumably caused by thickening with unmodified potato starch.

Retrogradation offers a third method for stabilizing the consistency
and texture of starchy foods, by encouraging the development of associative bonding within the pasted starch. Thus various cookbooks have
suggested that boiled potatoes be refrigerated overnight before preparation of potato salad, in order to impart a firmer structure to the potato
slices. Similarly, another process for making dehydrated mashed potatoes first precooks the tubers at a low temperature just sufficient to
gelatinize the starch, then allows this product to cool and stand, and
finally completes the cooking at full temperature (Cording *et al.* 1957;

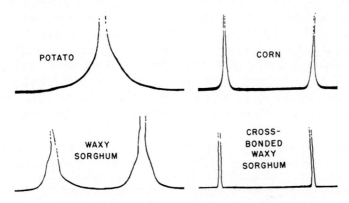

FIG. 129. VISCOELASTICITY OF VARIOUS STARCHES

Cording and Willard 1957). Retrogradation during the intermediate
cooling period stabilizes the granules against subsequent overcooking
and pastiness.

Starch can retrograde within the ungelatinized granule, particularly
in the case of the high-swelling root and tuber starches. A combination
of heat and moisture will cause further association of molecules within
the granule network (Sair 1964). Actually, this so-called heat-moisture
transformation represents a complete change in the crystal pattern of
the starch (probably by the elimination of a half mole of water of crystallization), since the altered material gives an entirely different X-ray
spectrum from that of the original starch. Other changes so induced
include a substantial increase in the gelatinization temperature of the
starch, and an enormous decrease in the swelling power (Leach *et al.*
1959). In the laboratory, we have heat-moisture treated whole potato
tubers by slowly heating them under water from 50° to 75°C at a tem-

perature rise of 1°C per hour. This treatment did not gelatinize the starch. The character of the treated potatoes was quite unique, since three hours' cooking in boiling water did not soften them to an edible state. Perhaps a somewhat less rigorous treatment might have some usefulness in improving the potatoes for some such specialized purpose as baking, potato chips, or "snack items." Some degree of natural heat-moisture treatment undoubtedly occurs during growth; thus Nikuni (personal communication) has found that sweet potato starch grown in southern Japan has a higher gelatinization temperature than that from the north.

One of the most important considerations in starchy foods is their stability under various conditions of storage. Retrogradation becomes much more important as the temperature is lowered and as the acidity increases. The most apparent changes in a starch-thickened sauce are the loss of surface gloss or sheen, increase in opacity of the paste, and separation of a watery phase by syneresis. And when syneresis develops in a starchy food, the housewife frequently jumps to the conclusion that spoilage by microbiological action has occurred. We have two methods for evaluating these changes: one by measuring the increase in opacity of the paste, the other by measuring the water separating by syneresis (Schoch 1967). If the opacity (by a light reflectance technique) is determined on 5% starch pastes over a period of 4 weeks, the curves shown in Fig. 130 are obtained. Corn starch retrogrades much more rapidly at 4°C than at 30°C. Potato starch deteriorates rapidly during the first week of storage, and much more slowly thereafter. Waxy starches and their cross-bonded derivatives show very little change over this entire period. Similarly, the relative retrogradation tendencies of the various starches are shown by plotting reflectance against starch concentration (Fig. 131). The increase in paste clarity of waxy sorghum at 2% concentration is very real and can be seen with the eye.

But this mode of evaluation by the slow development of opacity is much too tedious for routine use. Instead, equivalent results are obtained much more rapidly with a procedure involving "shock-treatment" by repeated freezing and thawing followed by measurement of separated water. One such freeze-thaw cycle is equivalent to about three weeks' cold storage at 4°C. In a freshly cooked paste, the water molecules are presumed to be distributed quite uniformly through the paste (Fig. 132). Freezing causes the separation of discrete ice crystals, and this water does not reconstitute uniformly when the system is thawed. Hence the starch paste loses its water-holding capacity and

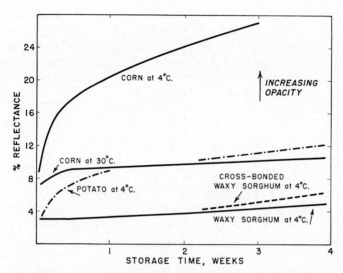

FIG. 130. CHANGE IN LIGHT REFLECTANCE OF VARIOUS PASTED STARCHES ON STORAGE

syneresis occurs (Schoch 1968). In Fig. 133, corn starch shows the maximum rate of syneresis. Waxy starch is relatively good, but cross-bonding seriously impairs its freeze-thaw stability. This fault is totally overcome by introducing ionic phosphate groups into the cross-bonded

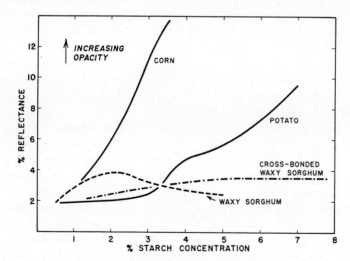

FIG. 131. EFFECT OF CONCENTRATION ON THE OPACITY OF VARIOUS PASTED STARCHES

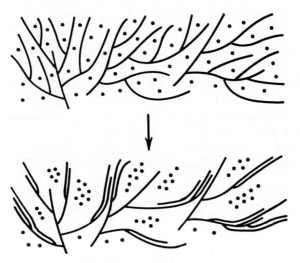

FIG. 132. SEPARATION OF WATER MOLECULES (black
dots) FROM A STARCH PASTE ON FREEZING AND THAWING

waxy starch, and indeed any degree of stability can be obtained by
sufficiently increasing the phosphate content. As a plausible explanation, the ionic sites mutually repel one another and thus prevent
association of molecules and consequent loss of water-holding capacity
of the paste. Instead of phosphation, cross-bonded waxy starches may

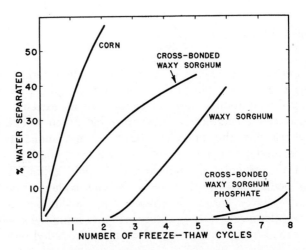

FIG. 133. EFFECT OF REPEATED FREEZING AND THAW-
ING ON SYNERESIS OF VARIOUS PASTED STARCHES

be acetylated; here the acetate groups act as mechanical obstructions to hinder the side-by-side association of linear chains.

And so we have considered the special properties of various natural and modified starches, the mechanochemical basis of these properties, and the specific functions which they perform in various foods. Perhaps it might also be appropriate to list some of our research failures, unsolved problems, unreported successes, and areas of potential usefulness of starches:

New Starches

Considerable exploratory research is needed to find new sources of starch, and to evaluate the possible unique qualities and specific applications of such materials. For example, rather extensive studies are presently being conducted on the starches from various Montana weed seeds (Goering 1967; Goering and Brelsford 1966; Goering and Schuh 1967). In particular, the starch of cow soapwort (*Saponaria vaccaria*) has granules of $1-\mu$ diameter, as compared with 15μ for corn starch. Hence this starch performs superlatively as a dry-dusting agent, for example in coating chewing gum and certain sticky candies. Also, because of its great surface area, it appears to have excellent dry-bonding qualities for such compressed tablets as aspirin. The starch from the dasheen tuber grown for food use in the southeastern United States likewise has $1-\mu$ granules, and perhaps might be easier to grow and process.

Controlled Swelling

We need to know much more about the architecture of the granule, and the internal forces which control gelatinization and swelling. We can readily decrease swelling by cross-bonding, by heat-moisture treatment, or by the addition of monoglyceride. As an extreme case, potato starch can be made completely ungelatinizable (even in boiling water), simply by refluxing the granular starch in methanolic HCl (Leach 1958). Seemingly, the acid promotes hydrogen bonding to such an extent that the granules do not lose their polarization crosses when cooked in boiling water. But there are no certain ways to loosen up the lattice forces within the granule and facilitate cooking by decreasing the gelatinization temperature and increasing the swelling. Such derivative groups as hydroxypropyl or ionic carboxyethyl will accomplish these purposes, but there may be some hesitancy in using these products in foods. The problem becomes particularly acute when cooking in high concentration sugar solutions as in canning, where the sugar sequesters most of the water and hence the starch granule does not swell sufficiently to have any thickening power. For example, the gelatinization temperature of waxy sorghum starch in 60% (w/v) sucrose solution is 87°–97°C, which does not permit enough swelling at boiling temperature to provide any thickening. As an interesting possibility, if an aqueous slurry of potato starch is freeze-dried, its gelatinization temperature is lowered 8°C (Leach 1959). Unfortunately, this treatment had no effect on the more difficultly cooked sorghum starch. Purely as glib speculation, freeze-drying may be a reversal of heat-moisture treatment, removing water of crystallization by ice evaporation and thus disrupting the hydrogen bonding within the granule.

Off-flavors and Off-odors

Certain starches such as wheat and tapioca are bland or even tasteless. However, acid modification or heat dextrinization of potato starch develops a strong cucumber-like flavor which precludes use of such products in foods. Corn starch will develop a somewhat unpleasant musty odor and flavor on storage; this is not rancid fat since defatted corn starch shows the same behavior. We would like to find ways of preventing or removing these off-flavors, particularly where the starch is used in more delicately flavored foods.

Flavor Loss During Staling

The delectable yeasty-buttery flavor and odor of freshly-baked bread disappear during staling and are completely regenerated by heating. I believe this may be due to the gradual complexing of trace flavor components by the linear fraction of the wheat starch. Might it be possible to prevent this by loading the bread with corn starch hydrolysate of 16–20 DE? The small linear fragments in the hydrolysate might complex with the flavor, but the bonding forces should be sufficiently weak to be dissociated with the warmth and salivary amylase of the mouth. Such low dextrose-equivalent material should likewise enhance the moisture retention and soft texture of the bread crumb.

Starches of Intermediate Linear Content

Where quick congelation is desired, as in gum confections, it is quite feasible to use a genetically-bred corn starch of intermediate 35–40% linear content, thereby avoiding the super-temperature cooking required for true high-amylose starches.

Heat Stable-complexing Agents

As previously mentioned, the complexes of linear fraction with such materials as monoglyceride dissociate at retort temperatures. The only known complex which survives at 123°C is that with cyclohexanol, and this of course is out-of-bounds foodwise. Perhaps a study of the relationship between dissociation temperature and the HLB (or hydrophil-lipophil balance) would reveal a food-permissible complexing agent which could withstand sterilizer temperatures. Such an adjunct would be of considerable value in such products as canned baby foods to maintain a high but nonelastic consistency and to avoid syneresis.

Starchy Matrix as Oxygen Barrier

If such oxygen-sensitive materials as carotene and the fat-soluble vitamins are emulsified into solutions of certain modified starches, and this emulsion then spray-dried, the sensitive material is protected within a starchy matrix which is impermeable by oxygen (Maywald and Schoch 1957). This encapsulation process likewise has application to the stabilization of flavors and volatile essential oils. Similarly, ground spices may be sprayed with an appropriate starch solution and the product then dried, to provide protection against loss of volatiles, oxidative deterioration or discoloration.

Edible Starch Films

It has long been a dream of the food chemist to formulate an edible starch film which was sufficiently flexible and transparent for use as a good packaging and wrapping material. For example, such a film might be used to package dehydrated onion soup; the film would totally dissolve when the packet was dropped into boiling water. While there are still many difficulties to overcome before this objective is completely realized, nevertheless a commercial film of hydroxypropylated high-amylose starch (Roth and Mehltretter 1967) is presently being marketed commercially. One successful use is for measured-dosage packets of enzyme for breadmaking (Anon. 1967).

Improved Emulsifier

As compared with such polyelectrolytes as carrageenan, starch is a poor stabilizer for oil-in-water emulsions. However, starch sulfate is excellent (Lloyd 1959), though not allowable in foods. Are there any food-permissible ionic groups which could be introduced into starch to improve its emulsifying power and protective colloid action?

Freeze Treatment

The Japanese make their "harusame" or "spring-rain" noodles by extruding and pasting mung bean or sweet potato starch, then freezing the cooked strands, and finally sun-drying the product. Freezing obviously induces retrogradation; without this step, the noodles disintegrate immediately when immersed in hot broth for serving. Perhaps the food industry should explore the possibility of modifying the textural character of starch-thickened foods by a freeze-thaw treatment (Schoch 1967).

Batter-bonding Starch

Cross-bonding corn starch is commonly used to improve adhesion of the batter coating to such fried foods as fish slices and chicken. Normally the unmodified cereal starches behave very poorly in this application, since the fried coating does not adhere, particularly to fatty fish such as cod. However, occasional instances have been encountered where certain factory batches of unmodified corn starch inexplicably gave truly excellent results. I suspect that these particular batches were accidentally heat-moisture treated by high humidity and high temperature conditions in the starch dryers during manufacture. In any event, a basic study is needed of the mechanics of this adhesion and the starch characteristics required.

Curdling of Canned Corn

Cream-style canned corn may contain a heat-sensitive protein which curdles to unsightly lumps during continuous retorting. Actually, this tendency to curdle is related to maturity of the corn (Maywald, et al. 1955). If the latter is picked a few days too early, or if the corn has not matured properly due to cold weather, this albumin-like protein separates out during processing. If allowed to mature, it appears to be transformed into a gluten-like protein which does not show a visible curd. Many attempts have been made to find a starch which would prevent curdling, but thus far without any obvious success. I believe that it is probably impossible to prevent heat denaturation and insolubilization of this protein, but that it may be feasible to produce a microcurd which is not of sufficiently large particle size to give an unpleasant appearance. A highly ionic starch might be effective, e.g., an oxidized product or a carboxyethyl derivative. Recent studies in the Food and Nutrition Department at Cornell suggest that the highly branched phytoglycogen in sweet corn may participate in the curdling reaction, particularly at low pH levels below the isoelectric point of the protein. Under these circumstances, a coarse curd is formed even at room temperature.

Protein-starch Reactions

Starch is almost never used alone as a food, but always in conjunction with various proteins and fats. With polar fats, it is the linear fraction which combines to give an insoluble helical complex. With proteins, it may be the branched fraction which is reactive, and the extent of the insolubilization may be directly related to the degree of branching of the polysaccharide. Thus only glycogen types give insoluble "coacervates" with concanavalin (jack-bean protein), and the branched starch fraction is inactive. This goes back to extensive but quite obscure studies by Koets (1936) and von Przylecki (1932), particularly on the coacervation between blood proteins and the starch fractions or glycogen. Certainly starch reacts with some of the proteins of milk (Osman 1967), but much more work is needed to clarify the mechanism of these reactions and its relationship to the character of starch-thickened puddings and sauces.

Cyclodextrins

The enzyme from *Bacillus macerans* converts linear starch molecules to crystalline cyclic dextrins containing 6 or 7 glucose units. This amylase likewise converts the outer linear branches of the branched fraction, but action is halted by the branch point. These cyclodextrins form protective complexes with certain volatile or oxygen-sensitive substances, in much the same fashion as the linear fraction helicizes with fatty acids. Research is needed to develop specific practical uses for the cyclodextrins. To improve yields, it should be possible to obtain almost complete conversion of starch to cyclodextrin by the joint action of *macerans* amylase and pullulanase (i.e., a starch-debranching enzyme).

Low Dextrose-equivalent Hydrolysates

Starch hydrolysates of 5–30 DE produced by acid or enzyme conversion of the waxy starches are finding increasing use as encapsulating agents and for enhancing the mouth-feel of clear soups. These lower polysaccharides are likewise excellent for preserving such fruits as maraschino cherries and candied citrus peel. These products are generally branched in structure, since the alpha-1,6 branch point resists hydrolysis by either acid or enzyme. Hence they are highly stable in solution and show little or no retrogradation. If darkening occurs as a result of a browning reaction between amino acids and aldehydic end-groups, this can be prevented by hydrogenation of the low-DE polysaccharide.

Linear Oligosaccharide

It is feasible to make a short-chain linear product of 25–30 glucose units simply by debranching a waxy starch with pullulanase. This material shows strong associative tendencies, and separates from concentrated solution as well-formed spherocrystals. It may have usefulness as a complexing agent and as a protective spray-coat.

Dextrose from Ungelatinized Starch

Granular starches are attacked only slowly by massive dosages of the common amylases. In contrast, the enzymes produced by microorganisms in the digestive tracts of ruminant animals rapidly attack ungelatinized starch. It seems plausible that a mixture of this rumen amylase and glucoamylase would give good yields of dextrose from high-concentration slurries of granular starch.

Extruder-gelatinized Starch

The high-shear supertemperature extruder is certainly the most efficient means for manufacturing pregelatinized starches, since the low amount of water to be evaporated greatly reduces steam costs. However, the extruded and expanded product tends to be somewhat horny and difficult to hydrate. While the commercial processing of starch has always been in water medium (with the single exception of defatting corn starch for gum drop use by continuous high-temperature extraction with 85% methanol), it may be quite possible to use another solvent medium in an extrusion process to improve the water-dispersibility of the final product. Solvents which might be worthy of investigation include liquid ammonia (which has the advantage of low temperature operation) and morpholine, either of which would overcome and prevent retrogradation. Perhaps the use of these solvents would also improve the texture and viscosity of the pregelatinized starches.

Low-calorie Starches

The last item is one which really has no legitimate right to be included on a program devoted to food carbohydrates—namely, low-calorie starches. In our American anxiety to stay slim and lovely, we have violated many of the principles of good nutrition and also consumed vast quantities of possibly harmful substances. This trend seems quite inappropriate at a time when the major anxiety of a good half of the world's population is centered on finding enough food to fill a starving belly. But we insist on a coffee whitener composed of nonnutritive fat dispersed in a matrix of nondigestible polysaccharide. Or a table syrup for pancakes must be sweetened with saccharine and thickened with nonnutritive starch. Some of the breweries have even test-marketed low-calorie beer where the "mouth-feel" is imparted by carboxymethyl cellulose rather than starch dextrins, but no one to my knowledge has suggested the ultimate step of removing the high-caloric alcohol from beer! If we must have nonnutritive starch thickeners, they can be made in either of two ways: (1) Pyrogenic dextrinization or graft polymerization to produce linkages which are not attacked by digestive amylases and which are more resistant to acid hydrolysis. (2) Hydroxyalkylation of starch to a degree necessary to stop enzyme attack. The second method is preferred because it is more effective and because high viscosity can be retained. Introduction of hydroxyethyl or hydroxypropyl groups at a level of 0.7–0.8 degree of substitution completely stops enzyme action. In collaborative studies with the Medical College of South Carolina, it was found that such a material used as a synthetic blood plasma was not degraded by amylases in human blood. In extensive tests on human subjects, the product appeared substantially better than clinical dextran. It likewise appears to prevent coagulation and "sludging" of the blood, and may perhaps be of value in preventing or even dissolving blood clots. Since the associative bonding within the granule is completely destroyed by this high degree of derivatization, this hydroxypropyl starch gives clear stable solutions of relatively low viscosity. Such products may be used for nonnutritive coffee whiteners and for pancake syrups. If high paste viscosity is desired, the starch must first be cross-bonded with relatively large amounts (0.1–1.0%) of epichlorohydrin before hydroxypropylation, in order to stabilize the granule against dissolution.

This list of food-starch problems and possibilities could be continued. However, the above examples serve to show the need for research to expand the types and markets for new starch-based foods and to suggest possible areas for further study.

DISCUSSION

A. Mustafa.—Concerning the retrogradation phenomenon, it seems to me that there are some glucose units from both side chains, linear and branched, which form hydrogen bonds. Which and how many carbon atoms participate in hydrogen bonding?

T. J. Schoch.—The hydrogen bonding is between hydrate water on carbon-6 of 1 molecule to the hydrate water on carbon-6 of an adjacent chain molecule. This would be the case in corn starch, wheat starch, and normal cereal starches. It is possible that the hydrogen bonding in the case of potato starch is through a water to water to water bridge. In other words, $1\frac{1}{2}$ moles of water per chain. This would still be through carbon-6. This is merely a plausible but unproven concept.

A. Mustafa.—Do you expect any difference in the enzymatic systems of the waxy and nonwaxy grains?

T. J. Schoch.—Yes, there is a definite difference with respect to the susceptibility of the ungelatinized granules to enzyme attack. Waxy maize starch is substantially more susceptible than is corn starch.

J. B. Sieh.—Is there any possibility of a complex similar to that you have proposed for the free fatty acids occurring with amino acids?

T. J. Schoch.—Yes, I think most certainly. The essential quality is a hydrophilic loading on a hydrophobic molecule. If you can get a hydrophobic molecule of an amino acid type, tyrosine for example, it should certainly complex with amylose.

J. B. Sieh.—I'm wondering if there has been any work done on this matter.

T. J. Schoch.—Not to my knowledge.

D. French.—It's well known that starch granules will absorb many enzymes of the amylase or phosphorylase type. By holding the pH at a suitable level, perhaps at 10, there's no amylase action on the starch granules. Furthermore Schram has shown that dextrins obtained from glycogen are capable of forming insoluble complexes with enzymes so that there is substantial work along this line.

BIBLIOGRAPHY

ANON. 1967. Packs enzyme for bread in new edible starch film. Food Drug Packaging, Nov. 9, 1967.

BROKAW, G. Y. 1962. Distilled monoglycerides for food foaming and for starch complexing. Can. Food Ind. 33, (4), 36–39.

CORDING, J., JR., and WILLARD, W. J., JR. 1957. Method for control of texture of dehydrated potatoes. U.S. Patent 2,787,553.

CORDING, J., JR., WILLARD, M. J., JR., ESKEW, R. K., and SULLIVAN, J. F. 1957. Advances in the dehydration of mashed potatoes by the flake process. Food Technol. 11, 236–240.

GOERING, K. J. 1967. New starches. II. The properties of the starch chunks from Amaranthus retroflexus. Cereal Chem. 44, 245–252.

GOERING, K. J., and BRELSFORD, D. L. 1966. New starches. I. The unusual properties of the starch from *Saponaria vaccaria*. Cereal Chem. *43*, 127–136.

GOERING, K. J., and SCHUH, M. 1967. New starches. III. The properties of the starch from *Phalaris canariensis*. Cereal Chem. *44*, 532–538.

GRAY, V. M., and SCHOCH, T. J. 1962. Effects of surfactants and fatty adjuncts on the swelling and solubilization of granular starches. Stärke *14* 239–246.

KITE, F. E., MAYWALD, E. C., and SCHOCH, T. J. 1963. Functional properties of starches of foods. Stärke *15*, 131–138.

KOETS, P. 1936. Coacervation of amylophosphoric acid and proteins. J. Phys. Chem. *40*, 1191–1200.

LEACH, H. W. 1958. Personal communication, June 13.

LEACH, H. W. 1959. Personal communication, June 23.

LEACH, H. W., McCOWEN, L. D., and SCHOCH, T. J. 1959. Structure of the starch granule. I. Swelling and solubility patterns of various starches. Cereal Chem. *36*, 534–544.

LLOYD, N. E. 1959. Determination of surface-average particle diameter of colored emulsions by reflectance and application to emulsion stability studies. J. Colloid Sci. *14*, 441–451.

MAYWALD, E. C., and SCHOCH, T. J. 1957. Protective starch matrix for xanthophyll oil and vitamin A. J. Agr. Food Chem. *5*, 528–531.

MAYWALD, E. C., CHRISTENSEN, R., and SCHOCH, T. J. 1955. Development of starch and phytoglycogen in golden sweet corn. J. Agr. Food Chem. *3*, 521–523.

MAZURS, E. G., SCHOCH, T. J., and KITE, F. E. 1957. Graphical analysis of the Brabender viscosity curves of various starches. Cereal Chem. *34*, 141–152.

OSMAN, E. M. 1967. Starch in the food industry. *In* Starch: Chemistry and Technology, Vol. 2, R. L. Whistler, and E. F. Paschall (Editors). Academic Press, New York.

PRZYLECKI, ST. J. VON, and DOBROWOLSKA, S. 1932. Combination between amylopectin and proteins or their products. Biochem. Z. *245*, 388–407.

RIST, C. E., DAVIS, H. A., and WOLFF, I. A. 1952. Method for defatting starch. U.S. Patent 2,587,650.

ROBINSON, J. W., and BROCK, F. H. 1965. Method for the production of starch base jelly candy. U.S. Patent 3,218,177.

ROTH, W. B., and MEHLTRETTER, C. L. 1967. Some properties of hydroxypropylated amylomaize starch films. Food Technol. *21*, 72–74.

SAIR, L. 1964. Heat-moisture treatment of starches. *In* Methods in Carbohydrate Chemistry, Vol. 4, R. L. Whistler (Editor). Academic Press, New York.

SCHOCH, T. J. 1941. Physical aspects of starch behavior. Cereal Chem. *18*, 121–128.

SCHOCH, T. J. 1952. Fundamental developments in starch for paper coating. Tappi *35*, (7), 22A–38A.

SCHOCH, T. J. 1965. Starch in bakery products. Bakers Dig. *39*, (2), 48–52, 54–57.

SCHOCH, T. J. 1967. Mechanochemistry of starch. J. Technol. Soc. Starch, *14*, (2–3), 53–78. (Japanese)

SCHOCH, T. J. 1968. Effects of freezing and cold storage on pasted starches. *In* Freezing Preservation of Foods, 4th Edition, Vol. 4, D. K. Tressler, W. B. van Arsdel, and M. J. Copley. (Editors). Avi Publishing Company, Westport, Conn.

SCHOCH, T. J., and MAYWALD, E. C. 1968. Some unusual properties of Pakistan shoti starch. Stärke *20*, 362–365.

CHAPTER 21

W. A. Mitchell

Carbohydrate Hydrophilic Colloid Systems

INTRODUCTION

The hydrophilic colloid holds a prominent position in food preparation and processing. So versatile are the physical properties of these materials that they have been applied to all phases of the food industry. Water is a most important constituent of foods and correct water management through the use of these "water loving" materials allows for the production of many interesting food textures. Use of the hydrophilic colloid in the food industry in the United States is counted in many millions of dollars annually.

Recent reviews have been written by Glicksman (1962, 1963) and Klose and Glicksman (1968) which give a good discussion of the general nature of these materials. This report will endeavor to point out some special areas which can be of importance to the understanding and application of the hydrophilic carbohydrate gums in the food field. The concept of the importance of the disperse state and the function of gums in disperse food systems will be developed. Suggestions will be made for creating a better vocabulary and descriptive system for diphase and triphase food systems. Data are presented showing the use of the carbohydrate gums in emulsions and dried emulsions; foams and emulsofoams; and liquid in solid—solid in liquid disperse systems. The importance of amylopectin in starch granule organization is stressed. The movement of amylose within and without the basic amylopectin granule organization is suggested as controlling many of the fleeting and varied physical properties of amylose-containing starches.

HYDROPHILIC CARBOHYDRATE GUMS—WHAT THEY ARE

The brilliant efforts of P. P. von Weimarn (1907) showed definitely that "colloids" are not materials but represent a disperse state of all matter. This viewpoint was supported by Arthur B. Lamb (1950). However, in spite of this, the concept of the "hydrophilic colloid" as material is well accepted among the scientific community as illustrated in the organization of Advances in Chemistry Series (1960). Generally, the carbohydrate hydrophilic colloid is looked on as being made up of polymeric materials derived from plants and are often referred to as polysaccharide gums. Among these are also included derivatives such as those derived from cellulose or starch.

A list of the most useful food polysaccharide materials is given in Table 31; these will be the chief materials to be considered in this report.

TABLE 31

MOST USEFUL FOOD POLYSACCHARIDE MATERIALS[1]

Natural	Derived
Starch	Dextrins
Guar	Carboxymethylcellulose
Arabic	Methylcellulose
Alginates	Hydroxypropylcellulose
Karaya	Hydroxypropylstarch
Locust Bean	Crosslinked starch
Pectin	Propyleneglycol alginates
Carrageenates	Low ester pectin
Agar	

[1] For a more complete list see M. Glicksman (1962, 1963).

The viewpoint that the above tabulated materials are "colloids" is not altogether wrong; as a rule, when these materials are dispersed in water, they usually and naturally, fall into the colloidal range; that is, the disperse dimensions obtained (1 mμ to 0.1 μ) have been arbitrarily assigned by the colloid chemist (Ostwald and Fischer 1918) as the colloidal range. It is this ease with which the colloidal state is achieved in water as well as their mutual solubility that justifies these materials being called "hydrophilic colloids."

A very important property of some of the polysaccharide gums is their ability to easily form a gel in a water system. Some that will do so are given in Table 32; those which can produce the clear elastic gel

TABLE 32

GELLING CARBOHYDRATE GUMS

Starch (amylose)
Pectin
Alginates
Carrageenates
Agar

(pectin, alginates, carrageenates, agar) must be considered the most elegant of the gums. They are the most difficult to produce and are the most expensive. Use is made of these gums in food systems requiring physical properties not obtainable by a lower priced gum. Of the remainder, two classifications can be made: those that disperse readily in a water system and those that swell but are dispersed with difficulty. The presence of impurities can influence the dispersibility or swelling. Industrial gums can have much impurity and can sometimes be the ground source material (such as ground locust bean).

The polysaccharide gums have been used since antiquity and many uses have been found over the years. The modification of the natural gums by the chemist has increased the use of these materials tremendously in recent years. In Table 33 is given a list of the more important food uses.

TABLE 33

CHIEF FOOD USES OF POLYSACCHARIDE GUMS

Food stabilization through water management
Emulsifying—dried emulsions
Gelling
Whipping—foaming
Texturizing—imparting body, mouthfeel, chewiness
Imparting viscosity—thickening
Making beverage clouds, lighteners and milks
Flavor and flavor enhancer fixation—encapsulation
Imparting freeze-thaw stability
Crystallizing regulator for sugars, salts, water and other food materials
Films—filming agents, glazing
Suspending—solids
Adhesives—food binder
Bulking agent—low calorie sugars
Antisticking—dusting—coating
Flocculating agent
Aid in food freezing and freeze drying
Pulping agent
Low calorie food

THE DISPERSE STATE AND THE FUNCTIONAL ASPECTS OF GUMS

In order to make the best use of the polysaccharide gum, it is necessary to get some understanding of their physical functions. In 1956 a symposium on physical function of hydrocolloids was held at the 130th National Meeting of the American Chemical Society. The papers presented discussed this difficult problem and made some progress in the understanding of these systems. It is hoped that a further attack on this problem can be made through a better understanding of the disperse state.

Colloidal State in Food Systems

The colloidal state is recognized as a metastable state (Ostwald 1937). Practically all of our food systems involve the colloidal state. As food scientists, we fully recognize colloidal instability, for we are always being called upon to do something about our changing food systems: the starch pudding gets grainy; there is syneresis in the pectin gel; the dry powdered whipping agent no longer can be whipped; the fat separates from our creams; fixed flavors leave the fixation medium; clear solutions become cloudy or cloudy solutions lose their cloud. When our food systems approach a condition of physico-chemical equilibrium they usually become technically useless; bread becomes stale and the emulsions "break." Even slight changes in colloidal aggregation or dispersion can alter markedly the textural qualities of food making them unacceptable.

Coarse Dispersions in Food Systems

Ostwald (1915) wrote *The World of Neglected Dimensions;* he chose to include the range of particles from 1 mμ to $\frac{1}{10}$ μ as the dimensions that were neglected at that time. In other words, the range of particles that were larger than those forming true solutions and those not being large enough to be seen through the use of a normal microscope. These dimensions Ostwald reserved for the field of colloid chemistry.

In 1968, for the food chemist, the realm of neglected dimensions must be considered as those particles falling in the range of about 0.1 μ to about 500 μ. It is in this range of particle sizes that we find the practical food emulsions, foams, raw and swollen starch granules, swollen protein and gum particles, and complex mixtures of these. These dimensions have normally been assigned to the physicist or the physical chemist. However, both the physicist and physical chemist have largely ignored the above area for food systems. Perhaps the reason for this was the complex mixtures involved; mixtures that could not easily be treated within the classical physical chemistry due to their constantly changing nature.

Recently, in recognition of the importance of nonequilibrium systems in technology, there has been more and more effort on the part of our workers in the universities (Mitchell 1967) to come up with mathematical expressions for handling such systems.

In the preceding sections the metastability of the colloidal state was discussed. It is this colloidal state that is usually called upon to stabilize the coarser dispersions of liquids, solids, and gases. It is through the use of gums that coarse dispersions are kept where they will function, for as long as needed, in making successful food products. It does, indeed, appear unlikely that a metastable state of matter should be called upon to stabilize a coarse dispersion. Because gums can have high viscosities, or even gel, larger particles, such as fat globules, can take much longer to move. Gas bubbles, too, can be easily trapped.

Modern requirements for convenience foods, which are foods sold ready for consumption or prepared with minimum effort, put more and more strain on the stabilizing agents. When the food is prepared in the home it is usually consumed in a day or two. If the food is prepared in a central location, it can take days, weeks, or months before the food is consumed. For the above reason the stability problem is becoming more and more severe.

The big question is—Can something really be done about these changes? Through the applications of the following it is believed at least something can be done: (1) better understanding and management of the dispersed state of matter; (2) correct water management thorugh the proper choice and use of the hydrophilic colloid system; and (3) proper choice and processing of ingredients.

Need for Better Vocabulary and Descriptive System

It is essential that we adequately describe the dispersed food systems in order to obtain understanding and control. For the present complicated food systems, the descriptive language which is available is not sufficient and leaves much to be desired. As an example, it is well accepted that cream is an emulsion and a whipped egg white is a foam; what should the cream be called after it is whipped? In addition to the fat being divided in the water phase (emulsion), a gas is also dispersed in the water or even in the fat phase making the system more than an emulsion. About the middle of the nineteenth century, Thomas Graham (the father of Colloid Chemistry) found it necessary to coin new words when confronted with phenomena which could not easily be described otherwise (Gortner 1934). It is here suggested that for a triphase system such as whipped cream (liquid–liquid–gas) the word EMULSO-FOAM or EMULSOGAS should be used. Cake batters and bread doughs would fall into this general category, although they are further complicated by the presence of solids.

There is a tendency for food workers to call almost any complex food system an emulsion. When this is done the term emulsion can lose much of its meaning. That the definition of the term "emulsion" is unclear is shown by Sutheim (1946). Becher (1966) expanding on the Sutheim listing, gives different definitions by 9 experts. Becher also gives a summary definition, close to that of Hatschek (1926), which is probably most complete and acceptable; this definition includes a lower limit ($0.1~\mu$) for the dispersed liquid particle and stresses the minimal stability of such systems. It should be noted that the diameter of the dispersed liquid phase for food emulsions generally exceeds 0.1 μ and usually they have a diameter between about 1 and 30 μ. Emulsions, then, should be classified as diphase coarse dispersions (as opposed to colloidal dispersions) from the viewpoint of the immiscible dispersed liquid. Buzagh (1937) does make such a differentiation for emulsions. However, even here there can be an overlapping. The so-called micro emulsions have a disperse phase of 0.05 μ or smaller and have a transparent appearance.

It is also necessary to consider the state of materials (gas, liquid, or solid character). Ostwald (1907), in looking at the dispersions of solids, liquid, and gases, prepared his classical table which has proved so useful over the years for diphase colloidal systems. Foods are usually polydispersed especially when there is liquid present. This means that the liquid (water for example) could have a salt (say, sodium chloride)

TABLE 34

SOME IMPORTANT DIPHASE COARSE DISPERSIONS FOR FOOD SYSTEMS

System	General Name	Example
Liquid in liquid	Emulsion	Milk
Gas in liquid	Foam	Whipped egg white
Solid in liquid	Suspension	Raw starch in water
Liquid in solid	Swelled particle	Water in swollen starch granule

dissolved in it, or a protein (say, sodium caseinate) dispersed in the colloidal dimension. In the water phase there could be dispersed an immiscible liquid (say, oil) in a coarser than colloidal degree of dispersion. It is in this light that the food emulsions and foams should be considered. Ostwald's chart was made chiefly for application to colloid dimensions, but one can choose diphase systems from his chart, which are applicable to coarse dispersions or all dispersions found in food systems. Some important coarse food disperse systems are shown in Table 34. The importance of the liquid phase (water) to food systems should be noted.

If diphase systems are looked on as being complex, the complexity is increased by adding another phase. Modern food requires all types of textures. A proper dispersion of the three states of matter can bring about many novel and interesting textures and forms. It is time that the diphase chart (Ostwald's) be expanded to a triphase one. Such an expansion is shown in Table 35. In making such an expansion, it should be pointed out that certain combinations of phases (such as gas–gas–gas) can only produce limited dispersion. Limiting the phases limits the number of physical properties obtainable. To cover the dispersed

TABLE 35

TRIPHASE DISPERSIONS[1]

So-So-So	Li-So-So	Ga-So-So
So-So-Li	Li-So-Li	Ga-So-Li
So-So-Ga	Li-So-Ga	Ga-So-Ga
So-Li-So	Li-Li-So	Ga-Li-So
So-Li-Li	Li-Li-Li	Ga-Li-Li
So-Li-Ga	Li-Li-Ga	Ga-Li-Ga
So-Ga-So	Li-Ga-So	Ga-Ga-So
So-Ga-Li	Li-Ga-Li	Ga-Ga-Li
So-Ga-Ga	Li-Ga-Ga	Ga-Ga-Ga

Solid—So
Liquid—Li
Gas—Ga

[1] In cases where the degree of dispersion is limited (such as gas in gas in gas) then only one of two phase systems can be obtained. Limiting the phases limits the number of physical properties obtainable. The continuous phase is given first followed by two phases which can be dispersed in the first. Also, the second listed phase can act as dispersant for the third phase.

state in food systems adequately it is necessary to include dimensions from the molecular to at least 500 μ.

APPLICATION OF THE CARBOHYDRATE GUMS IN DISPERSE SYSTEMS

Characterized by their affinity for water, resulting in viscous solution, and a tendency to gel, the polysaccharide gums are ideally suited for use as food texture modifiers and stabilizers. Furthermore, through the use of other food materials which can be dispersed (to all degrees as well as in the three states of matter) in the gum system, an increased number of textures are obtainable. One of the biggest advantages of the nonionic polysaccharide gums over other hydrophilic materials, such as proteins, is the relatively little influence that the presence of salts, acids, temperature or other food materials have on them.

Three rapidly growing areas in food technology concern the use of: emulsions and dried emulsions; foaming and whipping agents; and the use of the stabilized starch granule. Data has been collected applying the gums to these three areas.

Emulsions and Dried Emulsions

The carbohydrate gums have been used extensively since early history as a primary emulsification agent. Usually the concentration of the gum relative to the water used has to be high. The emulsifiers containing both hydrophilic and lipophilic groupings have an advantage because they can be used effectively at very low levels. However, textural qualities obtained from the gums make them desirable in certain food products. A series of gums as primary emulsifying agents have been investigated by Lotzkar and Maclay (1943), King and Mukherjee (1940), and others as given by Clayton (1954).

Very little has been published in the area of the dried emulsion. It has been stated by Fischer and Hooker (1917) that oil in soap emulsions as well as oil in hydrated protein emulsions cannot be dried, but that the colloid carbohydrates are more stable. Within recent years commercial dried emulsions have been made taking advantage of the revelation of Olsen and Seltzer (1945) when they fixed flavor oils in gelatin to produce a "fixed" flavor oil. General Foods Corporation soon thereafter spray-dried various oils and flavor material. Gum arabic has

TABLE 36

PARTICLE SIZE OF FRESHLY PREPARED EMULSIONS[1]

Sample	Diameter in Microns % of Total Particles				
Diameter	1.30	1.64	2.07	2.60	3.28
Gum arabic 10% cottonseed oil	51.4	34.5	10.7	2.5	0.82
Gum arabic 20% cottonseed oil	58.4	37.5	3.5	0.45	0.15
Acid dextrin 20% cottonseed oil	15.3	32.7	32.5	13.1	8.43
Acid dextrin 40% cottonseed oil	17.7	27.3	33.4	16.0	5.40
Gum arabic 10% laurate coating fat	19.2	38.3	31.9	8.5	2.08
Gum arabic 20% laurate coating fat	14.0	29.4	37.2	15.1	4.07
Acid dextrin 20% laurate coating fat	37.4	44.2	12.5	2.9	2.15
Acid dextrin 40% laurate coating fat	13.8	54.4	33.8	5.7	1.15

[1] Determined in 1% sodium chloride in the Coulter Counter.

proved to be an exceptional polysaccharide gum for fixing flavor material. A review of the use of gums in flavor fixation is given by Glicksman (1962), and Klose and Glicksman (1968). The gamut of oils and fatty material can be fixed in various gums to produce dry free flowing powders that find various commercial uses. In the production of dried flavors, it is important to capture the whole spectrum of flavor components and maintain these flavors in stable form under use conditions.

Dried emulsions have been made employing the Niro Spray Dryer with gum arabic and also an acid dextrin. Two fats (corn oil and a laurate coating fat) have been employed at two levels. Reconstituted dried emulsions have been compared with the freshly prepared emulsions by measuring optical density and by particle size determination through the use of the Coulter Counter. The data are given in the following tables and graphs.

Table 36 shows the particle size distribution of emulsions prepared with several gum and fat systems. By the presence of fat particles that are preponderantly in the 1-μ range these emulsions must be considered as fine emulsions. Table 37 shows the same emulsions after being spray dried in the Niro Spray Dryer and then reconstituted with water. By comparing the values of Table 36 with Table 37 it can be seen that the particle size distribution is shifted to a smaller dimension, apparently emulsification is increased during the spray drying process. From these tabulations it can be concluded that the polysaccharide gums are good emulsifying agents. The stability of these emulsions is such that they can easily be dried to produce stable dried emulsions. These dried emulsions can be reconstituted to produce usable food emulsions.

Dried emulsions hold their fat very tenaciously and special extraction

TABLE 37

PARTICLE SIZE DISTRIBUTION OF DRIED EMULSIONS[1]

Sample	Diameter in Microns % of Total Particles			
Diameter	1.3	1.64	2.07	2.60 and Larger
Gum arabic 10% cottonseed oil	80.0	17.9	4.6	1.15
Gum arabic 20% cottonseed oil	60.5	33.4	5.3	0.75
Acid dextrin 20% cottonseed oil	68.2	20.6	5.3	5.58
Acid dextrin 40% cottonseed oil	38.5	46.2	10.7	4.57
Gum arabic 10% laurate coating fat	65.7	28.3	4.5	0.71
Gum arabic 20% laurate coating fat	65.5	30.6	3.0	0.79
Acid dextrin 20% laurate coating fat	75.2	18.6	3.6	2.50
Acid dextrin 40% laurate coating fat	64.0	29.0	5.5	1.47

[1] Determined in 1% sodium chloride in the Coulter Counter.

procedures have to be used in order to extract all the entrapped fat. It is also very difficult to stain or locate the fat in the dried powders by microscopic means. Possibly glyco-lipids are formed or other semi-stable complexes to account for the stability of these dried emulsions. This is especially true when highly volatile materials are involved.

The relative opacity of the reconstituted dried emulsions was determined in a Coleman Nephelometer using the 40% laurate coating fat, 60% acid dextrin emulsion at the 0.1% level as standard (100% opacity). The results obtained are shown in Fig. 134 and 135. As expected the opacity of the reconstituted emulsion, above all, is dependent upon the fat concentration. The use of cottonseed oil appears to produce slightly more opacity for most of the emulsions tested; this was not necessarily reflected in the particle size distribution as determined through the use of the Coulter Counter (Table 37).

The stability of the reconstituted emulsion as determined by opacity

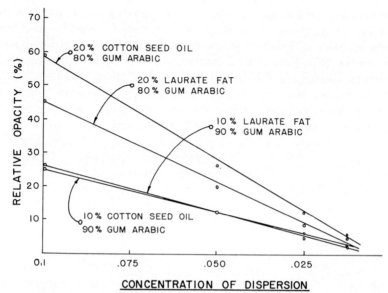

Fig. 134. The Relative[1] Opacity of Several Concentrations of Reconstituted Dried Emulsions Made with Gum Arabic

[1] Basis emulsion made with 0.1% dispersion of 40% laurate fat-60% acid dextrin, is equal to 100% opacity.

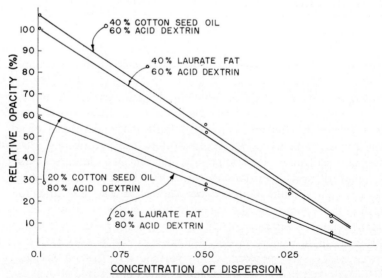

Fig. 135. The Relative[1] Opacity of Several Concentrations of Reconstituted Dried Emulsions Made with an Acid Dextrin

[1] Basis emulsion made with 0.1% dispersion of 40% laurate fat-60% acid dextrin, is equal to 100% opacity.

measurements is given in Fig. 136.   The stability of such gum stabilized emulsions is apparent, although there is a slow loss in opacity.   The relative stability of an emulsion is of importance for most technological uses.   Many of the dried emulsion powders can be stored for months under room conditions with little change.   However, some of the dried flavor emulsions can change rapidly on storage.   Here, too, an interaction of the gums with the flavor oil can take place.   With the more volatile flavors there is volatilization of these materials.

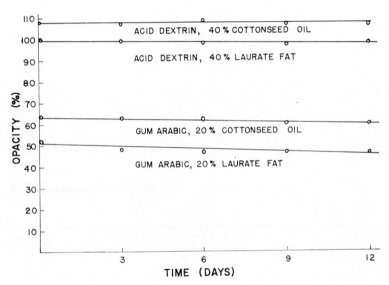

FIG. 136.   RELATIVE STABILITY AT 25°C OF RECONSTITUTED (0.1%) DRIED EMULSIONS MADE WITH GUMS

Foams and Emulsofoams with Gum Dispersions

Foams (gas in liquid) are produced with gum dispersion with little difficulty.   However, the stability of most food foams is much more difficult to maintain than the corresponding emulsions (liquid in liquid).   High viscosities or the gelled state of the gum water dispersion are necessary for any long storage of such systems (whipped gelatin dessert as an example).   There has been little work done on the effectiveness of various polysaccharide gums in producing stable food foams.   Berkman and Egloff (1941) conclude that high concentrations of substances that increase the viscosity of known systems increase foam performance.   In this connection polysaccharide gums have been used to stabilize foams produced by other "foaming" agents (such as those shown by Alikonis 1960, 1966).

We have employed the polysaccharide as a primary foaming agent and have measured the "overrun" with whipping time and stability as determined by drainage in a funnel. The data collected is shown in the following tables and graphs.

Figure 137 shows the influence of whipping time on the foaming properties of several concentrations of hydroxypropylcellulose. At the

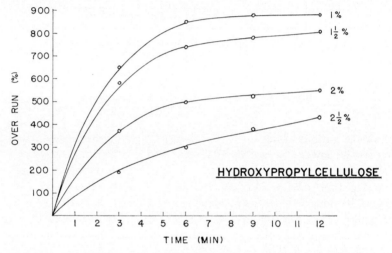

FIG. 137. THE INFLUENCE OF WHIPPING TIME ON THE FOAMING PROPER-
TIES OF HYDROXYPROPYLCELLULOSE AT SEVERAL CONCENTRATIONS

low gum concentrations the foaming properties are much better than at the higher concentrations. The "whip" volume increases very rapidly to produce a fluffy foam. Figure 138 shows the influence of concentration on the viscosity or body of the foam after 12 min of whipping. The viscosity increases with concentration. Figure 139 is a graph showing the "overrun" potential after 12 min of whipping when hydroxypropyl-cellulose is used. The graph shows that as concentration is increased, "overrun" potential is decreased. Finally, in Fig. 140 is shown the relative stability of the foam as determined by a funnel drain test and represents the time for complete collapse of the foam. From this graph it can be seen that foam stability increases rapidly with increase in hydroxyproplycellulose concentration.

In Fig. 141 is shown the whipping properties of an acid dextrin dis-

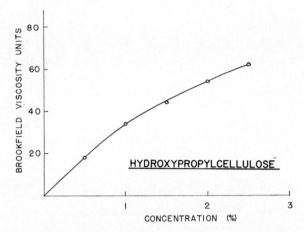

FIG. 138. THE INFLUENCE OF CONCENTRATION ON THE
BROOKFIELD VISCOSITY OF FOAMS MADE WITH
HYDROXYPROPYLCELLULOSE

persion at various concentrations.   As with the hydroxypropylcellulose
sample—the lower the concentration the more "overrun;" a straight line
relationship is shown up to about 15 min of whipping.   When the
whipping time of the sample containing 33.3% solids was extended
beyond 18 min, a decrease in "overrun" is observed much like that for
egg white.   However, no attempt was made to control moisture loss and
the lowered overrun could be loss of water resulting in a concentration

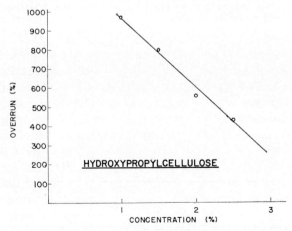

FIG. 139. THE INFLUENCE OF CONCENTRATION ON THE
FOAMING   POTENTIAL   OF   HYDROXYPROPYLCELLULOSE

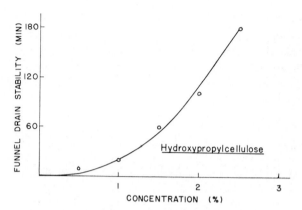

Fig. 140. The Influence of Concentration on the Stability of Foams Made with Hydroxypropyl-cellulose

effect. The whipping properties of 11.5% dried egg white sample is shown for comparative purposes.

The influence of gum concentration on the Brookfield viscosity of the whipped (12 min) foam is shown in Fig. 142. Here again the higher concentrations of gum, as expected, increase the viscosity of the foam. The viscosity can reflect the "body" and "mouthfeel" of the foam and can be a basic desirable property of edible foams.

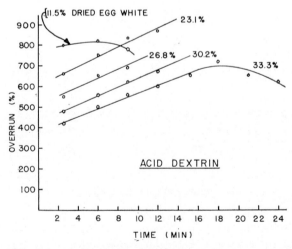

Fig. 141. The Influence of Whipping Time on the Foaming Properties of an Acid Dextrin at Several Concentrations

Table 38 shows that the "overrun" potential of the dextrin decreases as concentration is increased.

In Table 39 the stability measurements of the foam made with the dextrin gum are tabulated; these show again that increased stability is

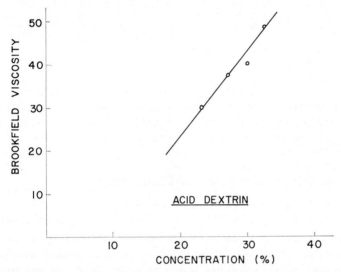

FIG. 142. THE INFLUENCE OF CONCENTRATIONS ON THE BROOK-FIELD VISCOSITY OF FOAMS MADE WITH AN ACID DEXTRIN

obtainable with increased gum concentrations. For reference, the stability of a whipped dried egg white reconstituted at 11.5% solids and whipped for 9 min is shown. This reference point shows that the stability of these foams is somewhere in the area of the egg white foams.

TABLE 38

THE INFLUENCE OF CONCENTRATION OF THE FOAMING CAPACITY OF AN ACID DEXTRIN[1]

Concentration in %	% Overrun
23.1	862
26.8	694
30.2	669
33.3	599

[1] After 12 min whipping.

The emulsofoams are more difficult to establish through the use of carbohydrate gums alone. However, when used in conjunction with the protein emulsifiers, such as sodium caseinate or gelatin, very definite advantages can be gained by using gums with the protein stabilized

TABLE 39
THE INFLUENCE OF CONCENTRATION ON THE STABILITY OF FOAMS
MADE WITH AN ACID DEXTRIN

Concentration in %	Funnel Stability in Minutes
23.1	20
26.8	27
30.2	50
33.3	95
11.5% Dried egg white	75

emulsofoams. These advantages are: faster whipping time; greater overrun potential (smaller gas bubbles); increased stability of the emulsofoam; improvement in body, mouthfeel and general textural properties; improvement in freeze-thaw stability.

Liquid in Solid Systems—Solid in Liquid Systems

The best example in the food area, and most generally used liquid in solid system, is starch. The most useful properties of starches are obtained by taking advantage of the granule structure to imbibe water. For starches of all kinds the amylopectin must be considered the basic organization material for the starch granule. For amylose-containing starches the granule is a solid solution of amylose in amylopectin; this is true even for the high amylose-containing starches. Furthermore, the granule solid solution must be considered as poorly organized or more or less in the glassy state. It is for this reason that amylose is soluble in water. Once the amylose is free of the entanglements of the amylopectin, that is, dispersed in water, amylose will crystallize in the normal course of events.

The amylopectin granule can be stabilized somewhat by the presence of the amylose, and the degree of stability is dependent upon degree of crystallinity of that amylose. Of course, the more crystalline the less soluble it will be and the greater the tendency to remain within the amylopectin matrix.

The physical properties (viscosity, gelling) of normal starches are in the main determined by the movement of the amylose within and out of the granule. It is in this light that we can interpret the one stage or two stage swelling patterns of different starches as given by the data of Leach et al. (1959), or most recently by Sandstedt et al. (1968) and also Goering et al. (1968). The waxy starch granules are relatively unrestricted to water imbibition while the starches containing amylose are more restricted until the amylose is made soluble.

If a complex with amylose (say, with fatty substance) is made within the granule, the swelling behavior and resulting viscosity of a slurry can be much different from that which could be expected if the amylose were allowed to move out of the granule. It is this behavior which could account for the difference in viscosity behavior of starches as reported by Mitchell and Zillman (1951) and that reported by Gray and Schoch (1962) where different procedures for adding the fatty materials were used. The heat treatment of starch in the presence of moisture (Sair and Fetzer 1944) (reviewed by Osman 1967) and its resultant change in physical behavior of the starch, again, concerns the movement and crystallization of amylose. The amylose is allowed movement by water entering the granule (much the same as the crystallization of candy glasses in the presence of small amounts of water) and arranges itself in the more stable crystalline form; in doing so the granule is stabilized depending on the degree of crystallization allowed the amylose; this could be amylose movement to the granule surface and a reentry into the granule. Such crystalline amylose in starch granules makes a cloudy paste with water. Taken generally, the starches producing more cloudy pastes are those containing more crystalline amylose (or amylose made insoluble by complexing with agents such as fat). No doubt the polymer size of both the amylose and amylopectin as well as the degree of crystallinity of both can have pronounced effect on movement of the amylose out of the granule as well as the movement of water into the residual amylopectin granule (swelling power).

Actually the stability of some raw starch granules is substantial as was shown very early by Tanret (1914) as well as the more recent work of Cowie and Greenwood (1957) and Adkins (1967). Fischer and Hooker (1917) note that greatly swollen starch granules may still be discerned microscopically even when a small amount of starch has been boiled for a long time in much water. The greater the imbibition of water the more susceptible the granule is to shear. The destruction of the granule structure can decrease viscosity or gel strength making unacceptable food products. Most recently there have been big strides made in the stabilization of the starch granule through the use of crosslinking agents such as epichlorohydrin as well as treatment with propylene oxide. Such stabilized starches have found extensive use in freeze-thaw foods, food systems having extended heat or acid treatment, and starch systems undergoing high rates of shear. These stabilized starch granule systems can be shown to retain much of their amylose during the cooking or processing step. On treatment with propylene oxide the amylose can be altered considerably, reducing gelling and retrogradation properties.

If the amount of water in a starch system is limited then the amylose cannot leave the granule (or leaves and then enters it again). Under these conditions there can be some but restricted gelling and crystallization of the amylose. If the swollen granule organization is destroyed by a shearing action, the total mixture will pass into the sol state of dispersion and the amylose is then free to crystallize. On the other hand, if water is plentiful the amylose does leave the granule and can gel or crystallize within a sea of swollen amylopectin granules.

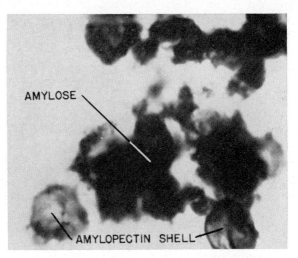

FIG. 143. BOILED REGULAR CORN STARCH SHOWING
STABLE AMYLOPECTIN GRANULE ORGANIZATION

By proper water management and shearing operation it is possible to have the following: (1) A swollen amylopectin granule with amylose dispersed within it. The amylose can slowly crystallize or gel. Also, the amylose can be so modified that no retrogradation takes place. On shearing the above, the granule organization is destroyed and both amylopectin and amylose are dispersed colloidally. (2) A swollen amylopectin granule and amylose in colloidal dispersion outside the granule with freedom to gel or crystallize. If this slurry undergoes a shearing action, the amylopectin granule organization is again destroyed and a sol dispersion of both amylose and amylopectin is obtained.

It must be clear that the physical properties obtained from any starch system is dependent on the final granule organization or disorganization.

In Fig. 143 is shown a photomicrograph of corn starch boiled for 5

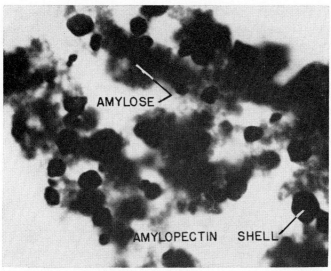

FIG. 144. BOILED HIGH AMYLOSE CORN STARCH SHOWING STABLE
AMYLOPECTIN ORGANIZATION

min in 2% slurry then stained with iodine. Note the free stable amylopectin granule (light) and the liberated and precipitated amylose
(dark).

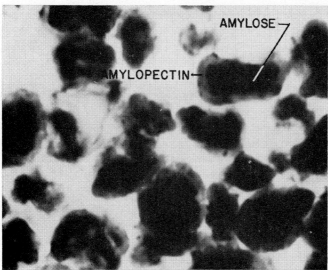

FIG. 145. BOILED CORN STARCH TREATED WITH 0.01% EPI-
CHLOROHYDRIN SHOWING AMYLOSE RETENTION WITHIN THE
GRANULE

In Fig. 144 is shown a high amylose corn photomicrograph. Note the granule organization of amylopectin (dark in photomicrograph) is intact even after extended extraction treatment in boiling water. Note also the copious amount of precipitated blue staining amylose (lighter color).

In Fig. 145 is given a photomicrograph of a corn starch treated with .01% epichlorohydrin. Note that the amylose is retained within the inner portions of the amylopectin granule organization even after extended extraction at the boiling water temperature (darker area). The outer portion of the granule (lighter area) has been freed of amylose.

SUMMARY

The versatile nature of the carbohydrate gums and the prominent position they hold in food preparation has been discussed. The importance of the disperse state in understanding the function of the hydrophilic colloid in food systems was elucidated. Suggestions were made for creating a better vocabulary and descriptive system for diphase and triphase food systems.

Data were shown in which gum arabic and an acid dextrin were used to produce emulsions with cottonseed oil and a laurate coating fat. Dried emulsions were prepared by drying in the Niro Spray Dryer. Particle size determination of the fat globule indicated that for both wet and dried emulsions the particle size fell between about 1 and 3 μ microns in diameter. There was a slight shift to smaller particles in the drying of the emulsions. Opacity measurements on the reconstituted dried emulsion showed that this property was dependent, above all, upon fat concentration. Relatively small differences in opacity were shown between the two gum systems as well as the fat system. The cottonseed oil reconstituted emulsions were slightly more opaque. Storage of the emulsions at 25°C showed only a slight decrease in opacity (stability) after 12 days of storage.

Two gums (hydroxypropylcellulose and an acid dextrin) were whipped into foams and the Brookfield viscosity, stability to drainage, and the effect of concentration on whipping and foam properties determined. At the lower concentration of the gum levels employed, the whipping was more rapid and the whipping capacity was greater. However, the viscosity and stability of the whipped foams increased with increasing concentration of the gums.

Starch as a liquid in solid—solid in liquid system was discussed. It was suggested that the amylopectin was the chief organizational material of the starch granule. The movement of amylose within and without the basic amylopectin granule organization is believed to be the controlling influence in the varied and fleeting physical properties of amylose-containing starches. Some photomicrographs were submitted in support of the amylopectin as the basic granule organization material.

DISCUSSION

Dr. T. J. Schoch.—You may be acquainted with the very nice method of measuring shelf-life of emulsions, as designed by Norman Lloyd (1959), which is a spectrophotometric light reflectance method. Essentially Lloyd was measuring the average particle size of the dispersed oil phase. In doing this work it was found that starting with a certain oil particle size, say a diameter of 3 or 4 μ, that there would be a rapid and linear aggregation of these particles up to let us say a particle size of 20 to 30 μ; thereafter the particles would stabilize, apparently reaching some sort of equilibrium state. Is this a surface tension phenomenon? Do these emulsion systems approach a certain equilibrium and thereafter do not change?

W. A. Mitchell.—No, I have not observed such rapid aggregation and reaching of an equilibrium at a particle size of 20 to 30 μ. It has been my experience that if the oil emulsion particle is of the order of 1 or 2 μ (smaller than 3) then aggregation changes take place very slowly. Apparently emulsions become more stable if one makes the oil droplets smaller. Actually many of the emulsions evaluated by Norman Lloyd had a globule diameter of 3 μ or smaller; they also aggregated very little on storage.

Dr. H. Schultz.—Dr. Mitchell, you gave information on a number of gums: there is one with which I've had a great deal of practical experience, but confess to knowing nothing about its structure; this is chewing gum. Can you say what chicle is?

W. A. Mitchell.—Chicle consists of oxidized hydrocarbons, a resinous hydrophobic material. Nowdays the term gum generally refers to carbohydrate hydrophilic materials not the hydrophobic chewing gum material you mentioned. It is unfortunate that the word "gum" has been applied to both types of material. The basic meaning of gum being—thick mucilaginous, adhesive.

BIBLIOGRAPHY

ADKINS, G. K. 1967. Starches and starch derivatives. Food Technol. Australia 19, 518–523.

ALIKONIS, J. 1960. Practical aspects of foam stabilization, In Physical Functions of Hydrocolloids, Advan. Chem. Ser. 25, Am. Chem. Soc., Washington, D.C.

ALIKONIS, J. 1966. Foam stabilization with carbohydrate colloids. Mfg. Confectioner *46* 9, 33–36.

ANON. 1960. Physical Functions of Hydrocolloids, Advan. Chem. Ser. *25.* Am. Chem. Soc., Washington, D.C.

BECHER, P. 1966. Emulsions: Theory and Practice, 2nd Edition. Reinhold Publishing Co., New York.

BERKMAN, S., and EGLOFF, G. 1941. Emulsions and Foams. Reinhold Publishing Co., New York.

BUZAGH, A. VON. 1937. Colloid Systems: A Survey of the Phenomena of Modern Colloid Physics and Chemistry, W. Clayton (Editor), translated by Otto B. Darbishire. The Technical Press Ltd., London.

CLAYTON, W. 1954. The theory of emulsions and their technical treatment. *In* Clayton's Emulsions, 5th Edition, C. G. Sumner (Editor). J. A. Churchill, London.

COWIE, J. M. P., and GREENWOOD, C. T. 1957. Physicochemical studies on starches Part VI. Aqueous leaching and the fractionation of potato starch. J. Chem. Soc., 2862–2866.

FISCHER, M. H., and HOOKER, M. O. 1917. Fats and Fatty Degeneration. John Wiley and Sons, New York.

GLICKSMAN, M. 1962. Utilization of natural polysaccharide gums in the food industry. Advan. Food Res. *11,* 109–200.

GLICKSMAN, M. 1963. Utilization of synthetic gums in the food industry. Advan. Food Res. *12,* 283–366.

GOERING, K. J., ESLICK, E., WATSON, C. A., and KENG JUIN. 1968. Barley starch III. A study of the starch properties of 30 barley genotypes. Paper presented at National Meeting of Am. Assoc. Cereal Chemists, Washington, D.C.

GORTNER, R. A. 1934. Colloids in biochemistry. An appreciation of Thomas Graham. J. Chem. Educ. *11,* 279–283.

GRAY, V. M., and SCHOCH, T. J. 1962. Effects of surfactants and fatty adjuncts on the swelling and solubilization of granule starches. Staerke *14,* 239–246.

HATSCHEK, E. 1926. Introduction to the Physics and Chemistry of Colloids. P. Blakistons Sons & Co., Philadelphia, Penna.

KING, A., and MUKHERJEE, L. N. 1940. Stability of emulsions II. Emulsions stabilized by hydrophylic colloids. J. Soc. Chem. Ind. *59,* 185–191.

KLOSE, R. E., and GLICKSMAN, M. 1968. Gums *In* Handbook of Food Additives, T. E. Furia (Editor). The Chemical Rubber Co., Cleveland, Ohio.

LAMB, A. B. 1950. Foreword to Colloid Science, McBain, J. W. Heath & Co., Boston, Mass.

LEACH, H. W., McCOWEN, L. D., and SCHOCH, T. J. 1959. Structure of the starch granule. Cereal Chem. *36,* 539–540.

LLOYD, N. 1959. Emulsion shelf life measurements. J. Colloid Sci. *14,* 441–451.

LOTZKAR, H., and MACLAY, W. D. 1943. Pectin as an emulsifying agent. Comparative efficiencies of pectin, tragacanth, karaya and acacia. Ind. Eng. Chem. *35,* 1294–1297.

MITCHELL, W. C. 1967. Statistical Mechanics of Thermally Driven Systems. Ph.D. Thesis. Washington University, St. Louis.

MITCHELL, W. A., and ZILLMAN, E. 1951. The effects of fatty acids on starch and flour viscosity. Trans. Am. Assoc. Cereal Chemists 9, 64–79.

OLSEN, A. G., and SELTZER, E. 1945. Preparation of flavoring material. U.S. Patent 2,369,847.

OSMAN, E. M. 1967. Starch in the food industry, In Starch Chemistry and Technology, Vol. II, R. L. Whistler, and E. F. Paschall (Editors). Academic Press, New York.

OSTWALD, Wo. 1907. Kolloid-Z. 1, 291–331. Cited in Soaps and Proteins (1921), M. H. Fischer, G. D. McLaughlin, and M. O. Hooker. John Wiley & Sons, New York.

OSTWALD, Wo. 1915. The World of Neglected Dimensions. Translated by M. H. Fischer (1917). Called An Introduction to Theoretical and Applied Colloid Chemistry. John Wiley & Sons, New York.

OSTWALD, Wo. 1937. Foreword to Colloid Systems, A. Buzagh, Technical Press Ltd., London.

OSTWALD, Wo., and FISCHER, M. H. 1918. Handbook of Colloid Chemistry, 2nd Edition. P. Blakistons Sons & Co., Philadelphia, Penna.

SAIR, L., and FETZER, W. R. 1944. Water sorption by starches. Ind. Eng. Chem. 36, 205–208.

SANDSTEDT, R. M., HITES, B. D., and SCHROEDER, H. 1968. Genetic variations in maize. Cereal Sci. Today 13, 82–94.

SUTHEIM, G. M. 1946. Emulsion Technology, 2nd Edition. Brooklyn Publishing Co., New York.

TANRET, C. 1914. Compt. Rend. 158, 1353–1356. Cited by R. P. Walton, (1928). A Comprehensive Survey of Starch Chemistry, Vol. 1. Chemical Catalog Co., New York.

WEIMARN, P. P. 1907. Kolloid-Z. 2, 76. Cited in Oedema and Nephritis (1921), M. H. Fischer. John Wiley & Sons, New York.

| Panel Discussion | Summary of Symposium |

S. M. Cantor.—In the deliberations over the past few days you heard Dr. Bollenback refer to this operation as a safari. I learned only recently what that meant. It is a Swahili word and it means long journey which kind of disappointed me. I thought it was something a lot more esoteric. This in many ways has been a long journey, a long carbohydrate journey and inevitably at this stage in a carbohydrate journey, certain things happen: Two sugars begin to appear, one in the audience and one in the speakers. The speakers is called verbose and the audience, comatose. That's an old sugar joke and I felt that since we have heard a lot of old sugar jokes we shouldn't miss that one. At any rate, we will try to keep those from crystallizing in the next few minutes as what we are concerned with is a hurried view of what has happened the last three days.

My particular view of the papers which I will give you in a hurried way of introducing a pattern or model for the panel to comment on is that we have had accentuated for us a vast storehouse of knowledge on the part of carbohydrate chemists and the ancillary disciplines that go along with carbohydrate chemistry. I do not mean to put those on various levels by saying ancillary. In this particular case we started out by talking about carbohydrate chemistry. It has been indeed a very complicated journey. We got various admonitions along the way. We heard Dr. Wolfrom right from the beginning admonish us about using the proper language and he got up repeatedly during the next two and one half days and did the same thing. Rightfully the language of carbohydrate chemistry is important to maintain in its pure state so that it can be understood and expanded. We heard Dr. French talk to us very wisely on not confusing widespread occurrence with simplicity, an important point, along with many other important points in his talk. Dr. Ward mentioned the same problem with respect to cellulose. The encyclopedic characteristics of gums and pectins was well brought out in the paper on that subject. The widespread occurrence of carbohydrates with proteins reported by Dr. Neuberger was an important contribution and a surprise to many of us. We heard on Wednesday morning three papers describing the analytical techniques, an important aspect of our Symposium. Then we went into a recitation of various reactions. The oxidative degradation studies reported by Professor Perlin was admirably done. The nonenzymatic browning story was thorough. We got an admonition from Dr. Horowitz in his paper on glycosidic pigments not to confuse fact and hypothesis. The effect of irradiation on polysaccharides was given thoroughly by Dr.

444

SUMMARY AND PANEL DISCUSSION

Massey. We turned this morning to another aspect of carbohydrates, the review of Dr. Harper and the attention to special diseases, obesity, and diabetes, and cardiac glycosides. Then we came down to the opposite side of the coin and we had in three rather hurried phases, application reports from experts in the sugar, starch, and hydrophilic colloids fields. At this point I suppose that you must feel a little bit like a badminton bird being batted back and forth.

On the other hand the dialogue which took place between the storehouse of information and those people who reported it and the applications I thought was the most effective one. One of the things it did do for me was to reemphasize what we've been talking about repeatedly throughout the Symposium, namely, the emphatic need for ways of breaking down communications barriers. Since most of the specific audience requests for panel discussion have been answered, we thought that we'd spend a little time talking about the communications problem.

In order to do this I would like to simply repeat in a way what I started out to talk about one day which was the large variety of information that is available. I thought I would put this down in the form of a process. The development of a new food product is a process dependent upon communication. Much information goes into the successful production of a new product. Take for example the utilization of a new agricultural crop. There must be all kinds of information pertaining to this commodity which must be fed into the process stream—research data, technology, economics, corporate organization, nutrition, production, financing, distribution, transport, and governmental standards and regulations—all of which pertain to the successful production, distribution and utilization of the commodity by a specific population. These are only a few of the items which ultimately wind up in products directed at the consumer. There is a concomitant type of activity going on in this process namely, market research and consumer testing, and all of the variables that go along with that. This information eventually gets fed back into the process and affects the product. Meanwhile all of these other aforementioned activities are feeding information into this process. Ultimately we get a product which should sell, and that is the whole business of developing a product. This is our system in North America and Europe, a rather small portion of the world. In other parts of the world, the system is complicated by political problems and adaptations. Each time you add a variable, you add a communication complication. Ultimately you are still trying to do the very same thing which is to get a product of value to the consumer.

All these variables can be organized under three headings; availability, acceptability, and value. The problem that we have starting with communication between technologists and food scientists is really an important one. How do you communicate all of these requirements so that you wind up with a reasonably high value figure going to the

446   CARBOHYDRATES AND THEIR ROLES

consumer not the least of which is nutritional value. This is the question
which comes naturally out of this Symposium although I must say the
kind of communication which we've enjoyed the past few days is ex-
emplary in many ways. Nevertheless it leaves problems. This is a
legitimate question to put to the panel. Let us see what happens.
I will call on Dr. Montgomery to lead off.

R. Montgomery.—The problem of communication is a most significant
one. I was struck by Dr. Cantor's remark that a tremendous amount
of knowledge has been offered during this program. As I look through
the audience and see many friends and colleagues in industrial en-
deavors, I would submit to you that the knowledge residing with you is
greater by many times than the knowledge shared with you by the
speakers on the program. This to me, in the academic field of endeavor, is the big problem. I don't know all of these problems. I
don't think that you know all of the problems, even within your own
company. This question of communication reminds me very much of
the advances that have been made in the field of biochemistry where
attempts are made to understand the biochemistry of the normal human.
If you look through the last 10, 15, or 50 yr of biochemical advances,
you find that many of the important leads in knowledge were derived
from reports made by clinicians, who write very good case histories.
This offers to the biochemist, who is concerned with the normal functioning of the human metabolic processes, a tremendous wealth of knowledge. I would suggest that if each time one of you go into the field
to study a problem, then this "case" could be written up and com-
municated to your own colleagues. The people in your research and
development sections might then have a better idea of the aims and
problems of your company than they do now. This could be developed
further by a bridgeman somewhere between your research and development and the outside sales team.

In the area of carbohydrates I think that we have better bridges
than I have seen in many other areas. If you think that a symposium
like this is the best opportunity for industrial and academic scientists
to share this knowledge, then I must point out that such symposia are
common in the field of carbohydrates. You could have come to this
meeting from the Gordon Conference on the Chemistry of Carbohydrates where similar problems are discussed. You could have been
to the Spring meeting of the American Chemical Society or to the
Annual Cereal Chemistry meetings. In the fall there will be another
ACS meeting and the Starch Round Table of the Corn Industries
Research Foundation. It is at times like this that communications improve. They improve not only in the lecture theater, but also around
dining and other tables. Industry cannot succeed unless it makes a
dollar into a dollar ten. You have to make it into a dollar ten and not
your competitor. It is understandable therefore that in an audience

such as this you are not going to discuss your problem unless you can do so in some sort of confidential manner. Communication must be somewhat restricted.

T. M. Reynolds.—I am going very simply to the purely personal level in this matter. A scientist is necessarily concentrated in a very narrow way on his own subject, otherwise he won't advance in it. This makes him perhaps a little impatient with what other other people are doing. We have even heard, in this meeting, that the carbohydrate chemists do not get on very well with other organic chemists. If this is so, how can we expect the carbohydrate chemist to get on with people in less related fields of pure or applied science? There is a great need for the scientist to have a sympathetic feeling for the work of other scientists, coupled, if possible, with admiration for the other man's courage in tackling his problem.

I think that there is also a great need for the scientist to have a feeling of personal responsibility towards the larger issues. It is easy to feel that the problems are too big and that one cannot make a personal contribution. This may be true, but the scientist who looks for opportunities may well find them.

G. E. Mortimore.—I would only second the suggestion made by Dr. Montgomery about elbow-bending. I must confess that I have had no experience in applied science, but I do believe that the model that Dr. Cantor used is quite applicable to all fields of science, whether it is basic research or applied research. It seems to me that there is a humanistic element that should not be overlooked in improving communication among individuals. In my experience at NIH, a very large institution, one often did not know what was going on in the laboratory next door until it appeared in print six months later. For this reason there did not exist an easy basis for communication. I noticed that people made a great effort to find some excuse to walk into someone else's office or laboratory to make small talk, primarily to overcome this barrier. It seemed to me that improved scientific communication was generally the result of such efforts.

A. Neuberger.—For some years I have tried to find out why academic scientists choose the career they have chosen. Some of us probably were not good enough to be taken up by industry, but I do not believe that this applies to all of us. The first reason is probably that the scientist has a need to fulfill himself in producing something which is esthetically satisfying in the intellectual field. The question is why should the community pay him to go on living this sort of life, satisfying his desires. I think this is becoming quite an important problem in all developed countries where science makes bigger and bigger demands on the national resources. The way scientists try to communicate their case to the public and politicians is based on their claim that ultimately they improve the living conditions of the community. Therefore it is the duty of all of us who are working in an academic atmosphere to help to see that some of the knowledge we have acquired can be put to practical use. I accept this as a social obligation. Then the question arises: how can this best be done? I think with the increasing specialization that is taking place in the world, you need people perhaps both in industry and nonindustrial organizations, who are essentially bilingual. You need someone whose main task it is to keep in touch with the developments which have a bearing on the activities of the particular firm or corporation and to choose the scientist who is most likely to be helpful and have him act as interpreter. There is an increasing need for such people in our civilization. For example, in my medical environment there is a need for a bilingual person who can translate both clinical and biochemical problems to people in the opposite field, and see how a medical or clinical problem can be attacked by biochemical means. If the person were not bilingual he could not do this. The same thing applies to science and industry. You need these people both within universities and industrial firms or government whose task it is to cultivate the bilingual character of their job, to be familiar with the industrial problems and at the same time keep up with scientific development, to know when to pose the problem to an academic scientist and to whom to go.

I would like to close by commenting on the success of the Symposium. I have been enormously impressed by the standard of the contributions and also by the interest shown by the whole audience in the problems, many of which are remote from the day-to-day problems they meet in their jobs. It is a unique experience to find such understanding and ready intercourse intellectually which has taken place here and outside. In many cases I found it hard to tell whether I was talking to an industrial chemist or to somebody holding a university appointment. This is the highest tribute I can pay.

G. N. Bollenback.—I would like to comment on the communication or conversation between technology and research personnel. If the research man and the technologist do not "sympathize" or "empathize" with each other, a gap remains between them. The information one man has never gets across the gap unless the research man appreciates what the technologist is trying to do and vice versa. The technologist must use the research man's facts in his processes. In the absence of a spontaneous exchange a liaison man is proposed who can translate research data for the technologist and who understands both ends of this setup. This type of liaison is a very effective technique. I think there are very few people who have the background to do it. Dr. Cantor is one but he is very exceptional. He was a highly regarded scientist who also worked in technology and now is in marketing. Such a background makes him especially attractive because it lends integrity to marketing.

If you think the dialogue between the technologist and the research man is poor, you should hear that between the marketing man and the research man. I will say this for the marketing man, he knows what to do with research data. For example, Dr. Emily Wick who is a very high caliber food technologist—objective and understanding of these problems—had a paper come out in one of the trade journals with this heading in large print: "Sucrose has no effect on Sweetness." Now, in very small and light print above that it says: "beyond a certain point." This is the sort of problem to expect when dealing with marketing people. The marketing man knows what to do with research data. The research man does not understand the marketing man at all. We undoubtedly need better communications here.

Between research and production the same sort of problem exists. The research man may have a series of data which he offers to the production man as a solution to a problem. The production man looks at the data and throws them away saying that he can't use them. Well, why not? Because the research man, having stayed in his lab, does not recognize what the problem actually is. His data may solve a problem, but not the one at hand. So in communicating, you should get out of the lab. If you don't get out you should have a representative who can describe problems for you and make use of your data in the prevailing process problems.

S. M. Cantor.—I too like to use artistic analogies like Dr. Neuberger. One of the things that occurs to you in listening to a program of this kind where models are referred to on a regular basis is that the rigidity of a model is sometimes self-defeating. We had many models displayed here in this program. One of the problems with the use of models, which artists seem to have less trouble with than other people, is that the model does not take on the characteristics of reality. Artists have a facility of taking liberties with their models that other people do not have. I recommend this sort of flexibility when dealing with models. It provides a kind of broad thinking that everyone can use.

One last thing in respect to that. If we take all of the pictures that come out of each of the disciplines that are represented here and superimpose then we begin to get closer to the reality of the situation, a resonance hybrid if you will.

H. W. Schultz.—It is my sad duty to bring this Symposium to a close. I would like to superimpose a comment on what the panel has said about difficulties in communication. In educating the food scientist or food technologist today, one of the greatest concerns is in providing young people with an appreciation for this matter of communications. We are faced with making "chemists" of them so they can communicate with the true chemists, on the one hand, about problems that involve chemistry, and with processing plant personnel on the other hand. Similarly they must become "microbiologists." Or they must communicate as "physicists" or "engineers."

Well, one can't be everything, but if we can convince our students

(I mean both undergraduates and graduates) to develop a desire and a language to communicate, they are going to be helped tremendously in solving problems. We emphasize, in our curriculum, that the food industry is built upon solving problems. A business that does not have problems is decadent.

I wish to thank each and every one of the speakers for making such fine presentations as representatives of those who are foremost in the knowledge of carbohydrates. An alert audience was present throughout the Symposium. It is sincerely hoped avenues of communication have been opened or widened for the ultimate benefit of all who study, process, or consume carbohydrates.

Index

A

Absorption, maximum, 283–284
Acetals, as chemical linkages, 13
Acetylcoenzyme A, 206, 308
Achroic point, in iodine-starch complex, 283
Acid treatment, for starch granules, 411–412
Acidosis, 306
Adenosine monophosphate, interactions, 368–370
    metabolic effects, 367
Adipokenetic factors, 329
Agar, source, 82
    structure, 83
Agaran, structure, 83
    sulfate groups, 85
Agaropectin, 85
Aglycone, location of free amino groups, 124
Aglycons, cardiac, 347
Albumin, egg, 116
    analysis by isotope dilution, 143
    sequence of amino acids, 120
Aldehydes, formation, 207
Aldoses, 13
    enzymic, 206
    oxidation by bromine, 205
Aldosylamines, rearrangement, 220–222
Algin, 82–84
    gelling, 84
    properties, 83
    structure, 84
Alginates, hydrolysis, 190
Allulose, 213–214
    synthesis, 209
Alpha particles, of glycogen, 31
    reactions, 32
Amadori rearrangements, of aldosyl-amines, 221–222
    mechanism, 222–224
Amines, aromatic reactions with uronic acid, 219
    in browning, 222
Amino acids, essential, 307
    sequences in blood proteins, 126
    sequences in glycoproteins, 118–119
    synthesis, 115
Amylase, 27, 416
    alpha on amylopectin or glycogen, 36–38
    on amylopectin, 39
    on glycogen, 37
    on raw starches, 43
    starch, 190–191

Amylopectin, 395, 421
    branching structure, 31
    definition, 26
    in water, 436
    starch, 29
    structure, 30–31, 396
Amylopectinosis, 314
Amylose, 395, 421
    definition, 26
    in water, 436
    starch, 29
    structure, 29, 47, 396
Amylose acetate, stereo structure, 49
Anomers, of glucose, 13–14
    of glycosides, 18
Anthocyanins, structure, 253
Antibiotics, occurrence in sugars, 15–16
Antigenic factors, 127
Apigenin, structure, 19
Apiin, from parsley 18–19
Apiose, structure, 19
Arabinan, structure, 77
Arabinogalactan, a hemicellulose, 81–82
Asparagine, in glycoprotein linkages, 118
    in imido groups in dissociation, 119
Atherosclerosis, 318
Avicel, 63

B

Baking industry, sugar usage, 382
Barry degradation, of polysaccharides, 212
Beta particles, from glycogen, 32
Betacyanins, 253
Betanin, structure, 253
Beverage industry, sugar usage, 379
Bisulfites, in browning, 231–232, 237–238
    reactions, with aldoses, 231–232
    with glucose, 237–238
Blood types, glycoprotein composition, 127
    specificity for synthesis, 127
Bread, processing, 9–10
    protein, 10
    role of carbohydrates to structure, 9–10
    use, 5–6, 9
Browning, factors affecting reaction, 225
    inhibitors, 231–232
    reaction in breads, 382–383

C

Calories, intake, 3, 6–7, 299
Canning industry, sugar usage, 384–385

Caramelization, by heat and acid, 17
Carbinol group, 207
    oxidation, 206–207
Carbohydrases, 178–196
Carbohydrates,
    analysis, 140, 147, 178
    as energy, 301
    available for metabolism, 179
    characterization, 133–134
    colloid, 421–442
    consumption, 6–7
    derivatives, 170
        in mass spectrometry, 152
    heterogeneity, 125
    in alimentary canal, 314
    in nutrition, 298–321
    in obesity, 322–346
    metabolism, defects, 312
        effects of insulin, 365
        hepatic, 359
        oxidation, 205–207
Carboxylase, 310
Cardiovascular disease, 317–318
Carotenoid glycosides, 254
Carrageenan, 84–86, 413
Catalysis, 5
Catecholamines, glucagon, 367
Cellobiase, from cellulose, 56, 67
Cellotriose, structure, 58
Cells, water content, 277
Cellulase, 189
    on cellulose, 62
Cellulose,
    as foaming agent, 440
    C₁ and Cₓ enzymes, 65–67
    effect of radiation, 279
    hydroxypropyl, 434–435
    increase in digestibility, 281
    microcrystalline structure, 69
    occurrence, 55
    properties, 55–57
    structure, 55–57
    synthesis, 67
Ceramindes, glycosyl, 107–108, 110
    hexosides, 107–108
Cerebrosides (ceraminde monohexo-
        sides), 107–108
Chalcone, flavor, 260–264
Chemical shifts, of carbohydrates, 168
Chitinase, 190
Chloroplasts, role of galactosyl diglyce-
        rides, 105
Chromatography, beverage analysis, 381
    column, of pectin, 75–76
    gas liquid, of alditol acetates, 153–154
        of carbohydrates, 147–148
        of glycolipids and gangliosides, 152
        of glycoproteins, 152–153
        of monosaccharides, 139–140
    temperature programming, 151

    trimethylsilyl derivatives, 148, 152–
        154
    of galactolipids, 105
    procedures for carbohydrates, 136
Chromogens, from carbonyl groups, 140–
        141
Chylomicrons, 326
Codons, in amino acids synthesis, 115
Coenzymes, 338
Cohesiveness, of starches, 406
Collagen, occurrence of hydroxy amino
        acids, 115
    soluble and insoluble, 122
    synthesis, 122
Colloid systems, hydrophilic, 421–442
    of carbohydrates, 148–149
Color, in confections, 386
Colorimetric procedures, of carbohy-
        drates, 136
Communication, 444–450
Confectionery industry, sugar usage,
        385–388
Conformational formula, fructose, 15
    glucose, 14
Congelation, 403
Convallatoxin, structure, 349–350
Conversion, carbohydrate to protein, 8
    hydrocarbon to protein, 8
    waste cellulose, 6
Convertibility, resources, 8–9
Coprophagy, 316
Corn curd formation, 414
    fiber, source of hemicellulose, 81
    starch, in confections, 403
Coulomb forces, in protein bonding, 115
    interactions in glycosidic linkages,
        123–124
Coupling constants, in carbohydrates,
        159, 165, 167–168
Covalent bonding, in glycoproteins, 115
Cyclamate, sodium, structure, 264
Cysteine carbazole reaction, 180

D

Dairy industry, sugar usage, 383–384
Dehydration, aldehydes, 17
Dehydrogenase, determination of glu-
        cose, 183–184
Dental caries, 315
Dextran, composition, 286
    degradation by radiation, 286
Dextrin acid, as foaming agent, 435, 440
    cyclic, 415
Dextrose, in corn sugar, 5, 11
Dextrose-equivalent hydrolysates, 415–
        416
    as sweetener, 388–389
Diabetes mellitus, 305, 312, 336, 359–
        372

Dialdehydes, production, 211
Difructoseglycine, decomposition, 227–
    228
  formation, 226
Digitoxin, 347–348
Digitoxigen, hexosides, 347, 352, 354
  lethal dosage, 353
  structure, 348
Diglycerides, digalactosyl, structure,
    103–104
  galactosyl, structure, 104
    role of chloroplasts, 105
    synthesis in microorganisms, 105–106
Dihedral angle, in NMR, 159
Dihydrochalcones, taste, 260–264
Diketoseamines, formation, 226
Dimethyl sulfoxide (DMSO), effect on
    mutarotation, 163–164
  oxidant of hydroxyl groups, 209
Disaccharides, glycoproteins, 122
Diseases, role of cerebrosides, 108
Dispersions, food systems, 424–427
  diphase systems, 426
  triphase systems, 427
DNA, 10
DPNH (NADH), 179

E

EDTA, in carbohydrate analysis, 139
Electron spin resonance, 280
β-Elimination, by bases, 17
  degradation of pectin, 79
  in glycoproteins, 121, 230
Emulsions, 424, 427–431
Emulsofoams, 425, 431–436
Emulsogas, 425
Energy balance, 335–336
  metabolism, 324
  requirement for irradiation degrada-
    tion, 281
  source, 323
  storage, 322–324
Enolization, 230–231, 238, 240
  basic, 17
  of ketoseamines, 234
Environment, for life development, 299–
    301
Enzyme, affected by radiation, 273, 284–
    285
  branching, specificity, 38
  C_1 and C_x hydrolysis of cellulose, 65–
    66
  commercial sources, 178
  conversion of glucose to fructose, 11
  deficiencies, 316
  in high carbohydrate intake, 310
  in lipogenesis, 311
  inhibition of glycolysis, 332
  isomerization, 5
  methods of analysis, 179
Enzymolysis, of cellulose, 64–66

Epichlorohydrin, as cross-bonding agent,
    399, 437
Epimers, of glucose, 15–16
Epinephrine, 329, 343, 359
Equilibrium relative humidity, 386–387
Equilibrium value, 377
Esterase, pectin, 273
Evatromonoside, 350

F

Fabry's disease, identification of trihex-
    oside, 108
Fats, in emulsions, 429
  in obesity, 322
  oxidation, 308
  synthesis, 308
Fatty acids, composition of galactosyl
    diglycerides, 102, 105
  composition of gangliosides, 111
  effect on starch, 404–405
  enzymes, 311
  free, 325–326
    concentration, 330–331
    in lipogenesis, 378–379
    in obesity, 325
    synthesis, 308–309
  hydroxy, in brain tissue, 102
  in starch, 404
  oxidation, 302–303, 305
  synthesis, 308–309
Fibrillar structure, of cellulose, 59
Fibrils, of cellulose, 59
Fibrinogen, role of carbohydrates, 116,
    129
Fischer formula, fructose, 15
  glucose, 13–14
Fish protein, in bread, 10
Flavanone, glycosides, of citrus, 254, 257
  structural properties, 257–260
  taste, 258
Flavors, browning, 219
  from sugar-amine interactions, 242–
    245
  loss, 412
  in starches, 412
  of pigments, structural requirements,
    255–260
Foams, 431–436
Food science, definition, 373–374
Food technology, definition, 374
Freezing point, ice cream, 384
Fringe micellar theory, of cellulose, 60,
    62
Fructose, from glucose, 5
  in gluconeogenesis, 311–312
  in lipogenesis, 311–312
  properties, 14–15
  structure, 15
Furanone, 242–243
Furanose, a ketose, 15

Furans, formation by dehydration, 17
Furcellaran (Danish agar), 86–87
  gelling properties, 86–87
  structure, 88
Furfurals, 244
  hydroxymethyl, formation, 234–236

G

Galactan, of pectic substance, 76–78, 81
  sequence, 78
Galactocerebroside, structure, 108
Galactolipids, in biosynthesis, 105
Galactose, properties, 16
Galactosemia, 314
Galactosidase, hydrolysis of lactose, 188
Galacturonan, from pectins, 76
Galacturonic acids, in pectin, 272
Gangliosides, composition, 109–110
Garan, structure, 93
Gaucher's disease, lack of glucosidase,
  108
Gelatin, protein emulsifiers, 435–436
Gelatinization of starch, 42, 397
Gelatinized (pre) starches, 402–403
Gelling of starches, 436
Gels, from pectins, 80
  formation, 422
Genes, allelic, 127
Gentiobiose, occurrence, 21
  structure, 21–22
  taste, 13
Geometrical environment, of protons, 158
Globulin, 116
Glucagon, interactions, 367–370
Gluconeogenesis, 303, 305, 324, 359
  enzymes, 311
  stimulation, 366–367
Glucopyranose, configuration, 167
Glucosamine, in glycoproteins, 117–118
Glucose, as precursor, 307
  degradation by irradiation, 280–281
  energy source, 306–312
  essentiality, 302–305
  in liver, 359
  in liver perfusion, 363–364
  isomerization, 4–5
  metabolism, 302–318
  oxidase, 178
  oxidation and reduction, 16
  properties, 13
  reaction, 16–17
  requirement in metabolism, 332
  stereochemistry, 13–14
  structure, 14
  structure of epimers, 15
  synthesis, 303. 305
  use by central nervous system, 304
Glucose-fatty acid cycle, 331–332
Glucose-insulin relationship, in fat me-
    tabolism, 343
Glucosidase, 186

Glucuronidase, 187
Gluten, in wheat, 9
Glycemia, hypo, effects, 302, 359
Glyceride, glycolipids,
  definition, 100
  from animals, 100
  of bacteria, 106
  of plants, 104
  structural requirements, 106–107
Glycerol, activated, in lipogenesis, 328
  enzymic determination, 191
  in fat synthesis, 309–310
  in gluconeogenesis, 360
Glycogen, branching, 34–35
  branching representation, 30
  definition, 27
  function, 26
  isolation, 32
  Meyer structure, 32
  on structure, 32–34
  particles, 32–33
  schematic illustration of particles, 33
  structure, 29
  synthesis, 360
Glycogen-storage disease, 313
Glycol, cleavage, 211
Glycolysis, 306–307
Glycone, location of amino groups, 123–
  124
Glycoproteins, classification, 128
  composition, 116–117
  structure, 115
Glycosides, cardiac, 347
  cardiotonic activity, 353
  structure, 348
  synthesis, 349. 355
  properties, 18–19
1,4-Glycosidic bonds, in glycogen, 29
  rates of hydrolysis, 137–138
  rotational angles, 41
  stability to radiation, 281
N-Glycosidic linkages, glycoproteins,
    117–118
  acid hydrolysis, 118
  specificity of attachment, 118–119
O-Glycosidic linkage, stability, 122
Glycosidic pigments, 253
Glycosphingosides, 107
Glycosuria, 306
Glycosylamines, formation, 220–221
Granules, starch, 438–439
  microappearance, 397
Grignard additions, conversion of ke-
    tones to amines, 209
Guar, 93
  use, 94
Gums, arabic, 81. 428
  properties, 89
  carbohydrate, 421
  colloid, 421

dispersions, 431
  enzymic hydrolysis, 191
  occurrence, 73
exudate, 88
from seaweed, 82
ghatti (Indian gum), 90
hydrophilic carbohydrate, 421
  application, 427
  gel formation, 422
  uses, 427
karaya, 90–91
locust bean, 92–93
tamarind, 94
Gyration, 80

H

Haworth arrangement of glycogen, 36
  ring formula of, fructose, 15
  glucose, 14
Helical configuration of starch, 404
Helix formation of starch, 40
Hemiacetal, oxidation by copper, 205
Hemicelluloses, definition, 81
Hesperidin, citrus, 254–257
  structure, 254, 256, 261
  taste, 255–257
Hexokinase, determination of glucose, 183–184
Heyns rearrangement, mechanism, 222–225
Homeostatic mechanism, 323, 327, 359, 361
Hormones, regulators, 360
Hydrocolloids, source, 80, 82
Hydrogen bonding, in starch, 396
Hydrogen peroxide, 181
Hydrolysis, glycoproteins, 124
  glycosidic bonds, 144
  starch, 4–6
Hydroxyl group, oxidation, 207–208
Hypoxia, 362, 364

I

Inborn errors of metabolism, 312–314
Infrared spectroscopy, in carbohydrates, 172
Inosinic acid, 17
Inositol, myo, oxidation, 192
Insulin, 305–306
  anti, 362, 367
  effects, 360–362, 365
  in adipose tissue, 326–328
  in fat storage, 336
  in glucose metabolism, 337, 365
  in growth, 336
  in synthesis of triglycerides, 325
  interactions, 366
  production, 312, 336
  transport of amino acids, 336

Intestines, 315–316
Invert sugar, 11, 20
Inversion of configuration, synthesis, 208–209
Invertase, 19
  hydrolysis, raffinose, 189
  sucrose, 188
Invertase-melibiase mixture, 189
Irradiation in sugar systems, 378
  on polysaccharides, 269
  on texture, 269
Isomaltol, 243
Isomaltose, occurrence, 21
  structure, 21–22
Isomerases,
  determination of arabinose, 180
  determination of xylose, 180
  glucose, 5
Isomerism, saccharides, 13
Isomerization, glucose to fructose, 4
Isotope dilution method, 361
  carbohydrates, 142

K

Karplus curve, in carbohydrates, 159–160
Ketone bodies, utilization, 333
Ketoseamines, formation, 220
  enolization, 234
Ketoses, 14
  role of carbohydrates, 305–306
Ketosylamines, rearrangement, 222–225
Kwashiorkor, 6

L

Lactate, determination, 193
  of uronic acid, 194
  oxidation, 193
Lactones, formation, 17
Lactose, in diet, 316
  reaction with lysine, 219
  reducing disaccharide, 20–21
  structure, 21
Laurate, in emulsions, 428–429
Lewis system, in blood types, 126–127
Lipase, 326, 329
Lipidemic, hyper, 317
Lipogenesis, 309, 328
  adaptive, 333
  enzymes, 311
Lipolytic, agents, 329
  anti compounds, 329
Lipophilic groups, 427
Lipoproteins, 326
Liver perfusion, effect of glucose, 362–364
Lumen, 315
Lysases (transeliminases), 190

M

Macrodextrins, 36–37
  from amylopectin, 38
  from glycogen, 37
Maltoheptaose, formation, 38
Maltohexaose, formation, 38
Maltol, 243–244
Maltopyranoside, methyl, structure, 41
Maltose, as reducing disaccharide, 21
  structure, 21
  rotational angles, 41
Maltotriose, 22
Mannan, 81
D-Manno-heptulose, occurrence, 17–18
D-Mannose, occurrence, 18
Marasmus, 6
Marketing, need, 2, 3
Mass spectrometry, carbohydrates, 169–170
  in combination with GLC, 170
Metabolic defects, 312
Metabolism, of carbohydrates, 301
  of glucose, 302–318
Metachromatic leukodystrophy, role of sulfatides, 109
Metal ions, effect in radiation, 278
Meyer structure, of amylopectin, 31
Microbial action, in sugars, 378
Microbiology, beverages, 381
Microwave heating, 383
Molecular geometry, 158
Molecular weight, determination for cellulose, 57–58
Monoglycerides, addition to starches, 142–143
  chemical analysis, 139–140
  derivatives, 139
  in glycoproteins, 122
Mutarotase, 181
Mutarotation, in NMR spectra, 163–164
Myelin, source of galactolipids, 102

N

NADH (DPNH), 179
NADPH (TPNH), 179
Nageli amylodextrin, 44
Naringin, conversion to chalcone, 260
  in citrus, 254–257, 260
  reaction, 255
  structure, 254–256
  taste, 255–257
Neodiosmin, structure, 259
Nervous system, glucose requirement, 332
Nitrogen dioxide, as oxidant, 206
Nitrogen-sparing effect, role of carbohydrates, 303
Nitromethane, in synthesis, 212
Nonenzymic browning, effect in foods, 219

Nuclear magnetic resonance spectroscopy,
  applications to monosaccharides, 161
  applications to polysaccharides, 165
  in structural determination, 164
  interpretation of spectra, 155
  parameter, 155–161
  of carbohydrates, 154
  spectrum, 166
    of monosaccharides, 157
  use in extent of branching, 165
Nucleic acids, 42
Nucleosides, 18
Nucleotides, cyclic, 367–371
Nutrition, caloric need, 6–7
  of carbohydrates, 298
  requirements, 300–301

O

Obesity, 322, 340–341, 331
Oils, in emulsions, 428–429
Oligosaccharides, from cellulose, 56
  from pectins, 77–78
  from starch, 38–39
  nonreducing disaccharides, 18–19
  properties, 12–14
  reducing disaccharides, 19–20
Opacity of emulsions, 430, 440
Optical activity, of monosaccharides, 13
Optical rotation of saccharides, 13
Oral cavity, 315
Organ perfusion apparatus, 363
Osmotic pressure, of carbohydrates, 315
Osuloses, formation, 227–229, 236–237
  properties, 228–229
  unsaturated, 227
Ovomucoid, 116
Oxalacetate, 360
  in metabolism, 305
Oxidase, galactose, 185
  glucose, determination, 180–181
    specificity of reaction, 181–182
Oxidation, in metabolism, 306–307
  of carbohydrates, 205–215
  overoxidation, 211
  periodate, 211

P

Pectin, degradation, 271–274, 276
  effect of radiation, 272
  formation of gels, 278
  hydrolysis, 77–78, 190
  properties, 73–80
  structure, 74
Pentose cycle, 307
Peroxidase, determination of glucose, 181
  of glycoproteins, 129
Phenylhydrazones, 212–213
Phosphate compounds, high energy, 306–307

Phosphorus oxychloride, as cross bonding agent, 402
Phosphorylase, degradation, 38
  occurrence in glycogen, 32
  of sucrose, 11
  stimulation, 367
Phosphorylation, 364–365
Photomicrograph, of starch granules, 438–439
Photosynthesis, role of galactolipids, 105
Pigments, formation, 236–237
  glycosidic, 253
Planck's constant, in NMR, 155
Polymers, cellulose, 57
Polysaccharides, 22
  alkyl derivatives, 142–143
  analytical procedures, 136
  chemical analysis, 137
  definition, 12
  derivatives, 95
  effects of irradiation, 269–270
    on texture, 269, 272
    susceptibility to damage, 271
    viscosity changes, 273, 276
  food uses, 422–423
  hydrolysis, 137
  in algal species, 82
  in plants, 81
  in wood, 82
Potatoes, instant, 406
Precursors, biological, 303–312
Product development, 2
Propylene oxide, crosslinking, 437
Protein, conjugated, 115
  consumption per capita, 8
  role in obesity, 322
  synthesis, 333
  Tamm and Horsfall (urinary glycoprotein), 117
Proteoglycans, 121
Protons, in NMR spectra,
    anomeric, 161–162, 165, 167
    axial ring, 158
    equatorial, 158
    hydroxyl, 158, 163–164
  resonances, in NMR, 156
Protonation, cleavage of glycosidic bonds, 124
Pullulanase, 38, 416
  action on, branched glycogen, 34–36
    dextrins, 36
    oligosaccharides, 35
  specificity, 34, 50
Pyranose, 13–14
Pyrroles, 244

R

Radicals, formation, 270, 275
Reactions, rate, 377
Reducing saccharides, 14
Reducing sugars, in breads, 383

Reductones, from glucose, 239
  from ketoseamines, 239
  mechanisms, 240
Retrogradation, mechanisms, 396
  of starch granules, 40, 43, 407, 438
Rhoifolin, structure, 259
Rhozyme-S, 179
Ribonuclease, amino acid sequence, 119–120
Ribonuclease B, 116
RNA, 18

S

Saccharides, mono, definition, 12
Saccharin, structure, 260, 264
Schiff's base, 223, 230
Serine, role in glycosidic linkage, 121
Serum albumin, 325
Serum lipids, 317
Sialic acid, in glycoproteins, 118
  in glycosyl ceramides, 110
Smith degradation, polysaccharides, 218
Sodium trimetaphosphate, as cross bonding agent, 402
Solubilities, limits, 377
Sorbitol, oxidation, 192
Spectrophotometry, of nucleotides, 183–184
Sphingoglycolipids, 100
  cerebrosides, 107
Sphingosine, in glycosphingosides, 107
Spin-spin coupling, in NMR, 158
Stabilization, of starch, 401–403
Stabilizer, in ice cream, 84
Starch, complexes, 48
    fractionation, 46
  conformation in solution, 40, 436
  cross-bonding, 399
  defatted, 424
  digestion, 27
  effect of radiation, 282
  film, edible, 413
  films and fibers, 48
  functions, 26, 395
  gelatin, 284
  granule, birefringence, 45
    crystal structure, 43, 44
    electron microscopy, 45, 46
    swelling, 398–399
  helical configuration, 404
  hydrolyzate, 5
  in foods, 395–420
  low-calorie, 416
  pastes, effect of freezing, 410
    effect of storage, 408–409
  reactions with fats and proteins, 415
  solutions, diagram of helices, 42
  structure, 28, 39, 48, 52
  synthesis, 26–27
  swelling power, 437
  viscosity behavior, 436

Staudinger arrangement, of glycogen, 36
Steroids, 347
Stoichiometric analysis, of carbohydrates,
    136
Stractan, a gum, 82
Strecker degradation, of amino acids
    244–245
Strophanthidin, 347, 354
    hexosides, 352
    lethal dosage, 353
    structure, 348
Structure determination, diagram of bi-
    polymer, 135
Sucrose, 5
    as nonreducing disaccharide, 18–19
    industrial use, 374–375
    structure, 20
    sweetness, 11
Sugar, analytical procedures, 136
    as leavening agent, 382
    corn, 5
    invert, 5
    irradiated, toxic effects, 289
    isomerization, 4
    liquid, 5
    phosphorylated, in browning reactions,
        247
    radiation effects, 285
    reducing, in browning reactions, 219
    standard curve, 141
    technology, 373–394
    usage, 374–375
Sugar-amine interactions,
    effect of oxygen, 238, 245–246
    flavors from, 242–245
    in dried foods, 219
Sulfatides,
    role in demyelinating disease, 109
    structure, 109
4-Sulfohexosulose, synthesis, 233–234
Sulfolipid, of plants, 106
Sulfur dioxide, browning inhibitors,
    231–232
Sweeteners, artificial, 264
    noncaloric, 3–4
    synthetics, 374
Sweetness, 376
    of invert sugars, 11
    tests, 388
Syneresis, of starch paste, 408–409, 423–
    424
Synergism, C$_1$ and C$_x$ enzymes, 66
Synthetase, of glycogen, 32

T

Taste receptors, of insects, 263
Tautomerism, of saccharides, 14–15
Tay-Sachs disease, changes in ganglio-
    side concentration, 110
    ceramide trihexoside, 108
Threonine, role in glycosidic linkages,
    121
TPNH (NADPH), 179, 309
Tragacanth, use as emulsifier, 91–92
Tragacanthic acid, structure, 92
Transeliminase, pectin, 273
Transferases, 50
    for fructose, galactose, 127
    in blood group specificity, 127
    in synthesis of ribonuclease A, 121
    specificity, 125–126
Tricarboxylic acid cycle, 305, 310
Triglyceride, precursors, 310
    storage of fat, 325
    synthesis, 328–329
Triglyceridemia, 317
Trimethylsilyl ether derivatives, of car-
    bohydrates, 148–149
    in analysis of sugars, 150–151
Trioses, 17–18
Trisaccharides, 22–23

V

Vicinal hydroxyl group, periodate oxi-
    dation, 214
Viscoelasticity, of starches, 407
Viscosity, breakdown in, 400
    changes in pectins, 80
    effect of acidity, 401
    instruments of measurement, 399
    starch pastes, 398
Vitamins, requirements, 315–316

W

Waste disposal, 68
Water, content in radiation, 275–276
    determination in starch, 172
    protective effect of, 275–277
    role in colloids, 421–441

X

Xylan, 81